BUILDING SORTED BY TYPE

CIRCULAR CONSTRUCTION METHODS

Edition **DETAIL**

Acknowledgments

We wish to express our deep gratitude to all the authors who contributed to this book: Werner Sobek, Thomas Auer, Andreas Hild, Hauke Horn, Christian Holl, Elena Boerman and Daniela Schneider. Their willingness to present their opinions and visions, their works, research and insight here made this book and the resulting discussions possible. We are also grateful to our students, whose contributions based on their participation in seminars and design studios laid the foundations of this book.

Further, we thank our respective teams at the professorships of Sustainable Construction and Building Construction at the Karlsruhe Institute of Technology (KIT) and, especially, Robina Behrendt, who was tireless in creating 3D illustrations, as well as Patrick Bundschuh, Luca Diefenbacher, Mattis Epp, Felix Caspar Jörgens, Sebastian Kreiter, Salesia Trenker and Amelie Vierhub-Lorenz for their valuable drawing contributions.

Our deep gratitude also goes to Andrea Klinge and Ulrich Röhlen, Riccardo La Magna, and Steffen Mayser and Karsten Jurk of the Munich-based office of PMI for sharing their expertise in loam construction, structural engineering and building physics, respectively. Our discussions with them allowed us to perfect the details, sections and 3D drawings presented here. We would like to thank, in particular, the Dean's Office of the KIT Department of Architecture and its Managing Director, Judith Reeh, for their support during the past years. Without their encouragement, creating this publication would have certainly not been possible. Very special thanks go to Sandra Hofmeister, Katja Pfeiffer, Jana Rackwitz, Cosima Frohnmaier, Barbara Kissinger, Steffi Lenzen and the entire team at DETAIL for their trust in us and their supportive help.

Dirk E. Hebel, Ludwig Wappner, Katharina Blümke, Steffen Bytomski, Valerio Calavetta, Lisa Häberle, Peter Hoffmann, Paula Holtmann, Hanna Hoss, Daniel Lenz, Falk Schneemann

Karlsruhe, May 2023

Contents

Acknowledgments	3
Preface	6
Foreword	9

INTRODUCTION

Building Sorted by Type	12

HISTORY AND STATUS QUO

History of the Building Culture of Reuse	30
More than a Mine – Existing Buildings as a Material and Cultural Resource	34
Vernacular Architecture	40
Learning from Temporary Buildings	48
Contemporary Examples of Circular Construction	56

MATERIALS – CONNECTIONS – LAYERS

Materials Selection for Circular Construction	66
Materials of the Circular Economy	70
Pollutants in the Cycle	80

Digitalisation in the Circular Economy	**92**
(Re)Building Simply	**98**
Reversible Assembly and Connection Methods	**104**
Principles of Joinery	**110**
Layering as a Circular Principle	**118**
Layer Compositions	**122**

DETAIL CATALOGUE

Focus on Timber	**130**
Focus on Masonry	**158**
Focus on Concrete	**174**
Focus on Steel	**190**
Focus on Loam	**206**

APPENDIX

Biographies	**222**
Image Credits	**224**
Sources	**225**
Index	**231**
Imprint	**232**

Preface

This book is a first step into the world of requirements, modes of thought and resulting architectural implementations of building sorted by type as a key element of a newly defined circular economy in the construction field. It can be understood as a contribution to a social paradigm shift that is becoming more and more tangible. Humankind must decide how to live together on this planet in a responsible manner and in harmony with natural circular flows and processes. This transformation needs to materialize in the field of construction in the form of rethought and differently implemented construction principles. As human beings, we must no longer tolerate emissions that are harmful to the climate. We must conserve resources and integrate them in cycles propelled by regenerative ideas. We must abandon the concept of "waste".

By way of introduction, the first part of this book deals with the urgency of the topic, underpinning it with numbers and facts. The second part presents historical perspectives of the topic, while the third features circular and recycling-oriented materials as well as methods of assembling, connecting and layering them. The important issue of pollutants in the construction field is also discussed. In the concluding fourth part of the book, we introduce a construction catalogue in the form of detailed architectural solutions sorted by type. They refer to the previously described principles and illustrate their exemplary implementation. The featured solutions comprise specifically selected combinations of built or planned cases of application. From the perspectives of architecture, construction, structural engineering and building physics, they establish a catalogue of opportunity that permits immediate implementation in moderate climate zones. For each construction project, for each site, for each regulatory, climatic or geological context, normative and legal requirements such as fire safety, soundproofing or moisture proofing demand independent examination and adaptation. Some of the construction types, products and materials presented here require additional deliberation on questions of building authority approval, in relation to the degree of public access and building class. Further, in certain parts of the book, the detail drawings feature a blue dot with guidance, intended to offer food for thought in reference to comparable topical issues.

While this book explicitly deals with questions surrounding building and construction sorted by type, what remains valid is that other questions, such as on the carbon footprint of construction materials, the availability of resources, or the possible omission of certain construction components should remain part of the thought process. In its year of publication, this book displays the state of the art of transformation that is highly dynamic. The intention behind the book is to show pathways and illustrate targets of socially responsible patterns of behaviour among actors in planning and implementation, instead of a fixed rulebook. Its guidance, thus, should be considered as critical commentary and enquiry. The aim is to inspire readers to apply a critical stance in further developing the presented solutions, within the scope suggested by the first three chapters of the book.

In Germany, approximately 19.4 million residential buildings exist to date. We should not forget that this existing housing stock comprises roughly 16.1 million detached and semi-detached houses, equalling a share of nearly 83%. This is why we intentionally show many detail sections of two-storey structures. We also refer to the topics of renovation and redensification, since they will increasingly gain relevance in the future.

With this book, we intend to show that already built, tried-and-trusted construction types, products and materials exist that support building sorted by type. Their application to our built environment, either within existing building stock or new construction, is both possible and sensible. We would like to encourage practitioners to use these examples and continue developing them, in order to support the paradigm shift towards a closed circular economy in the building sector. The approaches presented here are intended to inspire and change the construction industry in its entirety. As university educators, we consider this our obligation and mission.

Dirk E. Hebel, Ludwig Wappner, Katharina Blümke, Steffen Bytomski, Valerio Calavetta, Lisa Häberle, Peter Hoffmann, Paula Holtmann, Hanna Hoss, Daniel Lenz, Falk Schneemann

KIT Karlsruhe, May 2023

Foreword

Global resource consumption has already exceeded 100 billion tonnes per year and it will continue to rise rapidly in the future. The construction sector is responsible for more than half of this resource consumption.

Nearly 50 % of all construction materials used for building to date can be traced to industrialised nations. These nations are home to about 1.4 billion people. The estimated 6.6 billion people of the Global South need to make do with the other half. Often, they live in below-standard housing or lack sufficient utilities or waste management systems. Beyond that, the global population will further increase by two billion by 2050, reaching approximately ten billion in total. The resulting need for new buildings will trigger a demand for construction materials that can never be met. Why? Many traditional construction materials will no longer be available in the necessary quantities. Their production and transport requires energy that we are incapable of providing. Further, the resulting emissions will drive the global climate towards temperature ranges that are hardly supportive of human life, or even survival, for that matter. Building with deliberation, by using the materials available to us efficiently, needs to be the future standard. This is the only way to prevent the worst possible social disruptions and conflicts.

Building – deliberately, efficiently – means to avoid exhaustingly consuming construction materials and, instead, using them with the future in mind. The reuse or reclamation of already built-in construction materials will be the basis of future construction activity. How this can take place, from the big picture to the detail level, is presented in the book you hold in your hands. It is a pioneering achievement. In exemplary ways and rooted in practice, it addresses the requirements that building must meet in the future. I have been responsible for formulating related appeals for some time.

This publication is the first truly comprehensive professional reference book on construction that meets the demands of recycling and of building with recycled materials in the fields of design and construction. It will be important and valuable to students as well as architects and engineers already active in practice. I hope and wish that this book will soon be an integral part of the ongoing education of everyone involved in building.

Werner Sobek

INTRODUCTION

On Building Sorted by Type

Dirk E. Hebel, Ludwig Wappner

"Human beings are at the centre of concerns for sustainable development" [1]
United Nations, Rio de Janeiro, 1992

Established circular systems
In March 2020 the European Commission presented its action plan for the introduction of a comprehensive circular economy in the European Union by 2050. This is related to the ambitious goal of producing all goods and products and distributing them in a manner that permits full reuse and recycling. The result would be an economic system that completely abandons the concept of waste, including landfill and incineration. On our planet, natural circular processes have functioned according to similar systems for billions of years. Matter that exists on earth can't leave it (with the exception of space probes and satellites). Functioning closed loops have been established that continuously rebuild and reconfigure matter without loss. Plants use sunlight to produce oxygen and nutrients – in other words, through photosynthesis – that are absorbed by humans and animals. In return, the latter supply CO_2 as well as fertiliser (through natural excretion and biological compostation), which plants require for their own growth, aside from non-organic nutrients they extract from the soil. The space required for this form of metabolism in which the provision of biological substances takes place is defined as the biosphere. It refers to the part of the earth that is inhabited by all organisms, including humans. The biosphere is dependent on the lithosphere, the earth's crust. For instance, volcanic eruptions introduce substances into the biosphere that are reconfigured through photosynthesis and transformed into different material combinations. Correspondingly, substances are distributed from the biosphere into the lithosphere through mineralisation or sedimentation. These processes occur at different speeds and within circular systems that repeatedly interact and intersect, thereby existing in a condition of interdependence. Human beings, as elements of the natural environment, also constitute a factor within these established and complex systems with their perpetually changing and shifting closed loops. However, we happen to live in an era in which these natural cycles are impacted by humankind, exerting an influence so momentous that the circumstances of life on earth are subject to change to such an extent that it potentially calls survival in its current form into question. This era influenced by human activity is known as the Anthropocene.

The human factor
Until recently, layers of rock that offer information on the geological, biological and atmospheric conditions on earth served to define geological time scales in relation to geochronological units. The Anthropocene, however, describes the impact of humans on these conditions. Thus, it is the first geochronological epoch that isn't purely defined in geological terms, as the German geologist Manfred Menning has stated [2]. In the year 2000 the term appeared in an article titled "The 'Anthropocene'" and was, for the

first time, discussed in the context of designating a new geochronological unit succeeding the epoch of the Holocene [3]. In the article, the Dutch chemical and atmospheric scientist Paul Crutzen and the American biologist Eugene Filmore Stoermer describe a debate that had already began in the 19th century, dealing with the increasing visibility of human interventions in the natural environment, followed by a list of important, apparent and measurable impacts. As onset of the epoch, they propose the second half of the 18th century. Atmospheric concentrations of greenhouse gases were, for the first time, verified within ice core drillings, marking the beginning of industrialisation. This date also coincides with the invention of the steam engine by James Watt in 784. An official introduction (which has yet to take place) of the Anthropocene as a chronostratigraphic term was discussed by the International Commission on Stratigraphy in 2016, following a debate on the results of the Anthropocene Working Group convened during the 35th International Geological Conference in Cape Town. The conference proceedings indicate that human influence on the planet is so significant that the introduction of a new epoch denomination seems justified. The protocol of the working group was published on 21 May 2019 [4]. The majority of members – differently than encouraged by Crutzen and Stoermer – proposed the year 1950 as the beginning of the Anthropocene. This happened because sediment served to identify a so-called golden spike, indicating a demonstrable and verifiable geological sediment layer that enabled the new epoch in space and time to be defined. The nuclear fallout occurring from the mid 20th century onwards as a result of the first American nuclear weapons tests in 1945 in New Mexico was identified as the golden spike. Today, the radioactive deposits of this human activity can be proven without a doubt in Crawford Lake in Canada. This is why the lake was determined in 2023 as the reference point on earth to spatially define the beginning of the Anthropocene. Alternative proposals for a temporal marker have been made in the past and continue to be made. They reach as far back as the 17th century and the beginning of intercontinental shipping. This date coincides with an exchange of species, including pathogens, facilitated by humans and leading to subsequent and substantial impacts on existing ecosystems. Without a doubt, species dispersal and extinction, climate change, waste and scarcity of resources, or the increasing degree of environmental pollution measurable and verifiable within natural and artificial deposits are all indicators of human responsibility for massive terrestrial processes of change. Such is the consensus. We can also understand the Anthropocene as an epoch of human intervention in established natural circular systems without being able to comprehensively overview or even understand the resulting impacts. We may further assume that these patterns of behaviour did not begin in the 20th, 19th, 18th nor even the 17th century, but instead, constitute a human characteristic from the very beginning. Human life was always defined by a certain abundance of natural resources. A sustainable mode of thought oriented on the future was, thus, simply unnecessary. In this regard, defining the middle of the 20th century as a starting point also supports the conclusion that, due to the ever increasing global population numbers and the rapidly declining supply of resources, radical rethinking needs to occur while recognising that the preservation of our natural circular processes must receive absolute priority. The discovered golden spike of the Anthropocene, i.e. the unambiguous marker of human existence in geological history based on radioactive deposits, coincides with a different event that occurred at about the same time: The Apollo programme of NASA, the US National Aeronautics and Space Administration, was aimed at landing men on the moon. In 1969 this successfully took place. In 1972 the crew of Apollo 17 created a photograph of earth from a distance of about 29,000 km. It shows the planet as a highly vulnerable physical body drifting in the black void of space (Fig. 1). The official title of the photograph taken on 7 December 1972 by Harrison Schmitt is AS17-148-22727. It gained recognition under the name "Blue Marble", particularly due to its appropriation by the nascent environmental movement of the 1970s. In a blunt manner, it demonstrated to humanity both the uniqueness and the isolation of earth as a system. Suddenly, the planet was viewed as a closed loop in relation to matter and an open system regarding the exchange of energy between earth, the sun and the surrounding cosmos. It also became very clear how incredibly thin the biosphere – the layer inhabited by fauna, flora and humans, including the atmosphere – actually is and how deserving it is of protection. On this planet, the continuous reconfiguration of existing and finite matter in closed loops contrasts with the transformation of continuously newly introduced energy

1
Blue Marble, 1972, photo taken from the Apollo 17 spacecraft

within an open system, energy required to propel the closed loops. Recent calculations show that the sun will continue to supply the earth with energy in the form of radiation for another 5 billion years. Due to this extensive time horizon, we consider the sun as a regenerative or even infinite source of energy. Further, since the continuous reconfiguration of matter requires very long time frames, this begs the question whether it is correct to even speak of resource scarcity or even a crisis of resources, since nothing is actually lost in a closed loop. In past centuries, from a geological perspective, humans have intervened so deeply and so extensively in the cycles of matter that natural regeneration does no longer feel possible within a sensible time frame. For instance, sand deposits are rapidly shrinking – mostly driven by immense demands in the building industry. Within the technologically advanced systems of mining this resource, the results are major interventions in existing ecosystems and impacts on habitats for flora and fauna, and eventually, humans as well. Just consider the planet as a regenerative system spanning millions of years. Without a doubt, erosion will eventually turn stone into sand, to be swept into the oceans by rivers and streams. If, however, we understand sustainability as the ability to meet the demands of our own generation only to the degree that affords future generations precisely the same opportunity (which is the definition of sus-

Desert sand displays a great range of geological differences and variances across the globe. Desert sand is generally unsuitable for use in the building sector: Grain sizes of 0.1–1 mm are homogeneously small and their form is exclusively round. This is due to permanent movement caused by wind, in return resulting in a low packing density, meaning the number of gaps is too high when used as concrete aggregate. These preconditions are unfortunate, since the required greater amount of cement admixtures to fill the gaps leads to higher costs. As a further outcome, the mechanical performance of concrete decreases: Shrinkage and creep behaviour deteriorate in comparison to concrete with water-bound sand.

tainability according to Brundtland) [5], this also means that we need to think and act according to time horizons of 30 years. This is where the unlimited depletion of sand does indeed become a stress factor.

Linear systems

Sand continues to be the solid raw material that experiences the highest global demand, with the exception of fresh water, which is subject to even greater degrees of consumption. Sand is used for different purposes, such as concrete aggregate in construction components, for the production of glass, electronic components for computers and telephones, cleaning agents and even toothpaste. However, not all sand is the same. The construction industry demands grain sizes and shapes that can only be found in water-swept sand in rivers, lakes and oceans. Mountains and rock formations have been eroding for millennia, forming gravel, sand and dust. Precipitation sweeps it into waterways and, eventually, the ocean. Smaller particles continue to float, while grains ranging from 0.06 to 4 mm in size (what is typically described as sand) are predominantly deposited along coastal areas. This product of erosion is diverse in terms of its form and size. Sand particles of different sizes compact very well, display a large particle range distribution curve and, thus, only lead to small gaps between grains. Precisely these characteristics are why water-swept sand in combination with gravel types of different sizes is so valuable to the construction industry. It results in a lower demand for cement required to fill the gaps. Once it is used as concrete aggregate and, thus, bound within a so-called composite material, current technology is incapable of reverting it to an original form of the same quality. The only remaining option is so-called downcycling. This entails mechanically breaking up concrete into pieces of different sizes that are then reused as base courses for roads or other structures and, to lesser degrees, for new concrete mixes. In such cases, we should generally speak of linear systems or metabolisms (Fig. 2) that consume raw materials without any opportunity for their circular reintroduction or material reconfiguration. According to John Milliman, humankind is mining sand in quantities equivalent to twice the amount of sediment that all rivers of the earth are actually capable of transporting [6]. According to Schweizer Radio und Fernsehen, this is a quantity of 15 billion tonnes per year [7], with a trade volume of 70 billion US dollars. The United Nations Environment Programme (UNEP) even speaks of 30 billion tonnes [8], while the actual figure may in fact be higher than that. Fifty per cent of the sand that used to reach the oceans is currently and, in part, illegally extracted from rivers before arriving at its destination. This gigantic process of mining the material can't remain without consequences for the environment. Across the world, sand mining is assuming ever more drastic forms. Beaches along Africa's northern coastline are depleted under dubious circumstances. Rivers are excavated without any supervision. The sea bed is dredged, causing land masses to slide and entire islands to vanish. Impacts are globally palpable beyond the actual area of extraction and lead to far-reaching damage. Rivers in

2
Linear metabolism in the construction sector according to Richard Rogers: Cities for a Small Planet, 1996

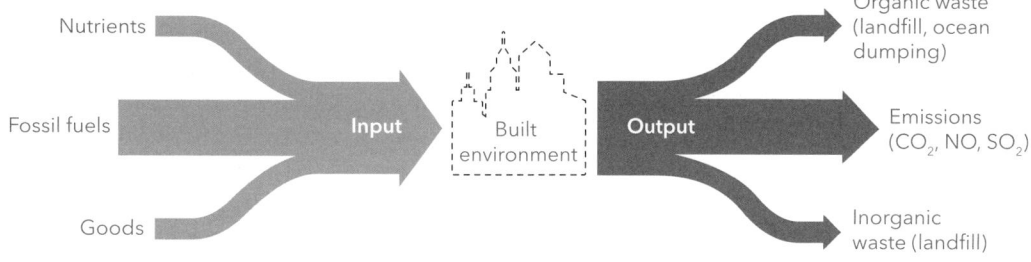

India, Thailand and Cambodia that have been dredged too intensively experience declining water levels, which result in the destruction of traditional settlements and cultures, to the detriment of the local population. Dredging the sea bed destroys the fragile basis of entire ecosystems. Sediment, once it is churned up, is distributed by currents into remote areas. Oceans are slow systems; their behaviour is difficult to simulate. The outcomes of sand mining will impact future generations and the consequences cannot be overseen.

The depletion of this resource – the Middle East and Europe have already begun to import large quantities of sand from Australia – is leading to an increase in illegal sand extraction and trafficking. The related news coverage even uses terms such as "sand wars" [9] or "sand mafia". In Singapore, for instance, this has led to clearly visible consequences. Here, sand is used for architectural construction, civil engineering and large-scale land reclamation projects. As an outcome, the island nation has increased its area by about 25 % in the past 100 years [10]. The sand export balance of countries neighbouring Singapore and the latter's import balance from these countries displays major discrepancies: While Singapore reports import quantities of 517 million tonnes in the past 20 years, the neighbouring countries speak of an export volume of 637 million tonnes to Singapore in the same time frame [11]. The origin of the material is something Singapore did not give much thought to, until vanishing Indonesian islands led to political tension and an officially issued export ban in the neighbouring country [12]. Malaysia, Thailand and Vietnam followed the example. Since then, Cambodia has become a major supplier of sand, likely illegally extracted from the country's rivers [13]. Even Dubai imports sand from Australia for its land reclamation projects and, increasingly, from Southeast Asian regions with low ethical and legal standards. This boosts the attractiveness of a market that is descending into lawlessness. Thus, if we speak of future urban design and consider the rapid rates of urbanization in Asia, Africa and South America, we also need to ask whether our established construction methods and materials are still sustainable. In all of these regions, gigantic construction projects in the form of future metropolises are looming over the horizon. Already today, nearly 90 % of cement (and, thus, twice that amount of sand as concrete aggregate) and 70 % of steel worldwide are used for construction projects in developing and emerging regions. If a raw material is missing and humankind is incapable of renewing it, which alternatives remain? And further, how can we plan and design future buildings that don't exhaustingly consume materials, but instead, store them for a certain amount of time and, when this period is over, allow their renewed deployment in a similar form and manner?

Take, make, throw
In her book "The Story of Stuff", Annie Leonard describes the previously mentioned linear systems (see p. 15f.) as aspects of a "take-make-throw" mentality. This mentality is characterised by consumers and producers who never waste a thought on where the employed product resources actually come from and/or were they go once products are expended and eventually discarded [14]. She writes that 99 % of all consumer products are in use for less than half a year before they are dumped in the waste bin (Fig. 3). In order to comprehensively understand this unsustainable product world, the book presents five chapters describing the extraction of raw materials, the production of goods, their distribution, their consumption and disposal. This way, it offers insight on the related consequences for the environment and the immediate impacts on health, culture and society. The author explains this using examples of daily consumer items and the resource consumption they are based on, beyond the actual, visible object. In addition, she describes certain excesses of human action, for instance planned obsolescence. This process never even aims at sustainability, but instead, at an intended and prematurely effected reduction of prod-

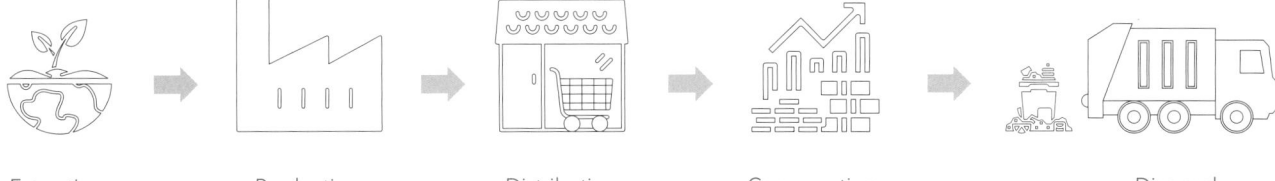

Extraction → Production → Distribution → Consumption → Disposal

3 Linear system: The majority of products are currently dumped in a landfill or incinerated after their use has concluded

uct life cycles. Its aim is to give priority, over anything else, to the repeated economic gains of a linear throw-away system defined by ever shorter intervals. The price we pay for this course of action is the exploitation of natural resources and the unfettered pollution of the natural environment (see "Pollutants in the cycle", p. 80ff.). This type of environmental stress increasingly impacts population groups that are already socially and financially marginalised, comprising individuals who can't afford to live in an ecologically conscious way and, thus, free of health hazards. The ecology and social justice are deeply intertwined. This circumstance has been subject to intense debates in the USA since the 1980s in the context of the environmental justice movement. This also raises the question of how we can design and produce items and goods in a manner that transcends the concept of waste or the discharge of materials and, instead, aims at establishing a completely closed circular system. Nothing that humans extract from existing and established natural cycles should be considered valueless after consumption. Once we accept this position and begin to think in terms of sensible and actionable circular systems, we can gain a completely new perspective on planning, design and function, as well as reuse, repurposing, reprocessing and recycling.

Resources, reserves and ratios

Linear thought and action become particularly apparent when we consider the remaining time frame within which primary material resources will remain available in their original form. Raw material resources are commonly understood as the totality of verified or potential geological deposits of a certain material within the lithosphere (metal or mineral), the biosphere (biological), the hydrosphere (dissolved in water) or the atmosphere (gaseous). Raw material reserves, on the other hand, comprise the verified deposits that permit extraction in technological, ecological, economic and ethical terms from today's viewpoint. Based on this definition, production forecasts (mining and/or extraction) can be quantified in terms of how long we will continue to have access to non-renewable resources. This is where ranges and ratios come into play. Calculations are based on expectations of ongoing consumption being related to deposits that are referenced in currently known data, without adding new reserves or considering changes to consumption patterns. The German term for this is "fixed range" (statische Reichweite), commonly known as reserve-to-production ratio (in short, R/P).

It describes the ratio between the reserves and the existing annual production output, measured in years. Neither the recycling of materials nor the sourcing of materials from already existing stock are included in such calculations. This aspect is of paramount importance: The reserve-to-production ratio of fossil fuels, for instance, is based on a completely non-regenerative process related to their combustion and successive chemical reconfiguration. This is different in the case of metals or minerals: In theory, non-energy raw materials – given intelligent planning and use – can be fully reclaimed. Here, the concept of a circular economy comes into play: Raw materials are supposed to be used in such a manner that their successive deployment remains possible within a consumption chain that spans many life cycles. Equally so, the cycle itself needs to be propelled by regenerative energy sources, in order to avoid the chemical transformation of fossil fuel into emissions harmful to the climate. In 2005 the German Ministry for the Economy and Technology (BMWi) published an assessment on the global reserve-to-production ratio of select materials (Fig. 4). The numbers show that the global reserves, in part and for certain metals, will only be available for another two, three or at most four decades, if the consumption of primary materials from the known deposits remains unchanged.

Material ratio for copper as example
In the case of copper – a metal of paramount importance to the building sector, used in large quantities for pipes, cables and sheet metal applications – the numbers are a cause for alarm. Forecasts indicate a reserve of only 470 million tonnes and a resulting reserve-to-production ratio of 32 years. In 2021, 16 years after the German assessment was published, the United States Geological Survey released updated numbers in their Mineral Commodity Summaries [15]. They estimate a global reserve of copper of 870 million tonnes with an annual production of 25 million tonnes of refined copper for the year 2020, leading to a reserve-to-production ratio of 34.8 years. The ratio seems to have increased in the years that passed between the publication of the two assessments. The consideration of a reserve-to-production ratio is, thus, insufficient for substantial forecasting. In order to more effectively display factors of perpetually changing markets, both in terms of production and consumption of raw resources, dynamic calculation methods are more appropriate. They allow integrating predicted changes, such as technological progress in mining, increases in efficiency in production itself, discovery of new reserves, substituting particular resources within goods, political changes and/or variants that otherwise defy calculation. Nevertheless, reserve-to-production ratios of copper and other metals are an unambiguous indicator that we need to find and use other methods and sources in order to meet our demand for raw materials. The quantities of soil extracted in order to mine particular metals or minerals are continuously increasing [16]. The mining process is, thus, becoming less and less productive – in the top 10 km of the earth's crust, 1 tonne of rock contains about 33 g of copper. This coincides with the use of various chemicals and huge amounts of water, as well as major excavations. The results include land degradation, soil erosion, the loss of flora and, thus, the loss of wildlife habitat – with the well-known consequences for the biodiversity that our lives depend on. Further, the energy required to produce 1 tonne of copper from primary sources is about five times that of the amount needed to recycle already existing copper [17]. Millions of tonnes of CO_2 emissions could be avoided. The transformation of our society into a community of ecological responsibility specifically and emphatically calls for technological solutions that demand an increased use of copper for electric power lines, electric motors and electromagnetic coils, heat pumps and related cables, as well as other electronic fields of application. In order to guarantee this, architects need to successfully deploy such essentially important materials in a more intelligent manner and carefully plan and ensure their renewed deployment. Roughly 55 % of copper worldwide is used for buildings [18]. The construction sector must responsibly install the material in such a way that future generations will be able to salvage and use it for different purposes.

Time and time again
Keeping materials in a continuous cycle without loss in quality is the prerequisite for change in terms of action, planning and, hence, building. Often the term recycling serves to describe the underlying idea. However, it is more or less collectively used for different types of strategies. The German language provides more pre-

	Extraction [in MT]	Reserves [in MT]	Resources [in MT]	Reserves [in years]	Resources [in years]
Bauxite	159	25,000	> 55,000	157	> 346
Lead	3.15	67	> 1,500	21	> 476
Iron ore	1,340	160,000	> 800,000	119	> 597
Copper	14.6	470	> 2,300	32	> 158
Nickel	1.4	62	140	44	100
Zinc	9.4	220	1,900	23	202
Tin	0.26	6.1	> 11	23	> 42

Sources: USGS (2006), BGR (2005)

4
The calculated linear reserves of select construction materials (metals), compiled in 2004 by the German Federal Ministry for the Economy, published in 2005.

cise terminology. The renewed deployment of materials without change of form (Gestalt), function (Nutzung) and material composition (Zusammensetzung) is considered reuse. An example for this practice is a door leaf that is reused as a door leaf in a different building. If form and material composition remain unchanged while the function changes, we speak of repurposing. Again, the door leaf serves as example, now used as a framing element within a wall construction. If the form changes within a mechanical process (e.g. shredding and/or melting) and the function changes as well, yet the material composition is maintained, we speak of reprocessing. In this case, a plastic yoghurt cup is turned into panelling for kitchen cabinetry (see "RoofKIT", p. 60ff.). If the form is changed while the material composition and the function are maintained, the term recycling is here correctly used. The yoghurt cup enters a new cycle as a container for groceries.

Reuse and repurposing of copper

Copper is a material that permits very good reuse and reprocessing. There is no difference in quality between primary and secondary raw resource. Even after multiple cycles of use, there is no loss in quality. Currently, the assumption is that two-thirds of copper mined globally since 1990 are still in circulation [19]. When comparing the production of copper in Europe with reclaimed quantities, the share is 50 % [20]. However, this figure is of limited significance, if the time the copper remains bound within a particular application is not considered. Copper stays in a residential building for 60–80 years and for 10–15 years in an electric motor. The method of "real recycling" (not downcycling) relates the assumed average 35-year period a material remains bound within certain items to the rate of production at the time. This results in a reclamation figure of approximately 80 % [21]. While this figure is impressive, it also indicates the need to deploy materials in a manner that enables better demolition and easier reclamation. Our current built environment is, however, not prepared for this.

The urban mine

The existing anthropogenic (human-made) material stock is often described as an urban mine (Fig. 5 and 6). In Germany, this stock contains 28 billion tonnes of bound material. Per year, the construction industry utilises 534 million tonnes of mineral-based raw construction materials [22]. Urban mining is rooted in the idea that existing buildings and infrastructure can be used in order to extract and produce construction materials for other tasks and, thus, minimise the consumption of primary materials as far as possible. The amounts of energy and labour required for this purpose are large. The term mining in this context is, therefore, appropriate. Similar to, for instance, coal mining, it is dangerous, labour-intensive and requires tremendous effort to extract materials and construction components from buildings that were actually manufactured and integrated for precisely the opposite purpose – long-term use. Initially, this practice seemed to be an adequate solution to the resource problem: We simply use what already exists and is no longer in use. We also perpetuate common linear thinking and action. On the path to a completely circular system for the building economy, the major effort urban mining requires given its low effectiveness can only be an intermediary step. We can slightly improve the existing linear system, yet we can't fundamentally change it. Many materials that are considered sorted by type in terms of their characteristics cannot be reclaimed in a way that preserves their quality, due to impurities and assembly methods in existing buildings that contradict circularity. This is why they are subsequently used at lower levels of quality.

Loss in quality in recycling
Glass as a material that technologically permits recycling is a fitting example for the loss in quality in recycling. In the case of glass used for consumer products, German industry production benefits from return and collection methods with strict colour separation and optimised products sorted by type. A recycling quota of up to 90% is achieved [23]. The average recycling share of flat-glass products in the building sector consists almost completely of production waste and is currently at 20% [24]. Due to impurities that occur during the construction and the reclamation processes, glass salvaged from the demolition of buildings fails to achieve the required threshold value for its reintroduction into flat-glass production [25]. As a consequence, only lower-grade reprocessing for use as foam glass or infill material is possible, and eventually, landfill at the end of the supply chain [26]. The same is true for the so-called cascading use of wood. Initially processed into a high-quality product, the material rapidly declines in quality in its second, third or fourth life cycle. Eventually, it is used as chips, shavings or sawdust for wood-based material panels or as admixture for other composite materials. Finally, however, the majority is incinerated or goes to landfill. This is also related to legislation, which partially prohibits declaring salvaged materials or construction components as material resources. If the material is salvaged in the process of a demolition and was not previously removed and recovered within a separate procedure, it is declared as waste and inevitably ends up in the previously described cascading use with a rapid loss in quality.

Downcycling and upcycling
Mixed or compound building components are particularly unsuitable for circularity if they include both a biological and a mineral, metallic or synthetic component. This prevents biological compostation (carbon cycle). Further, incineration leads to a product that needs to be treated as hazardous waste. Its chemical composition even prevents using the ashes as fertilizer for natural systems. It is no longer suitable for such purposes and requires waste separation. This process is known as downcycling. Wood, for example, loses its original quality or workability. Upcycling is the opposite of this process. Here, previously used materials are once more utilised in order to create high-quality products. Wood is also an example for this. Beams or boards salvaged for the manufacturing of exclusive pieces of furniture or facade elements tell stories of their previous use and, thus, display their unique identity. For this purpose – aside from the previously described processes of recovery that require tremendous amounts of effort – it is also necessary to maintain

5
The existing urban mine was not designed for the circular economy.

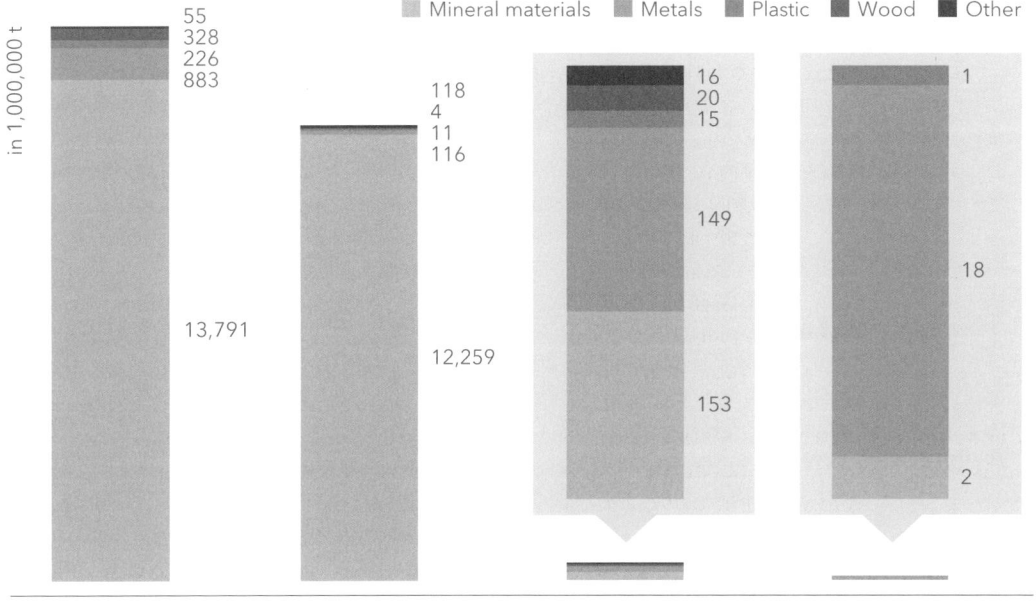

6
Anthropogenic material storage according to grade and material, in Germany, as of 2010 (ICT = information and communications technologies)

the original form of the specific materials or construction components. Upcycling, repurposing or reprocessing should be supported in the planning and construction phases by avoiding the application of coatings, adhesives or inseparable connections, or those that can potentially destroy materials.

Sorted by type

Maintaining the original condition of materials and ensuring during construction that they are not affected by impurities, mixing, damages or insolubility is summed up under the term sorting accuracy. It can serve as a standard for materials and construction. Materials sorted by type display their original and basic configuration. Neither are they mixed, amalgamated, coated or connected to or merged with another material with other characteristics. In the building sector, establishing methods of construction and assembly sorted by type is the indispensable precondition for the circular reclamation of materials while maintaining levels of quality. Many material fractions that are considered sorted by type regarding their characteristics are inadequate for reclamation in the manner of a comprehensive circular economy, due to connections or assembly methods that are non-circular [27]. Adhesive connections, foaming, wet sealing, synthetic mortar types, impregnations and grouting can prevent damage-free recovery sorted by type. They lead to the contamination of materials, with landfill or incineration the eventual consequence. Currently, only very small quantities of secondary raw resources extracted from buildings are suitable for reuse, repurposing and reprocessing at identical levels of quality. Most of them become so-called downcycling products, serving for the construction of roads, landfills or mines [28]. Published recycling quotas should be viewed critically, such as data on the use of construction and demolition waste released by the German Environmental Agency. Such materials and their use don't represent closed cycles, due to

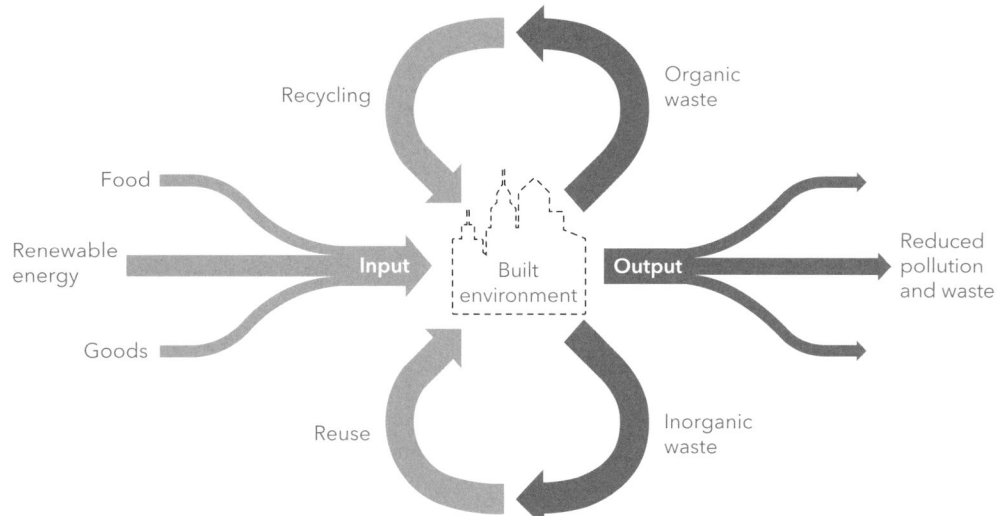

7
Circular metabolism in the construction sector according to Richard Rogers, Cities for a Small Planet, 1996

the dominant role of downcycling and since they contain residual material for energy recovery or byproducts from other waste streams [29].
Nevertheless, an increasingly extensive system of digital platforms is currently emerging that offers pre-used construction components or materials displaying great potential for simple reuse or repurposing according to current norms and standards. Challenges remain, including accessibility, liability, certifications and, often, complex norms and standards for application as required by the current market situation. The central task is to develop new construction methods and technologies in order to transform deployed primary and secondary materials into construction materials that are endlessly recyclable in significantly greater quantities than before, based on maintaining quality levels, preventing ecological harm, technologically enabling sorting by type and economic attractiveness. However, we need to begin with the creation of a new type of material stock that guarantees continuous and easy access and digital management, based on material passports and digital twins of our buildings (see "Digitalisation in the Circular Economy", p. 92ff.). This future-oriented perspective can serve to transform our buildings into opportunities for mining a diverse set of raw resources – a tremendous challenge that holds immense innovation potential.

The circular system
In principle, sustainable action encompasses three different strategies defined by three terms: Efficiency, sufficiency and consistency.

Efficiency
Efficiency is at the root of a strategy that aims to improve a known system within its own boundaries. This approach is very popular, because a certain spirit of innovation is ascribed to it, in terms of increasing the economic profitability of existing systems and, thus, boosting their yield or their productivity. Consumers often hear promises stating that given circumstances can be maintained, because they will be consuming less. This mostly entails technological process that soon reaches physical limits. These limits need to be negotiated on other levels, which mostly leads to impacts on the environment, such as in the case of the Dieselgate scandal of the 2010s. Efficiency, therefore, aims at the reduction of negative effects, either in economic or ecological terms.

Sufficiency
Sufficiency defines strategies that aim at doing less. It is, thus, rather unpopular, since it can coincide with abstinence, in order to achieve a lifestyle change and a reduction of resource consumption. However, what often tends to be misinterpreted

is that this "less" holds major potential for innovation. Light switches without the need for cables, portable lighting devices or even flexible heating elements based on the idea of creating comfort only in those places that are actually in use belong to this category.

Consistency

Consistency points out the need to harmonise the sum of our activities with natural circular principles and, thus, make them compatible or "friendly". As a strategy, it entails doing things differently by looking for new pathways that permit acknowledging the consequences of our economic activity within established circular systems, without disturbing the latter or negatively impacting them. Eco-effectiveness is how Michael Braungart and William McDonough [30] describe this approach in the context of their cradle-to-cradle principle. It refers to a mode of thought that doesn't prioritise the quantity of materials, products or items people consume, but instead, asks whether they are established within a circular system. The intention is to prevent their destruction through temporary usage and support their continued use within a circular system. The symbol that Braungart and McDonough chose for this idea is the cherry tree. It produces vast quantities of flowers, the majority of which never produce fruit. In systems optimised for efficiency or sufficiency, this would be considered outrageously wasteful. In a consistent system, however, the leaves falling to the ground contain nutrients for the soil that supplies the tree. As a result, year after year, cherry trees can grow new flowers, supplied by the open system of solar energy – according to Braungart's and McDonough's definition, a case of "intelligent waste". Therefore, consistency aims at reinforcing positive effects, which explains the term of eco-effectiveness. "Life under a cherry tree" is what the Belgian startup RotorDC called an exhibition of used construction materials in Brussels in 2019. The firm is dedicated to the reuse of construction materials and products. The notion that the building sector can operate within such a model aimed at consistency is very appealing. Germany in particular, a country with raw resources available only to a limited degree, could benefit from this enormously and substitute currently necessary imports, at least in part. However, current linear economic models (see p. 15ff.) are in absolute contradiction to the ideas circular systems represent (Fig. 7). We extract incredibly huge amounts of materials from the natural environment (primary materials). We process them in order to create products and items. Under current circumstances, mostly fossil fuels are used for this purpose, resulting in major emissions harmful to the climate. We exhaustively consume these products and dispose of them in landfills or burn them once they become so-called waste.

Circular economy

A circular system, unlike a linear economic model, is oriented on reuse, repurposing, reprocessing and recycling from the outset (Fig. 8). This implies that goods and products should be designed and manufactured in a manner that they never become waste and, thus, never lose quality or value. In this case, the value of the materials to be reclaimed plays a decisive role. The greater the profit outlook, the greater the possibility of reclamation. Therefore, building technology and political decision-making become important aspects in order to ensure and, at the same time, prioritise the concept of circularity. If societies across the globe succeed in the task of recognising every existing and every newly erected building as an available material mine and as a future circular material depot or bank, the concept of real estate could receive a completely new connotation and evaluation. Real estate could become a continuously changing, temporary aggregate state of reversibly assembled raw materials – a means of storing materials, the value of which can be measured and balanced in the long term. It is, however, necessary to shift the focus of interest from investment costs to life cycle costs. Long-term value chains need to be tied to the provision of loans and funding: The greater the potential for reclamation is, the more favourable the financing becomes.

Incentive systems for recycling-friendly action

Political decision makers can offer important and game-changing incentives and accelerate the related processes. For example, reclaimed or circular construction materials could be exempt from a carbon tax that will become necessary in the future. We need to discuss whether the same taxation system applies to secondary as well as primary raw resources. It is entirely conceivable that future remodelling and new construction will require the submission of applications for building permits in conjunction with demolition and reversibility plans and material data to be documented in central databases. Further, maintaining and preserving buildings instead of demolishing them needs to be prioritised and embedded in legislation (see "More than a Mine – Existing Buildings as a Material and Cultural Resource", p. 34ff.). Currently active political decision makers are increasingly focusing on the idea of a circular economy. However, they remain stuck on the beaten track of efficiency (making the existing system a little bit better). An example for this is the Resource Efficiency Program (ProgRess), enacted in 2012 in Germany. It only benefits lobby organisations and special interest groups that follow outdated notions of the economy as a linear system.

Approaches to a sharing economy

New approaches are required that are oriented on use in the context of a sharing economy, instead of placing emphasis on the ownership of materials and products. Initial models of such innovative circular economy approaches are already under development. For instance, carpet manufacturers have begun to adapt their product in such a manner that it consists of only one completely recyclable material which, in addition, no longer requires an adhesive connection to the floor construction below. Users pay for this carpet flooring based on how long they use it. If it needs to be replaced, the old carpet flooring is "collected" by the manufacturer and becomes subject to 100% recycling. Manufacturers are increasing their independence from the market of required raw materials and are becoming more resilient to fluctuations. Customers receive an "all-around carefree package", which is very appealing to wholesale consumers most of all, considering the time span from installation to replacement of surface materials such as flooring in buildings (see "Layering as a Circular Principle", p. 118ff.).

The so-called "pay-per-lux" concept is a further example of the notion of product-as-service. Here, the provider does not sell a lighting fixture, but, instead, a service: a guaranteed illumination quantity in a particular place. The selection of the lighting fixture, the lamp and even the energy supplier remain within the purview of the provider, just as the liability for and maintenance of all components. The aim of such a company is, thus, to deploy products that are as robust, energy-efficient and durable as possible, in order to pay as little as possible for repairs, outages or waste. Such economic considerations support the development of a diverse range of new models that all have one idea in common: We need to fundamentally rethink the construction of our products and adapt them accordingly.

Circular construction

Building today is an extremely challenging endeavour. There is a plethora of directives and standards that are intended to uniformly govern the technical and functional quality of buildings. This includes requirements for structural stability and thermal building-envelope properties with regard to fire safety, heat protection and soundproofing. In Germany, absolute insulation hysteria has led straight into an impassable jungle of product applications, the exact construction and composition of which remain largely unknown. For years, the building industry has been implementing a strategy of offering new and ever more complex product ranges and system solutions for entire application fields such as air tightness, heat-loss prevention, durability or soundproofing. As a result and in return, dedicated norms, codes and standards are established in order to meet specific

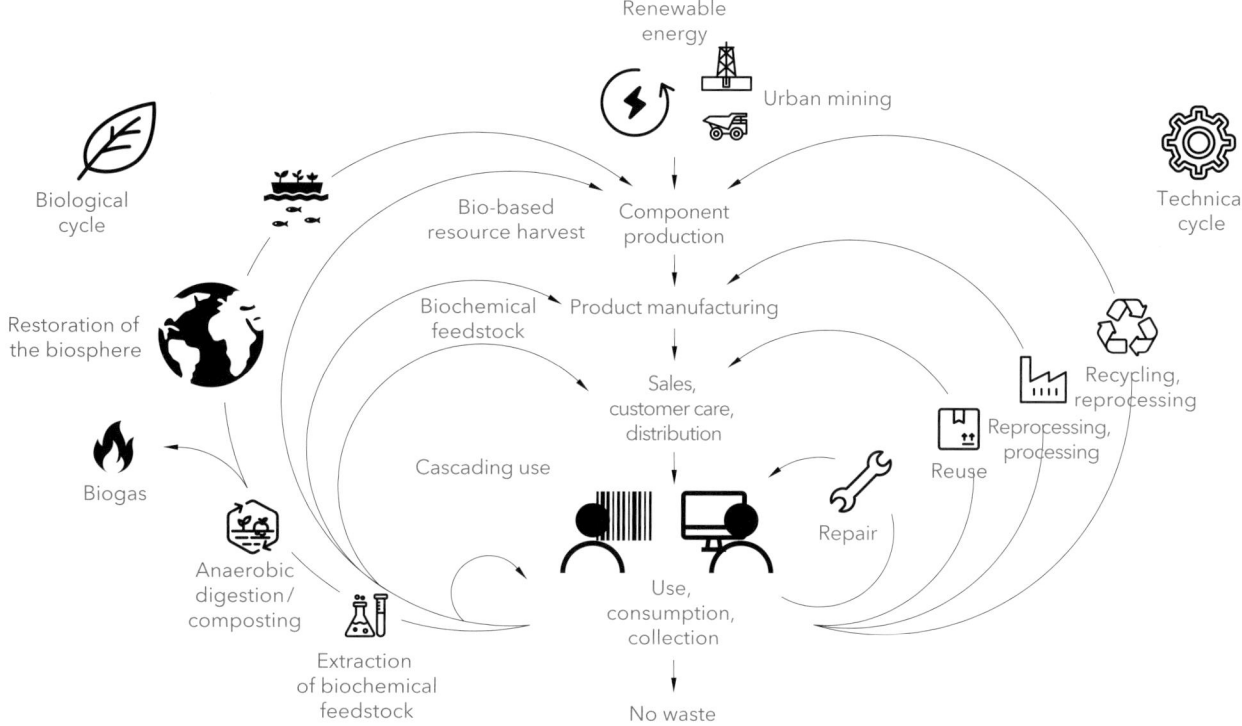

8

8
The circular economy butterfly diagram by the Ellen MacArthur Foundation proposes to control and maintain materials within separate technological and biological cycles, similar to the cradle-to-cradle principle according to McDonough and Braungart. This principle was, however, amended by the factor of time: The diagram contains cycles of different length and their characteristics. The name originates in the symmetrical image that resembles a butterfly.

liability requirements that each application of a particular system solution demands. A key reason for pushing this standardisation of building products to ever greater heights is legislation: In legal disputes, obligations stated in contracts for architectural services aimed at building according to the technological state of the art mostly amount to referrals to DIN norms. They are, for the most part, however – as previously mentioned – co-authored and promoted by the industry itself. In neighbouring countries, instead of standardised system solutions, state-of-the-art technology is understood as solutions proposed by professional experts that meet all demands placed on security and protection – planned, calculated and certified for a particular kind of application. There is a significant difference between the two that indicates how the desire for simple and sensible construction types for circular buildings can be fulfilled in a financially feasible way. Unfortunately, in the current contractual context, incentives for proposing the best possible solution in terms of circularity are yet to become relevant. Currently, they are still considered equal to the previously described requirements for protection and safety. However, the issue of providing protection and security for natural environmental systems should find expression just as strongly as those related to providing protection and security for human life. The consequence of thinking in terms of standardised building components is that there are still too many composite materials and system solutions lacking transparency in terms of their constituent parts. Simply put, their reuse, repurposing, reprocessing and recycling are impossible. Demolition is guaranteed to produce large quantities of materials that are, eventually, dumped or incinerated. Our very own architectural culture of tectonics and design has, for many years, led us to render connections and joinery invisible or conceal them as thoroughly as possible. Architects rarely make an effort to develop simple and self-evident detailed solutions motivated by the logic of construction anymore. Instead, details are culled from product catalogues to be assembled quickly, without thought, without making a professionally qualified estimation on which consequences this assembly process, based on predetermined connections and conglomerates, will actually entail while a building is occupied and beyond. This has increasingly led to adhesive applications, interconnected cladding and other multifunctional smoke-and-mirrors applications. All of them are representative of a throw-away mentality. They will inevitably force us into a dead end of resources that society can and must no longer tolerate and accept.

An interesting approach was presented by the Bavarian Chamber of Architects in 2022: A new building class titled "E" was introduced. The letter "E" refers to "experimental building" or "building simply", in German "Einfaches Bauen". The Chamber states: "This classification does not entail a strict catalogue of requirements. All approaches are considered experimental that aim at cost-efficient construction, exploring new residential types, or something different, something completely new. The important thing is that an architect is responsible for a qualified planning process. Simplifying codes enables clients and architects to coordinate standards, materials and details in such a manner that sensible, sustainable and affordable buildings are the result. Sustainability also calls for appropriate design and its coordination with user requirements. Limiting the project to the essential protection goals as stated by building law is one possible option. Grading according to Building Class "E" communicates this to provisional renters or owners, without any false pretences" [31]. This common sense, back-to-the-roots approach aimed at simple and comprehensible basic rules of design and construction can harmonise the importance of circular structures with the basic requirements of durability, permanence, security and protection. The great responsibility all those active in planning, architecture and design need to bear is in how we can design future buildings and how we can plan both their assembly and their reversibility. The aim is, on the one hand, to enable buildings to function as future raw material storage in a verifiable manner. On the other hand, such new developments can also establish a new aesthetic that corresponds to stated challenges and informs the production of architecture. For this purpose, common connection and assembly concepts and the related use of materials need to be consistently and comprehensively tested. They should also be oriented on the question of how we can build with materials that allow future generations to reuse them while maintaining their original quality. Further, architects need to develop technologies and easily comprehensible details while maintaining economic feasibility and aesthetic attractiveness. These additional challenges will leave their mark on the design of new and transformed buildings. It is an aspect that played a decisive role in the tectonic character and performance of construction components in the days of Carlo Scarpa and Karl Josef Schattner, especially with regards to building in an existing context. This further includes the effort in terms of time, energy and technology required to reclaim materials and construction components already. Is it only a promise on the material level of construction that, perhaps, materials used in the future will eventually display circularity? Or do we use materials right away that have already been verified as circular? We can already use materials today that are experiencing their second, third or even

Notes
[1] United Nations 1992
[2] www.helmholtz.de, 29 August 2016
[3] Crutzen/Stoermer 2000, p. 17f.
[4] Subcommission on Quarternary Stratigraphy 2019
[5] Hauff 1987, p. 46
[6] Milliman/Syvitski 1992, p. 525–544
[7] ECO Spezial. TV documentation, SFR 2014
[8] Peduzzi 2014
[9] Delestrac 2014
[10] Hassler/Topalovic/Grün 2014
[11] see note 8; United Nations Statistics Division
[12] see note 7
[13] Boden/Courtney 2010
[14] Leonard 2010
[15] U.S. Geological Survey 2021
[16] ISE 2019
[17] Deutsches Kupferinstitut Berufsverband e.V.
[18] Gäth/Eck 2018
[19] ibid.
[20] see note 17
[21] ibid.
[22] Umweltbundesamt: Bauabfälle 2021
[23] Bundesverband Glasindustrie e.V. 2014
[24] Neroth/Vollenschaar 2011, p. 555–635
[25] European Commission 2012
[26] Heisel/Hebel 2021
[27] ibid.
[28] Kreislaufwirtschaft Bau: Monitoring
[29] see note 22
[30] Braungart/McDonough 2009
[31] Architektenkammer Bayern 2021

fourth life cycle. In addition, architects need to deliberate on how to assemble and connect these materials and construction components. Usefulness during operation as well as reversibility need to be equally guaranteed.

Reversible, future-proof connection details
An entire range of connection details that were designed with reversibility in mind are already available on the market. This includes detachable, undercut dovetail joints in timber construction or metal hanging systems with bolted connections, which both guarantee reuse sorted by type for solid timber members. This further includes stacked timber systems that avoid the use of adhesives and are only connected by temperature-treated wood dowels and the friction they produce, in order to avoid mixing biological and technological cycles. New connection methods exist for masonry walls without mortar, prestressed construction systems for masonry block, reusable prefabricated concrete elements, dry seal compression systems, window construction systems without wet seals, light switches without the need for cables, wall cladding without colours, painted finishes or other forms of coating. Sealant systems for buildings that contain plastic sorted by type or insulation systems in the form of loose infill that can cover a building and its construction are already available on the market, just as dry construction systems for heating coils, interlocked raised floor systems, bolted installation systems sorted by type or removable roofing and framing. The range of solutions that can be easily acquired today or have been in use for centuries is extensive. Corresponding demand could lead to an expansion of this range in a targeted and rapid manner. However, thus far, planners have rarely tended to focus on these solutions.

The book in your hands is intended to inspire a discussion on a new understanding of building. It aims at providing a basis for creating circular material storage capacities respectful to the materials used for all future generations. It is our duty to critically enquire on how our discipline operates and to rethink and redesign tried-and-tested, self-evident construction solutions of the past. This approach can also generate a new type of aesthetics by combining the related necessities on the one hand and the important aspects of usability and adequacy in terms of architecture and construction on the other. Sustainability and simplicity in building, in the spirit of this book, should go hand in hand as a matter of course. Their interrelation can contribute to a circular building practice of the future. This publication presents theoretical, historical and visionary basics and principles. In its catalogue section, it includes initial solutions that support simple realisation. We would like to call for a reorientation of architectural teaching. Change is urgently required. The related development of intelligent, sustainable and well designed solutions on the levels of projects and their realisation can take place in cooperation with professional organisations, building and construction trades, as well as the industry.

HISTORY AND STATUS QUO

History of the Building Culture of Reuse

Hauke Horn

Preservation and continued building

For a long time, a linear model was dominant in architectural historiography, similar to the construction industry: New buildings most of all received acknowledgment and awareness. To a major degree, new forms and styles defined the image of architectural history, aside from technological progress. What was overlooked from such perspectives was that reuse and continued construction were important historic forms of building culture. In terms of culture, religion and politics, many important buildings of the past, such as Aachen Cathedral (Fig. 1), Trier Cathedral or the Church of the Holy Sepulchre in Jerusalem are not constructions adhering to a pure style. Instead, they are structures that developed over the course of many centuries, where old and new parts were intentionally connected and combined. In the construction of medieval castles, development occurring over time was the rule, not the exception (Fig. 2). Significantly, the picturesque appeal of castles that is so popular today can be traced back to successive steps of addition, alteration or continued construction of these edifices. The historiography of modernity negatively evaluated the preservation of older components. By doing so, it misjudged the value and meaning of buildings that developed over the course of history. In many cases, as recent research shows, the history and traditions of a place were intentionally and visibly preserved [1]. This is why buildings were incrementally developed around an old core that fostered a sense of identity, never to be demolished. The Palatinate Chapel of Charlemagne in Aachen (ca. 800), the legendary palace of Empress Helena in Trier (4th century), or the legendary Tomb of Christ in Jerusalem (upon which the first building was erected in the 4th century) were literally the measure of all things in terms of additions, alterations or reconstructions. Intentional contrasts between the old and new building fabric were employed in a targeted way. The aim was to distinguish the old according to the image of the new as something decidedly "ancient", and vice versa. In this context, "old" was understood as "time-honoured". Viewed from today's perspective, such concepts of preservation and continued building are attracting renewed interest in relation to societal demands for sustainable architecture. On the one hand, traditions of place are preserved that also foster identity. On the other hand, there is virtually no better strategy for waste prevention. In the case of buildings that are not demolished, we do not have to consider recycling at all. Continued use is the most efficient form of reuse.

Spolia

Aside from concepts of preservation and continued building, the reuse of old construction materials, elements and components within new contexts plays an important role in architectural history. In the city of Rome, the medieval and early modern practices of building can be viewed as prominent examples for a

1
Building parts from the early Middle Ages to the Baroque era (core, 9th century), cathedral (former minster), Aachen

2
Building fabric developed across many centuries (oldest existing elements 13th century, newest 20th century), Schönburg, Oberwesel

3
Medieval house with numerous antique spolia, Casa dei Crescenzi, Rome

4
Antique spolia columns in the 12th century central nave, Santa Maria in Trastevere, Rome

5
13th century portal zone with Byzantine spolia, St. Mark's Basilica, Venice

historical form of urban mining (Fig. 3). For more than a millennium, Roman builders repeatedly reclaimed resources from the colossal legacy of classical antiquity. There is hardly a medieval building in Rome that does not feature antique spolia [2]. On the one hand, old stones and fragments were pragmatically reused for the masonry walls of new buildings. In part, especially relevant components became visible spolia integrated in the architectural concept. For instance, when the early Christian Basilica of Santa Maria in Trastevere was rebuilt from the ground up in the 12th century, the colonnades of the central nave were intentionally erected using historic columns that were significantly different from one another in terms of colour, proportion and sculptural capital form (Fig. 4). This approach clarified to observers in an awe-inspiring manner that valuable construction elements of antiquity were reused here. At the same time, using them referred to the early Christian tradition of the site. Hence, reuse was a central element of the architectural concept. The 20th-century compendia of architectural history avoid medieval Roman buildings precisely for this reason: They did not have enough in common with the new edifices of central and northern European Gothic style – a problem of perspective in terms of scientific theory. However, one of the most famous examples of the extensive use of spolia is located in Venice: The entire facade of St. Mark's Basilica is decorated with columns, incrustations and sculptures seized by the Venetians during battles for dominance over the eastern Mediterranean region against the Byzantine Empire (Fig. 5). This means they were appropriated as trophies, to be put on display on Piazza San Marco in the heart of the city, in order to visually represent the victories and substantiate the related political claims of the maritime republic. In the present age, this kind of reuse would be considered problematic, at least in ecological terms. The spoils would have to travel long distances across the sea, from Constantinople to Venice. In this case, preserving natural resources is counteracted by the significant amount of

6
Aedicule (16th century) with asservatia from the old St. Peter's Basilica (4th century), St. Peter's, Rome

7
Magdeburg Cathedral
a 13th century cathedral choir, featuring column shafts in second reuse: asservatia from the old cathedral, brought to Madgeburg as antique spolia in the 10th century.
b Cathedral cloister with antique capitals from the old cathedral, second reuse as column bases in the refectory

power required for transport. For medieval sailing ships, this power was still completely climate-neutral and renewable (wind power). Today we might fear that a significant amount of fossil fuel would need to be consumed. We can conclude that the shorter the transport routes, the more energy-efficient and sustainable the practice of reuse will be. The best possible energy balance is, thus, achieved by the reuse of building materials and components that are acquired in the vicinity of the actual construction site. In such cases, opportunities most of all exist in the remodelling and conversion of existing buildings and contexts.

Asservatia

Building materials and components can be reused within the same structure, yet in a new context. They will be described as "asservatia" in the following [3]. Due to their origin, they can be distinguished from spolia, which were taken from other buildings. Architectural history contains countless examples of employing asservatia. It hardly comes as a surprise that churches that developed over time, such as Trier Cathedral or the Church of the Nativity in Jerusalem, feature many asservatia. Yet, many edifices that are considered new also include reused parts of precursor buildings. For example, during the construction of the new Basilica of St. Peter in Rome in the 16th century, many of the columns of the elongated nave of the early Christian church, Old St. Peter's Basilica, were reused to frame aedicules and portals in a reverential manner (Fig. 6) [4]. Even the famous spiralling columns of the legendary tomb of Saint Peter, which had already been introduced in the early Christian church as spolia, were re-reused in the aedicules surrounding the Renaissance dome. Thus, Old St. Peter's Basilica became part of the new church. The new construction of Magdeburg Cathedral in the early 13th century is a further interesting example of the repeated reuse of building components (Fig. 7). The splendid and colourful natural stone columns, on display in the choir and placed on pedestals, were acquired from Italy in the 10th century, at the behest of Otto the Great. Originally, these valuable spolia likely supported the walls of the central nave. They became asservatia following the new construction of the cathedral beginning in 1207, when the magnificent column shafts of the demolished cathedral were impressively integrated into the new cathedral choir, thus denoting the founding of the church by Emperor Otto. The column shafts still stand there today, demonstrating their second reuse, at least. The capitals that proved unsuitable for placement in front of a wall were reused again as bases within the refectory of the cloister (Fig. 7 b). In any case, reuse in a new context in the case of capitals and other sculptural building elements that no longer permit use according to their original function constitutes a common practice. One example is the holy water font in 14th-century Essen Cathedral following the new construction of the early Romanesque long nave (while preserving old building elements). An early Romanesque capital from the old church nave was repurposed as a column base (Fig. 8). The reuse of building materials, components and elements was, however, not limited to highlighted artefacts

Notes
[1] Froschauer 2020; Horn 2015; Horn 2017; Kappel/ Müller 2014; Hoffmann 2005; Albrecht 2003; Meier/Wohlleben 2000
[2] Spolia are considered as "building elements that are reused intentionally and, thus, generally put on display", Meier 2020, p. 9
[3] Horn 2017, p. 61-77
[4] Bosman 2004
[5] Gruber 1937

8
11th-century capital from the old central nave as pedestal for the 14th-century holy water font, cathedral (former minster), Essen

9
Floor surface including reused rubble, St. Kolumba, "Madonna in the Ruins", Cologne, 1950, Gottfried Böhm

10
Wall with visibly exposed aggregate consisting of crushed rubble stones, St. Anna, Cologne-Ehrenfeld, 1956, Dominikus and Gottfried Böhm

that observers were able to recognise as "old" within a design concept that intentionally put historic layers on display. During times in which building materials were more expensive than labour, their reuse on site was, simply stated, a matter of course in economic terms. It is now common knowledge that many historic buildings feature materials that were retrieved from older layers. A descriptive example of this practice is the construction of a new long choir in the Freiburg Minster in the 14th and 15th centuries. At first, perimeter walls were built around the existing Romanesque apse in order to be able to continue religious service without interruption. The old apse was eventually demolished, yet the masonry stones were used to build the new choir walls. It is hard to imagine a shorter building material cycle.

Spolia and asservatia in the modern age
Established practices changed on a large scale once industrial production methods emerged in the 19th and 20th centuries, including efficient means of construction combined with inexpensive transport opportunities. Karl Gruber, professor of architecture in Darmstadt, expressed his criticism in 1937, stating that "brickworks located next to the construction site were closed in order to use giant concrete slabs that were driven to the site from production facilities many kilometres away" [5]. Yet, modernism was far more open to the option of reuse than its proponents are willing to admit. The non-load-bearing exterior walls of Le Corbusier's masterwork in Ronchamp, for instance, consist of the old stones of the destroyed preceding church. The difficult accessibility of the site, in earlier times described reluctantly as the motivation for the reuse, can hardly be the actual reason, most of all considering the sheer mass of concrete that had to be delivered in order to create the spectacular roof. In Germany, the reuse of old building materials – as spolia or as asservatia – was a self-evident practice in the years of rebuilding after World War II. Again, remarkable examples can be found in sacred architecture. In 1950 Gottfried Böhm, for the chapel "Madonna of the Ruins" in Cologne, his debut building, not only integrated the masonry wall of the Gothic church in his new structure, but also used the rubble and debris on site in a creative manner for the plinth walls and the floor mosaic [Fig. 9]. The Gothic buttresses even received new architectural relevance as ornamental decoration of the eaves. In the meantime, the chapel itself has been integrated as an older historic layer in the Kolumba, the new museum for the archdiocese, built by Peter Zumthor and completed in 2007. If old building materials can no longer be reused, they can still be recycled. Once more, a church in Cologne, designed by Gottfried Böhm together with his father Dominikus in the post-war years, offers an instructive example. The aggregate used for the concrete walls of St. Anna consists of crushed stones from ruins [Fig. 10]. By exposing the aggregate on the visible surfaces, the ruins were made evident and elevated to the level of a design concept.

Summary
Clearly, reuse and continued building represent long-standing traditions in architectural history. Yet, for a certain period of time, this fact was acknowledged only reluctantly. Against the background of current global challenges such as climate change, environmental pollution and the depletion of natural resources, these old concepts are once more coming to the attention of architects. They comprise a rich repository of approaches and inspirations. Great potential for architectural design is offered by the double sustainability of reused building materials, components and elements: Aside from the ecological sustainability of recycling, the cultural sustainability of preserving history and identity plays an equally important role.

More than a Mine – Existing Buildings as a Material and Cultural Resource

Christian Holl

Sometimes it seems as if the construction and real-estate industries were experiencing last-minute panic. As if they were guided by the notion that it was necessary to demolish buildings as quickly as possible, before those actors prevail who demand ever greater obstacles to allowing demolitions from happening. All the while, current German funding policy still gives preference to new construction over the existing context. When it comes to new buildings, support is given to what happens to be the current technological state of the art [1]. On federal levels and from 2023 onwards, it is also true that the provision of funding grants for new construction will be reduced in scale for the benefit of renovation and conversion. However, this is simply not enough [2].

In the future, demolition should be subject to approval in every and all instances. This is what Architects for Future demand, since: "Demolition has until now and in most cases been exempt from approval. No assessment takes place whether valuable building fabric – which supports renovation – should be demolished or not. Under consideration of the energy demands and the emissions spanning the entire building life cycle (construction, operation, demolition), renovation is preferable to demolition and new construction without exception" [3]. Instead, we are confronted with daily news reports on the demolition of valuable buildings that has already been decided or is still being debated, such as the Billwiese student dormitory in Hamburg, completed in 1965, historically listed and designed by Heinz Graaf and Peter P. Schweger [4], the Braunschweig city hall from 1965 by Heido Stumpf and Peter Voigtländer, also listed [5], and the Cantian Stadium from 1951 in Berlin by Rudolf Ortner – the latter scheduled for demolition after all [6]. On the other hand, there is encouraging news of successful interventions on behalf of, e.g., the 1971 Potsdam computing centre by Architekturkollektiv Sepp Weber, which will likely be preserved [7]. The fact that we consider this a success indicates the minor degree to which the preservation of buildings is self-evident. Similar to Potsdam, examples of successful preservation endeavours are often based on the significant effort of individuals in preventing the demolition of buildings, equally often without acknowledgement.

Circular economy means: Leave it be
Beyond historic preservation and the struggle for the continuation of non-commercial uses, we have not even begun to talk about what else is scheduled for demolition (Fig. 1). Architects for Future investigated that, in Germany between 2015 and 2019, "per year on average around 1.9 million m^2 of residential space and 7.5 million m^2 of other usable space were demolished – without assessment or whether the existing buildings in their entirety or, at least, individual components permitted ongoing use in the future. Existing potential for continued building and use was not exploited" [8].

1
Everyday life in German cities: Valuable building fabric is demolished. Shown here: Stuttgart

Even the new ideal of circular economics can only help if a clear distinction is made between recycling and downcycling. The high rates of recycling that the EU envisions are, thus, to be viewed with caution. In her research, Annette Hillebrandt, who teaches at the University of Wuppertal, focuses on urban mining and material cycles in architecture. She was asked about her thoughts on the fact that all non-residential buildings covering more than 2000 m² have to meet a recycling quota of 70% from 2027 onward: As long as no distinction is made between recycling and downcycling, "we will easily meet the 70% requirement" [9]. Hope appears justified when considering that the first new buildings have been completed in a circular manner. However, they at best comprise a small part of a possible solution, because they leave the question unanswered on how to deal with existing structures: 85% of buildings existing today in the EU will still exist in 2050 [10]. The glory of new circular edifices reinforces the belief that starting to build differently at some point is enough, in this case, for a change, in a circular manner. Without a doubt, it is an important step forward. Still, we should be realistic about it and contextualise it accordingly. Compared to the sheer mass of new buildings, the impact of such projects has the appearance of a plea. They merely offer potential, making promises that others will have to keep. It will be decades until the new buildings of today are demolished and their construction components are reused, if at all. If we really want to take the circular economy seriously, we need to understand that it isn't only about building new houses in a circular manner. Neither does this suggest that we simply have to examine the existing context and its building fabric in such a manner as to identify how best to utilise it to develop something new. Nor does truly thinking in terms of circular economics mean to turn demolition into a sort of "tetris project" that demonstrates how material from a demolition can be reused in a new building to the greatest possible degree. Instead and simply put, it means simply leave as many buildings as possible standing for the time being. Many of the materials and products used within buildings, most of all concrete, can only be maintained on the same levels of quality – meaning without the renewed application of water, energy and other materials, such as binding agents – if the buildings they were used for are maintained. In other words: The new ideal of circular economics specifically prevents us from conceding further to the artificially accelerated demand for new items. In the context of building, it does not liberate us from affording the existing context a much greater degree of attention than is currently the case. When it comes to building in an existing context or fabric, circular thinking will need to – and can – prove its merit, through additions, conversions and adaptations. Yet, it is only an element of the great challenge we are currently facing: Truthfully, in 2050 more than 85% of buildings existing today should still be standing.

2
In 1953 the Federal Court of Auditors in Frankfurt a.M. was completed. In the 1990s it was historically listed (as opposed to an addition from 1955). For the conversion into a hotel, large parts of the existing building were demolished, only the facade remained of the historically listed building.

Desperately seeking creativity

We are called upon, at least compared to the practice of the past decades, to actually achieve something new. More precisely, to understand the new as a continuous form of appropriation and adaptation of the existing building fabric. This applies in particular to underappreciated buildings: Often disregarded everyday structures, from detached homes to car parks, from supermarkets to shopping centres, from simple warehouses to factories and even residential construction from the era of postwar modernism, including large-scale housing estates. We can only succeed in this task, if preserving the existing context isn't understood exclusively as a question of technology, once the reasons for maintaining existing structures are no longer an admission that new construction is the actual object of desire. All too often, as in the case of the Billwiese student dormitory, the reason given for the demolition of existing buildings it that they are "simply no longer up to date" [11]. At the end of the day, lazy minds and lacking imagination are the major forces driving a demolition, aside from costs. We therefore urgently need to formulate a "rebuilding code" in order to propose a different and more generous kind of funding policy, to develop simpler methods of amortisation for renovation projects and, hence, counter the trend.

The task at hand is also a cultural challenge that should not be underestimated. To appreciate what seems to be insignificant – this actually is a megaproject, particularly because it shouldn't be understood as a schematic "one size fits all" solution that draws from standardized models, while neglecting what actually constitutes the existing context and its social relevance. Instead, we need something else. Something that the architect Niloufar Tajeri has described as "precisely developed design tactics that address the construction-related idiosyncrasies of the existing context and reveal its capacity for adaptation – instead of schematic execution, an analysis of the object and the financial resources of the residents in line with their housing needs" [12]. Only this way can renovation be reconciled with a socially responsible rental practice, sensible approaches to energy renovation and the provision of affordable residential space.

The aim is to acquire an understanding of the existing context no longer contingent upon juxtaposing old and new or measuring one against the other, or, as Boris Groys states, to facilitate the new by preserving the old [13]. Viewing the existing as a testimony worthy of preservation should remain an important exception. Once we view the entire existing stock according to whether it should be historically listed or not only leads to the devaluation of the existing: It is too easy to grade it as a testimony that isn't necessarily worthy of preservation (Fig. 2). We should therefore raise the question whether assistance from historic preservation authorities can be expected or

3
Conglomerate of old and new:
a Split, Croatia
b Alte Pinakothek, Munich, reconstruction and alteration: Hans Döllgast, 1952–1957

even hoped for. Not only because these authorities are under pressure [14], also because their instruments and their mission can neither capture nor protect the entire mass of existing structures, nor should this even be considered their mission. However, what we should be able to expect is a critical discussion on how historic preservation authorities excessively establish the original condition of a building as the measure of their evaluation. A position that, by the way, actually prevented placing the Marlene Poelzig house in Berlin or the Schmitthenner villa in Stuttgart on the historic preservation list. Both had been remodelled. Regarding the Poelzig house, "this exemplary building of architectural modernism was reshaped according to the Heimatschutzstil" [15]. In the case of Schmitthenner, the architect himself was responsible for the changes, putting his different creative phases on display in a condensed manner. This was eventually considered unsatisfactory and viewed as an "interesting, yet not necessarily a representative example of the architect's creative oeuvre", as stated by the historic preservation authority [16]. This is why it is important to integrate continued building in the logic of historic preservation, not least with reference to particular examples. The aim is to make comprehensible the effects of processes of reshaping, what they demonstrate and, thus, what they communicate, thereby creating new pathways to an intrinsic understanding of history (Fig. 3).

This does not mean that the vast majority of existing architecture must no longer be altered in any way, form or manner. Neither should we view the existing context exclusively in terms of whether it is of value to the present, simply because it tells us something about the past. In a more substantial manner than is currently the case, architecture should be understood as an open system comprising elements that permit repeated and renewed arrangement, rather than an immutable opus to which any successive change in principle results in a perceived decline in quality. We should no longer ask what we want to have, but instead, what we can achieve with what we already have. Small, precise interventions are required, as well as inventive proposals. What we need is real creativity.

We have much to gain

Understanding architecture in this manner is also a challenge to the way architects see themselves. Similar to the modernisation of residential buildings, it is a challenge that cannot be addressed by schematic models. Once architects no longer aspire to create seemingly "timeless" edifices aimed at avoiding the need for successive alteration, amazing opportunities arise. Dealing with the existing context permits utilising the richness of the existing building stock as a means of expression, deriving from it forms of ornamentation, or risking new combinations that designers deny themselves based on self-imposed demands of needing to

4
Conversion, Werk 1, factory district (Werksviertel) near Munich East station, 2019, Hild und K

create timeless designs, simply because they would otherwise be considered faddish. Such approaches become meaningless once we begin to work with the existing context. What we need for this purpose are "architects that match the type of a tinkerer who defies conventions" [17]. Placing emphasis on the existing context can enliven architecture itself, allow it to gain new possibilities of expression. Changes undertaken by users shouldn't be viewed as damages, but instead, as explorations of the potential that architecture offers them (Fig. 4). Eventually, we need to abolish the tether that binds form and quality, a tether that has made discussions on architecture and urban design so agonising and exhausting for quite some time. Either/or discussions that invariably deny particular forms of architecture any appreciation while affording it to others with similar prejudice miss the challenges of the existing context by a wide margin. In which style and with which formal preference a building is erected, according to which idea of the city a district is planned, cannot be the basis for asking how to assess their quality. The basis for a future-oriented, direction-setting design approach to existing conditions and their multifaceted characteristics is to encounter the built fabric impartially and curiously, in order to secure existing qualities or establish new ones with a minimum of material usage. It is no longer sensible to give preference to one model of the past above others. Whoever continues to hold on to such notions, for instance by reducing the European city to its urban core and the Wilhelmian-era districts, meaning a very narrow part of the overall city, fails to understand the challenges posed by the entirety of the existing fabric: The major share of the existing stock is denied its urban character. This limits dealing with existing conditions to responses of individual preference. It also prevents understanding the logic and the processes that have led to alternative forms. As a result, the range of options required for deliberating on the existing fabric productively and creatively is severely constrained. Hence, in such cases, demolition seems the only justifiable option. Finally, people, the environment, the buildings they inhabit and which they are familiar with are also devalued in a sweeping manner (Fig. 6). However, there is an opportunity to discover new qualities by reflecting on the existing stock. Using the existing context as the basis for developing something new requires the cooperation of many actors and demands skilled craftsmanship. The results originate in the process and whatever happens to be available. To maintain what exists means to accept how it was appropriated over time. Additions and expansions are also part of the existing conditions. Establishing architectural uniformity is more often than not declared desirable, yet neither does it do justice to the existing fabric, nor to its history. Architects should envision a moment in time when their new construction has become

5
"A story that never ends!"

the existing context. The challenge of dealing with this context is merely shifted in time, into the future. Controlling what happens to and within these existing buildings is only simple in the short run: Buildings and sites are often optimised in relation to a certain use that take place there. People become subordinate to notions of action and movement that are structured by architecture. In the majority of cases, this is intentional and has been so from the beginning. However, such notions are valid only to the point when buildings begin to display patina, when the framework conditions change within which buildings were erected, and – this is precisely the quality of the existing building stock – a new use or purpose is layered upon the original use and design intention, creating those productive voids that planners were not able to foresee. To interpret this circumstance as a quality and respond to it accordingly is a benefit that architects should not underestimate. They could, in greater ways than functionally optimised and specialised new buildings permit them, focus and deliberate on developing space as a quality in itself. Once this quality has been recognised, we can move forward in a manner that, in terms of building, values the existing building stock more than the new. An existing building would no longer be a deficient new construction, but instead, new construction would be the unfinished existing condition – because it has not yet proven its value as a resource, not yet been able to show that it can cope with change and that, based on mutations and transformations, it is able to enrich everyday life (Fig. 5).

Notes
[1] Deutsche Umwelthilfe 2022
[2] Schröer 2022
[3] Architects for Future 2021
[4] Berkemann 2022a
[5] Berkemann 2022b
[6] Dittrich 2022
[7] lernort garnisonkirche 2022
[8] see note 3, p. 11
[9] BDA-Denklabor 2022
[10] European Commission: European Green Deal 2021
[11] Berkemann, Braunschweig city hall/town hall 2021
[12] Tajeri 2018, p. 206
[13] Groys 1992
[14] Deutsche Stiftung Denkmalschutz 2022
[15] Kasparek 2021
[16] cited from Sellner 2021
[17] Confurius 2017, p. 102

6
Reanimating interventions undertaken by residents and users

Vernacular Architecture

Peter Hoffmann

A discussion of traditional or vernacular forms of construction and architecture is of significant importance to building sorted by type. This is particularly the case due to the seemingly advanced character of multilayered and complex building methods employed in contemporary projects. In his book "Vernacular Architecture", Christian Schittich defines the term used in the title as "the way simple people build without the assistance of professional planners" [1]. Ever since the so-called Neolithic Revolution, ways of creating things started to emerge as people began to live in permanent settlements. Across the millennia, people also developed methods of building and construction. From the Neolithic period onward, these methods were part of an exclusively oral tradition and continuously improved throughout history. A look at traditional buildings in the context of the recently arising sensibility for how we deal with the environment astonishingly reveals that the principles of reduction and reuse formulated in the German Closed Substance Cycle Management Act were already fully self-evident millennia ago.

Reducing building materials and land use
Before people began to stay in permanent settlements, they lived as nomads, moving from one place to the next (Fig. 1). Dwelling in tents and cooking over an open fire, they left scarecly any traces behind. Building materials were a valuable resource and would routinely be disassembled and reused. Today, however, many buildings are supported by pile foundations that will continue to exist long after humanity itself expires. The traces that people leave behind in their buildings are a testament to cultural developments. They are fascinating from an archaeological perspective, yet they should also inspire reflection in terms of what future generations will discover. Against the background of the increasing accessibility of inexpensive construction materials in the context of progressing industrialisation since the previous century, it became less economically feasible to reuse building materials embedded in existing structures. Instead, in the majority of cases, they are either reprocessed or dumped in a landfill. In 2019 construction and demolition waste accounted for about 55 % of overall waste accumulation in Germany [2].

A further field of activity in which vernacular architecture can serve to inspire is how we deal with built-up areas. Nomads and early settlers alike lived together with their families and animals within the tightly confined space of tents and, much later and for the most part, in single-room dwellings. The reasons for the limited size of these dwellings were, typically, poverty and scarcity of resources. From today's point of view, the question on how to sensibly deal with spaces for living is gaining relevance once more. The per capita living space in Germany in 2020 was 47.4 m², compared to 34.8 m² in 1990 [3]. The reasons for this are multifaceted and, most of all, rooted in social and typological structures. Using built-up space sparingly, however, automatically leads to the reduction of material and energy resources required for the production, operation (mostly heating and cooling) and demolition of buildings. Traditional houses occupied by members of local communities were mostly subject to the economic necessities of resource minimisation. Other aspects of sustainability – such as social and ecological demands – were mostly considered subordinate to the financial aspects. For all those involved in building to become interested in learning from the traditional ways of dealing with resources, we should first deliberate on the efficiency and sufficiency of architecture in relation to questions on how ecologically friendly, how supportive of equal opportunity, how relevant in economic terms it is. We can answer these questions only by rejecting linear notions of the economy (waste creation) in the building sector.

Accessibility and change of local building cultures

Prior to the onset of the Industrial Revolution in the 18th century, the appearance and image of houses and cities, as well as the underlying methods of construction were defined by building materials available in nature and the resulting achievements of local building cultures. Following the emergence of related tools, materials and technologies, construction methods were advanced and adapted to local conditions throughout history. The design-based integration in the local context was self-evidently facilitated by available materials and traditions. People learned early on to orient their dwellings in relation to the topography and, for instance, created cave villages, such as in Matmata in Tunisia (Fig. 2) or in Shaanxi in China. The aim was to reduce exposure, to consume as little material as possible and to preserve available arable land. In cold regions, people quickly understood how to use solar power. In warm regions, they built dwellings in a manner that allowed adequate shading and ventilating interiors. The examples of the cave villages show how similar methods of building developed completely independently from one another in very different regions of the world – as long as geographic and geological conditions were similar. To this day, in many less affluent parts of the world, people still use natural and locally sourced materials for construction. In part, they are less expensive and available to a greater degree than synthetic construction materials or those from remote sources. Nevertheless, industrialisation and globalisation contributed to a significantly accelerated form of change and adaptation, compared to earlier ways of building. Prior to the onset of industrialisation, fewer construction materials were available in the required quantities and varieties. Increasingly economically feasible production and transportation methods led to the global distribution of industrially produced construction materials, which increasingly replaced traditional materials and craftsmanship-based means of building. In particular due to their superior weatherproofing characteristics, the use of more durable construction materials such as sheet metal or concrete also became desirable in less affluent regions. Building with natural construction materials required and continues to require relatively high maintenance. Otherwise, the materials used, in an undesirably rapid manner, return to their natural cycles, due to rot (e.g. wood, reeds) or erosion (loam) (Fig. 3, p. 42). Structural building protection measures aimed at waterproofing play an important role for all construction types. Steep roof surfaces with waterproof roofing materials and deep overhangs, for instance, offer protection to the construction components they cover. A plinth (e.g. made of

1 Yurt in western Mongolia. The structures were designed for disassembly and renewed assembly.

2 Berber underground dwellings, Sidi Driss, Matmata, Tunisia

natural stone) prevents rising moisture and splash water from impacting walls from below. Curtain-wall facade construction types can protect a loam wall from erosion. The guiding idea remained the same across the centuries: The intelligent layering of materials based on their climate- or construction-related characteristics found repeated use where it was sensible to do so, given they were reversible and locally sourced (see "Layering as a Circular Principle", p. 118ff.).

However, the growing demand for comfort and durability in the construction field, most of all in relation to thermal building envelopes, generally leads to ever more complex components. This triggered massive processes of change in the industry. From the early 20th century onwards this has also contributed to the development of composite building components that cannot be dismantled sorted by type. Eventually, once the use of these components terminates, the demonstrable results are gigantic quantities of hazardous construction waste. Thermal insulation is only one example. In order to reduce thermal loss and power demands in buildings, their envelope is increasingly becoming the focus of attention. The use of economically practicable thermal insulation composite systems is, at the same time, the least circular method of insulating a structure and, thus, the most intensive in terms of waste creation. This is often overlooked. Without needing to discuss the technical and construction-related issues of many of these systems in depth, obviously most of them can hardly be dismantled sorted by type. In the case of curtain wall facade systems with multiple layers and back ventilation, reversibility is indeed possible. On the one hand, the omnipresent question on how to consume resources in a responsible manner requires us to minimise the energy footprint of buildings. On the other hand, we need to understand insulation as an element of a cycle of reversibility and reusability. The examples presented on page 130ff. show approaches that meet both requirements in terms of materials and connections. However, future research and innovation remain indispensable.

Vernacular building materials

The design and construction methods for vernacular buildings, aside from traditional craftsmanship that has been passed down through the ages, substantially depend on local accessibility to building materials. In return, this defined the appearance of individual structures, even entire cities. The materials have been known for centuries and can be distinguished according to categories of renewable or mineral building materials.

Renewable building materials

Without a doubt, the most well-known renewable building material is wood. By depositing lignin within the cell walls of plants (lignification), these turn into wood and provide the material with its compressive strength. As a result, wood became the most universally used renewable building material. Wherever wood was available, timber buildings dominated the regional image of settlement spaces for a long time. The material can be used for load-bearing structures, as facade siding and cladding, as wall and roof sheathing, or for interior outfitting. The example of wood shows very well how regional accessibility and the appearance of vernacular architecture are interrelated. Wood grows significantly better in areas with high precipitation than in those with low precipitation. Log cabins, for instance, were built where long and straight growing wood was available. Such cabins are still built today in the form of modern log structures, most of all in Alpine regions (Fig. 4). These examples show well how technological progress influences the traditional appearance of buildings: Prior to the development of large saws, log cabins were, for the most part, built by using entire

3
Mud huts in Ghana, exclusively built of natural materials
4
Traditional log construction, Grisons, Switzerland

3

4

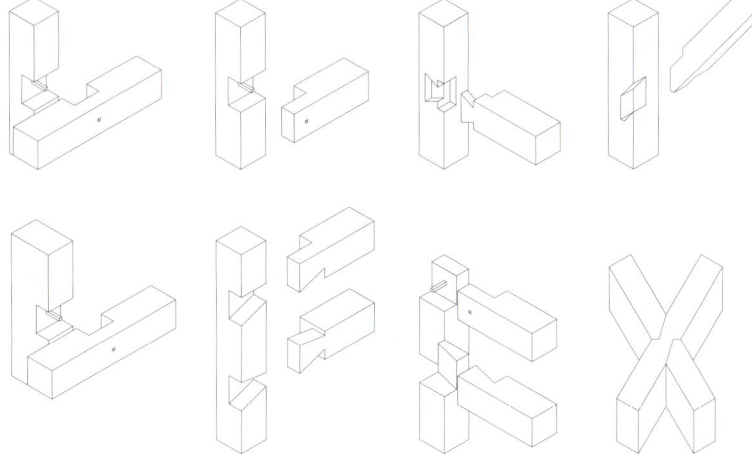

5

uncut logs. Today, the individual timber members of modern log structures are set in computer-controlled CNC machines with millimetre-level accuracy. The material is used in a resource-saving manner according to structural engineering and building physics requirements.

Since time immemorial, in regions with high rainfall, wood was employed extensively for construction. In a corresponding way, structures required some form of waterproofing. Sloped roofs dominated the architectural image, such as in the form of gable roofs. Large overhangs and canopies were created and, as a result, structural timber protection. In this manner, particularly durable and, thus, sustainable typologies were developed, such as the typical Black Forest house with its striking roof shape. With a view to building sorted by type, these examples show that early timber construction, with its mechanical joining methods, often consisting of a single material, always demonstrated reversibility (Fig. 5). It was possible to disassemble buildings and reassemble them at a different site, or reintegrate the beams of old, demolished buildings into newly erected structures. Such traces remain visible to this day in many historic houses. The fact that wood is of low weight and can be easily transported contributes to this.

Compared to other load-bearing elements, wood has a low mass. However, it also requires particularly diligent planning in order to establish a balanced indoor climate. Construction components providing additional thermal mass (e.g. loam panels, types of render, layers of interior walls or floors) can help achieve required climate conditions in timber frame structures and meet given specifications given in legal regulations.

In regions were wood is available to a limited degree, the material is used more sparingly and only for load-bearing structures. In times in which wood may becoming scarce once more due to climate-related changes, a view to traditional building methods appears worthwhile. In the half-timbered houses typical of Central Europe (Fig. 6 and 7), the Hanok of Korea, the Minka of Japan, or the timber frame structures of Oceania, Southeast Asia and Latin America, all of them similar in terms of construction, wood was used sparingly for

5
Traditional timber joints, shown here: Lap joints, cf. Krauth-Meyer 1895

6
Traditional half-timbered construction as bar-type frame. All joinery details comprise timber sorted by type without any additional materials.

7
Close-up view of a traditional half-timbered structure with loam infill, the latter 100% circular.

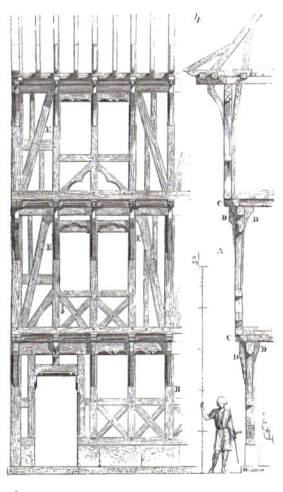

6

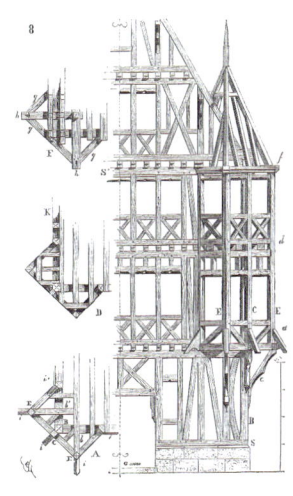

7

Vernacular Architecture 43

frame structures. Depending on the climate, hybrid construction types with timber primary structures and infill were developed, for instance by using loam blocks in colder regions, or building envelopes consisting of bamboo panels or reed mats in warmer regions. Based on the use of mineral binding agents such as mortar, separation sorted by type in the case of half-timbered houses is slightly more difficult than in the case of solely timber houses or frame structures clad in (reed) mats. However, infill materials of traditional half-timbered houses can mostly be returned into an ecological cycle, as long as they consist of naturally occurring loam, without synthetic additives.

Mineral-based construction materials
The typical mineral construction materials used in traditional architecture include natural stone, loam and fired brick. These materials serve as infill in the previously mentioned hybrid half-timbered buildings, as well as for monomaterial-based construction types. Similar to other construction materials, buildings consisting of natural stone mostly exist where stone is available in nature and permits extraction. Due to the effort required for processing, the material was preferably used for representative architecture, either secular or sacred. However, in the Ticino Alps or the mountain regions of Central Asia, such as Tibet, houses built with stone also reflect a long tradition of creating structures for dwelling and agriculture. Buildings featuring mortar-bonded or dry-set stone walls are extremely resistant to weather. In such cases, timber was only used sparingly for horizontal construction elements and components, such as ceilings or roof structures, while stone tile roofing was highly common. Particularly dry construction types, e.g. masonry walls without mortar, permit reversibility in the manner of sorting by type exceptionally well. The same is true for returning them to a cycle of reuse, since they are merely layered and not joined with a bonding agent. The layering of stones into nearly jointless masonry walls can achieve stability based on their weight and being canted

8

into each other. However, this method is highly sophisticated and requires a significant amount of experience. Unlike wood, stone is a material that is difficult to process. Yet, it is also very durable (Fig. 8). Many traditional houses with stone walls and horizontal components consisting of timber show how both materials can complement each other and how hybrid structures sorted by type can be developed by employing different building materials in a manner that takes their respective characteristics into account.

Working with loam, just as with stone, represents a rich tradition of construction (Fig. 10). It is one of the world's oldest and most frequently used building materials. Estimates state that about 8–10 % of humankind live in houses that are partially or completely built of loam. In developing countries, the share is about 20–25 % [4]. Loam is available as a building material in a nearly unlimited manner and can be fully returned into biological cycles The material consists of clay, silt and sand. It can be found in nearly all soil layers in different types of composition. Humankind has benefitted from the natural cohesiveness and strength of loam since millennia in order to build houses. The material mostly finds use in the form of rammed earth walls or as air-dried loam brick. Due to its sufficient formability and durability, loam is very often employed as infill material for frame construction, including half-timbered struc-

8
Stone structure, Casa d'Estate, summer residence, Linescio, Switzerland, 2019, Buchner Bründler Architekten

9
Mud/adobe construction, Great Mosque of Djenné, Mali

9

tures, or for render layers. In addition to many ecological advantages, loam construction elements and components support maintaining a constant indoor climate. Based on their large thermal mass, humidity and heat are stored and released after a certain delay. Loam – compared to other building materials displaying similar formability, such as concrete – does not require chemical curing processes and, instead, cures physically. As a result, it is sensitive to erosion, such as caused by rainwater. So-called erosion proofing (e.g. trass lime layers or larger pebbles) can control such processes. The alternatives are particular protection measures against moisture on the exterior of buildings, in addition to recurring maintenance and care. In consequence, such buildings tend to be located in regions in which precipitation is typically low. For instance, in Yemen or Mali, impressive loam (or mud) architectures have been created that are considered cultural and technological masterworks (Fig. 9). These regions experience low rainfall and, to a large degree, feature architecture with roofs that are flat instead of sloped. These examples once more show the interrelations between geographical conditions, the accessibility of materials and the appearance of traditional forms of architecture.

On its quest for ever more robust construction materials, humankind discovered about 5000 years ago that the water resistance of loam or clay brick can be increased through firing. High temperatures above 1100 °C lead to sintering of the material and its clay content, making it particularly resistant against moisture. Both unfired loam blocks and those created by firing, commonly known as brick, display high compressive strength. Kiln-fired brick demonstrates increased resistance against water and improved solidity. Traditionally, brick was set in lime mortar beds to build walls or vaulted ceilings. It was also frequently employed as infill material for frame structures and, later, as roofing material in the form of thin tile.

In the 19th century, use of the material was limited to representative architecture and the houses of wealthy individuals. This increasingly changed following the introduction of industrialised production methods. "Fired brick finally gave simple people the opportunity to build weatherproof and stable replacements for high-maintenance air-dried loam block walls" [5]. However, the history of the building material also shows how the improvement of durability and comfort immediately influenced the consumption of resources for manufacturing purposes. Clear felling countless square kilometres of forest worldwide in order to generate the power needed to fire brick is one of the most striking examples for how the success of a single building material impacted entire natural and cultural landscapes in a sustained manner. This development is neither overly dependent on local climate conditions, nor the accessibility of construction materials and energy sources. The example of brick manufacturing clearly illustrates social, cultural and contextual consequences: Had all the wood consumed for firing brick for masonry wall buildings instead been suitable for construction, had it been used for building timber rather than masonry structures, vast quantities of the material could have been saved and carbon could have been sequestered instead of emitting it.

Metal based construction materials
To this day, metals constitute a significant portion of the most common building materials. Traditionally and most of all in classical antiquity, metals were used to create connectors for construction purposes, such as clamps, clips, pins, bolts or nails. In the 18th century and following the emergence

10
Loam construction, Sustainable Urban Dwelling Unit, Ethiopian Institute of Architecture, Building Construction and City Development (EiABC), research project on the use of loam as a local resource in two-storey residential construction.
a Rammed earth technique using timber sliding formwork
b Timber sliding formwork for the exterior building walls

of early industrialisation and new coking methods, the cost-efficient production of cast iron began. This also enabled the manufacture of metal sections on a greater scale. Rolled sections were, from the mid-19th century onwards, a common product of ironworks, primarily applied within engineering and industrial construction tasks. Initial applications benefitted from the high compressive strength of the material. However, the tensile strength and the bending stiffness of iron products during that era were low. This is also indicated by structures that often resembled vaults. Elements were connected with rivets or, later, with bolts, which simplified the assembly and disassembly process. An impressive example of this practice is the Crystal Palace built for the Great Exhibition in London in 1851 (Fig. 1, p. 49). Following the conclusion of the Exhibition and the complete disassembly of the structure, a large share of the material was reused for a museum and exhibition hall in Sydenham, where it remained until 1936.
In the late 19th and early 20th centuries, based on new manufacturing processes, industrial production facilities increasingly improved material properties by reducing the carbon share of iron-carbon alloys. This also resulted in higher tensile strength and bending stiffness. The new alloy known as steel supported mass production and far-reaching applications in the construction field. It also paved the way for the global success of steel construction in the building boom of the 1920s and 1930s. However, in this context, other connection methods increasingly found use, such as welding sections together, which in turn made the reuse of elements and components difficult or even impossible.
In the early 20th century, building with reinforced concrete became a dominant practice. Concrete turned into the most extensively used material in the building sector, mostly for reasons of durability, costs and workmanship. The steel embedded in the material performed very well, due to its improved qualities in terms of tensile strength and efficient handling. Yet, here as well, monolithic reinforced concrete construction types made later separation impossible and, thus, prevented reuse. As a result, in the mid-20th century, a different pathway for reclaiming these construction materials emerged: Recycling. In 2019 the share of recycled material in German crude steel production comprised 44.6 % [6]. Due to increasingly strict political specifications and legal amendments to the Closed Substance Cycle Waste Management Act, it is estimated that this figure will further increase, since the carbon footprint of recycling is significantly lower than for primary production.

Vernacular building joinery

The traditional methods of connection, assembly and joinery of construction materials are of considerable importance to building sorted by type and illustrate one of its key aspects (Fig. 11). From this viewpoint, mechanical connections are preferable to chemical-based methods, since they permit much easier disassembly. One example of this is the force-fit and form-fit connections employed for dry-set stone walls or dovetail joints in timber construction (Fig. 7, p. 43). With monomaterial connections (e.g. carpenter-style joinery in timber construction, Fig. 5, p. 43), metal-based connectors such as screws or nails can be completely omitted. Due to requirements of building air-tightness becoming

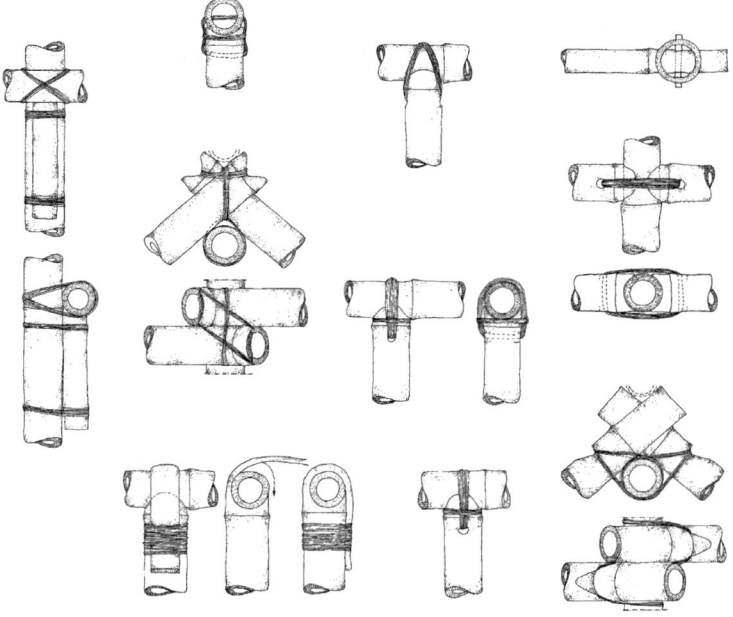

11
Traditional joinery, shown here: Bamboo construction. It allows the consistent application of mechanical connections that permit complete reversibility sorted by type. Illustration cf. Gernot Minke

increasingly strict, material-locked connections are also becoming increasingly important. These connection types, which include mortar joints, are not easily disassembled and reused. Further, chemical or material-locked connections need to be distinguished according to natural and synthetic bonding agents. Natural bonding substances can be returned into biological cycles in a problem-free manner. Natural organic adhesives, so-called biopolymers, have been in use for millennia. However, they play scarecely any role in contemporary construction, since their durability compared to synthetic bonding agents is mostly low, with limited moisture resistance being one reason. For instance, bone glue is completely biodegradable and has been in use for centuries – long before the development of synthetic glue or adhesives – as a chemical connection method based on processed animal bones. Currently, timber members and wood components are mostly bonded by the use of synthetic adhesives, which corresponds to surface bonding. Glueing is, instead, understood as anchoring an element to a surface [7]. Research-based approaches are increasingly emerging that aim at developing a new class of synthetic adhesives that permit microbial biodegradation in order to return them into biological systems [8]. However, here as well, extensive research is required, as well as the political will to support the broad application of results.

From the perspective of mineral chemical means of connection with regard to vernacular construction types, particularly air-dried, non-hydraulic lime mortar is of interest. Burnt lime was discovered as a mineral bonding agent thousands of years ago. By adding water, a formable type of mortar is created that permanently hardens by absorbing CO_2 from the atmosphere.

Learning from traditional methods for the future

How can architects learn from vernacular architecture in order to design more sustainable structures for the purpose of building sorted by type? Aside from commonly known principles such as effectiveness (improving the use of resources) and sufficiency (reducing the consumption of land, building dimensions and deployed resources), most of all consistency of operation within existing natural and technical cycles is relevant for the future of building. This relevance results, on the one hand, from the societal task of building better, with greater social impact, by saving resources and by avoiding waste. On the other hand, it is the outcome of political decision-making that no longer allows simple "business as usual". In order to introduce construction materials and, thus, important resources into existing cycles, planners and actors in the building sector should develop and create parts of structures and entire edifices in a manner that permits problem-free reversibility into their constituent components, while maintaining original quality levels. Mechanical connectors that play a much more prominent role in traditional building than in contemporary construction are preferable to current synthetic bonding types. Deliberating on mechanical and reversible connections between elements and components means taking a giant step forward towards a circular form of building sorted by type. Synthetic adhesives that permit disassembly despite material locking become acceptable once the materials used are biodegradable. Rather than consuming globally distributed products, a return to using locally sourced construction materials is desirable, also with regards to limiting transport routes and strengthening local value chains. A look back to traditional construction types can be helpful in taking a step forward into the future.

Notes
[1] Schittich 2019, p. 18
[2] Umweltbundesamt 2022
[3] de.statista.com
[4] Marsh/Kulshreshtha 2021
[5] see note 1, p. 16
[6] de.statista.com
[7] Habenicht 2009, p. 142
[8] ibid.

Learning from Temporary Buildings

Falk Schneemann

Temporary buildings are understood as buildings and constructions with an atypically short duration of use. They are designed and built in a manner that permits easy disassembly, either to rebuild them in the same form at another site, or to reuse elements, components or materials. Unfortunately, this also allows simply discarding them. The motivation for reuse in the same or a different form was, until recently, nearly exclusively considered a question of economy. In the meantime, this has changed. For instance, a market stall, simply to save costs, should not be created anew for every site or for every market, but should permit disassembly and reassembly at a different location. In order to ensure this, disassembly must be possible with limited effort and without causing damages. These requirements correspond precisely to those relevant to designing and building sorted by type. This raises the question whether temporary or mobile structures of the past offer lessons to learn for the present and the future. However, temporary buildings aren't automatically synonymous with circular construction – just think of mobile site containers consisting of adhesively bonded sandwich elements.

Innovation as driver

Examples of temporary buildings go back a long way. The assumption is that the first dwellings were mobile and, thus, temporary buildings, either to meet the demands of a nomadic lifestyle, or to remain mobile when hunting. Of greater interest to this chapter are edifices and structures that were created at the onset of industrial mass production in the building sector. Here, material usage and construction methods were already very similar to current building practices.

The Crystal Palace: An early industrial circular building type?

The Crystal Palace was planned and erected for the Great Exhibition of 1851 in London (Fig. 1). The intention was to create a suitable building given a very tight budget. Further requirements included, from the very beginning, short construction time and the ability to disassemble the structure, principally because the building was located in Hyde Park. The London public therefore would not have tolerated it permanently occupying the site. An initial competition failed to produce interesting results and, hence, British all-rounder Joseph Paxton was asked to propose a design. Developed in cooperation with engineer William Barlow, the design still exceeded the budget. However, a solution was found: The contractors were promised that they could take back the components of the building for a fixed sum. This implies notions of reusing and recycling materials or construction elements and components. Yet, this actually wasn't the first step. The building was disassembled after the exhibition concluded, the components transported to a different London location, reassembled and

1
The Crystal Palace consisted of millions of individual standardised elements. Crystal Palace, Great Exhibition, London, 1851, Joseph Paxton, William Barlow

2
The Lustron houses were designed by Carl Strandlund as construction kits that permitted rapid and easy assembly and disassembly by two individuals.

amended by an expansion [1]. As early as 1852 the architect Charles Burton published a design for a 198-metre-tall skyscraper. For the project, he intended to reuse the frame construction of the Crystal Palace [2]. This would also have implied a material cycle. However, the project remained unbuilt. Thus the structure itself was nearly completely destroyed in a fire in 1936.

The dimensions and the construction type of the Crystal Palace were, at the time, unparalleled. The building volume was 563 m long, 124 m wide and 33 m tall in the central hall. The construction was strictly modular and based on the dimensions of the largest glass panes available on the market at the time. More than 1000 cast-iron columns together with beams of the same material formed a skeleton frame that stiffened itself and did not require load-bearing masonry walls. All elements were prefabricated. As a result, the construction time spanned only eight months. The combination of early industrial means of building, Victorian style and ornamental loans from the plant world seems, from today's perspective, rather surprising. In terms of construction and from the perspective of circularity, the edifice set new standards that are valid to this day. Particularly the fact that such a highly specialised structure consisted of many smaller and "neutral" elements and components that permitted new combinations in different forms, similar to interlocking building blocks, remains impressive to this day. We may remain sceptical whether the idea of reshaping temporary exhibition architecture into a skyscraper was a realistic one or not. However, the visionary intention is highly appealing.

The Lustron houses: A serial interlocking building system

Ideas for industrially produced buildings that can be freely assembled based on serial elements were developed time and again, most of all during eras of great social crisis. The so-called Lustron houses were prefabricated steel-element structures with enamel-coated components. After World War II, the industrialist and inventor Carl Strandlund from Chicago developed them for soldiers returning from the battlefields as a response to the housing shortage in the United States. They were considered low maintenance and extremely durable, intended to attract modern families who possibly didn't have the time or weren't interested in the significant effort required to maintain common timber or stone houses. Already in the 1940s Strandlund had managed to build gas stations according to a similar principle. Thus, following World War II, he entered the housing market with the same idea. The houses featured an extremely simple design. Two people were able to assemble the correspondingly dimensioned, small and lightweight parts in only a few days. A handbook that resembled a technical manual from the home appliance sector was supposed to offer assistance.

3
The detachable and prefabricated Maison Tropicale established a language of details and materials that was derived from machine, ship and aircraft construction, West Africa, 1949–1951, Jean Prouvé

Production was halted in 1950 after only one year of business. The company was unable to repay the loan it received to launch its operations. During the short production time of the Lustron house, however, more than 2000 houses were built, many of which are still occupied.

Jean Prouvé's principles of detachable construction

French architect Jean Prouvé followed a similar visionary-industrial approach. In his architecture, many details and production methods were inspired by automotive or aircraft construction. A significant part of his work also comprises furniture demonstrating production and assembly methods that also had a significant impact on his buildings. Prouvé developed construction methods, organisational principles and details with simple assembly and disassembly in mind. To him, an individual house always represented a temporary developmental stage of his own research or the adaptation of an existing principle to, for instance, the related particular climate zone. Correspondingly, deliberating on Prouvé as a pioneer of circular building does not lead to focusing on a singular project, but instead, a continuously developed set of construction principles. The point of origin was a patent for a "detachable steel frame construction" from 1939. The principles can be retraced in different iterations and among others in the Maisons démontables (1944), the Maison Ferembal (1948), the Maison Métropole (1949) and the Maison Tropicale (1949–1951) (Fig. 3). While the buildings appear very different at first glance, the principles of interest here remain the same [3]. The houses always consisted of elements which were assigned a clear primary function: Foundations, floor slab, load-bearing structure, facade, roof. All elements featured dimensions that enabled their transport by truck or plane. They displayed clear principles of assembly and were easy to detach. Screws and bolted connections were predominant. Strip foundations consisting of masonry or concrete were used. Variations existed, where stacked concrete pavers served as the foundations. Upon them, for the most part, lattice grids were placed that, in return, supported the floor slabs. This led to creating a platform that was slightly elevated above the landscape and upon which the load-bearing structure of the house was set. The structures consisted of steel or also timber. They functioned similarly to gantry cranes or larger-than-life carpenter's trestles. Based on their geometry, they were capable of bearing vertical and horizontal loads. These trestles were arranged in an exposed and expressive manner within the building volume. Facades were divided into elements matching the floor grid dimensions and were delivered to the construction site complete with windows, sun protection and opaque panels. The final assembly step comprised the roof elements. Similar to the floors, they were supported by joists.

All elements and construction components were minimised in terms of material usage. The language of forms and construction methods resulting from the approach is highly reminiscent of machine, ship or aircraft construction. As a result, Prouvé's work features a highly individual and striking expression. He repeatedly worked with canted sheet-steel elements that were often connected by screws or bolts. His preferred material was steel. If necessary, he combined the material with wood, for instance for walls and floors. For the joints between these basic elements, Prouvé often used very simple sections into which elements were placed or inserted.

In his work, Prouvé achieved a certain degree of complexity by use of simple connection details and effective basic elements, recognisable in terms of their construction. We can therefore consider his work a guiding example for our discussion on circular building methods. He planned his system to be modular, in order to ensure that the multiple use of prefabricated construction components supported problem-free remodelling as determined by the construction type. The basic precondition was to be able to simply detach the individual elements, despite their complex overall composition. This permitted repairing them upon demand or, towards the end of their life cycle, detaching them sorted by type and returning the materials to the corresponding cycles.

The Paper Log House: Waste as a resource

According to the thesis that crises, time and again, motivate research on adequate and intelligently constructed temporary buildings, the Japanese architect and Pritzker laureate Shigeru Ban developed the so-called Paper Log House in Kobe in 1995. The aim was to create emergency shelter structures following the massive earthquake that occurred in the region (Fig. 4). The resulting requirements included short construction time, simple transport and assembly, as well as rapid availability of building materials. The houses featured a 4 × 4 m footprint and a gable roof with a low slope. Shigeru Ban, in the design of his building, was repeatedly drawn to the notion of repurposing materials: He used cardboard tubes, discarded as waste in large quantities by the paper industry and other industry branches, as a simple and cost-efficient material resource. The cardboard elements he employed possessed an exterior diameter of 106 mm and a thickness of 4 mm. They were filled with newsprint in order to improve their insulating performance and, additionally, covered in transparent polyurethane in order to protect them from moisture. This, however, disallowed sorting them by type.

Beverage crates filled with sand served as foundations. They were simply placed on the ground, similar to strip foundations, to elevate the houses in order to protect them from rising moisture or splash water. The floors were comprised of a cardboard-tube sandwich structure set between two layers of plywood. This resulted in creating a flat box that was able to span the distance between the beverage crates. The positions of the walls were marked by small plywood crosses along the floors. The cardboard tubes of the walls were mounted onto them. The airtightness between the individual, vertically arranged tubes was achieved by the use of sealing tape. Wood windows and a door were set into the walls. A plywood ring beam of sorts formed their upper edge. A simple roof structure consisting of bars,

4
The Paper Log House was constructed of cardboard tubes and beverage cases. Emergency shelter, Kobe, Japan, 1995, Shigeru Ban

Learning from Temporary Buildings 51

5a b

5
People's Pavilion, Dutch Design Week, Eindhoven, 2017, bureau SLA and Overtreders W
a Exterior view
b The pavilion was, for the most part, built from loaned materials in a manner that allowed returning them after use in the same quality as before, such as the construction grade timber members connected by metal straps.

struts and a ridge beam was set on top. These elements were also created from cardboard tubes, into which plywood connector nodes were set. A roofing layer consisting of simple tarp served to cover the roof structure.

Even if covering the cardboard tubes with polyurethane, similar to the use of adhesively bonded plywood, is objectionable in terms of circularity, the Paper Log House nevertheless displays many construction principles and details that can serve as a reference for simple, circular building and construction today. All construction components allow simple assembly and equally simple disassembly. Details, such as the nodes of the roof structure or the interlocking connection between the cardboard tubes and the floor elements are exemplary in this regard. In addition, only a small variety of materials was used to create the houses, many of which were previously deployed in a very different context. The unorthodox construction principles and details, such as foundation walls made of beverage cases or walls consisting of cardboard tubes establish novel sources for building materials and generate opportunities for reuse and reprocessing.

The People's Pavilion: Building material on loan

The People's Pavilion, built as a temporary festival centre on the occasion of the Dutch Design Week 2017 in Eindhoven, represents a similar, yet even more radical approach (Fig. 5a). The pavilion was designed by the Dutch architecture offices bureau SLA and Overtreders W. It comprises a roughly 7-metre-tall event space with a cross-shaped floor plan reminiscent of sacred architecture and offering about 250 m² of space. It was in use for only one week. What made the design unique was that all construction components and materials required for building the pavilion were merely loaned from existing cycles and systems and, after conclusion of their use, returned to the original owners. The temporary function was planned in a manner preventing changes to the individual elements that would result in a loss of quality and, thus, disallow later continued use or their return to the loaners as agreed upon by both parties (Fig. 5b). In order to meet these goals, the pavilion displayed atypical material usages and connections that all aimed at avoiding drilling, glueing or sawing. Materials were not only provided by or loaned from hardware stores or directly from the manufacturers, but also from residents of the city.

The lower part of the load-bearing structure consisted of prefabricated concrete elements that typically serve as foundation piles. Here, they became columns. Set on top of them, the upper part of the load-bearing structure featured timber beams – only connected by steel ties and belts and without cutting them – that also determined the dimensions of the building. The glazed ground-floor facade elements had been

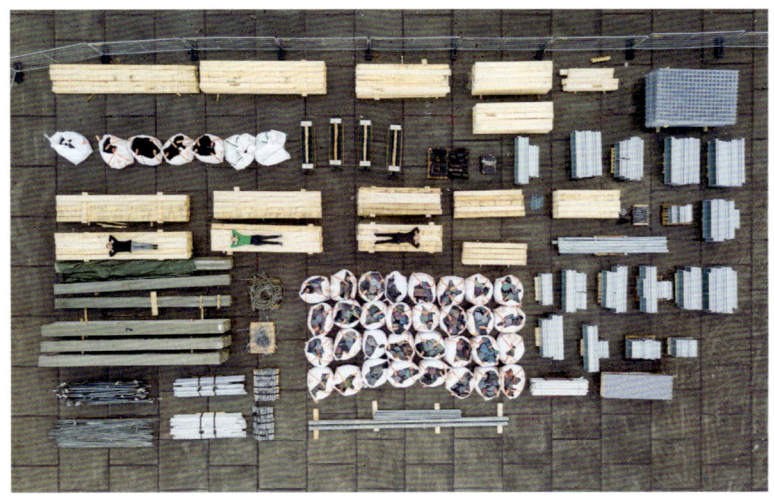

6
People's Pavilion after disassembly

retrieved from the renovation of an office building and, following their use in the pavilion, were built into a different office building. The facade above consisted of plastic shingles in different colours that were sourced from household waste by the architects themselves and manufactured using specifically created machinery. These shingles or tiles have, in the meantime, received building law approval and are available on the market under the name Pretty Plastic. For the roof of the pavilion, a building system for greenhouses was selected that is in widespread use in the Netherlands. After the pavilion closed its doors, it was returned to the loaner (Fig. 6). For the furniture, loaned material was employed as well. The podium consisted of loaned concrete pavers. The benches were provided by a church for temporary use [4].

With its radical approach and atypical construction methods, the People's Pavilion shows how completely closed material cycles can function today. It is not just about the right choice of materials or the development of suitable joinery and connection methods. Questions on the logistics, availability and use of materials as opposed to owning them are increasingly becoming the focus of attention. In the broadest sense, we can understand temporary architecture as an overarching term for new business models for a circular economy. A Dutch company, for instance, offers carpet tile under the premise of "product as service" (see "RoofKIT", p. 60ff.). Clients do not pay in order to own the carpet. Instead, they pay for using it. This model became possible because the company changed the composition of its product in such a manner that enabled recycling it to 100% and transforming it into new carpet flooring. Why sell products and acquire new raw materials, again and again? Hence, the company merely loans the product and requests its return once it needs to be exchanged, after which it becomes a new resource.

Looking at these exemplary temporary products and buildings from the perspective of circular building sorted by type, we could keep on going, ad infinitum. Further examples include the aeroplane hangars by Konrad Wachsmann, the IBM Traveling Pavilion and Diogene by Renzo Piano, the small-scale architecture by Richard Horden, nearly the entire oeuvre by Buckminster Fuller, the Swiss Sound Pavilion for the Expo 2000 by Peter Zumthor, the Benthem House by Benthem Crouwel or, just as well, anonymous structures including scaffolding, market stalls or exhibition stands and, eventually, the above mentioned products and business models. Even if few of the presented buildings meet all requirements for thermal insulation, soundproofing or fire safety common in the Central European climate zone or, for that matter, of the legal framework of Germany or the EU, their potential for circular construction is very clear.

Basic principles for circular construction
As temporary buildings show, the following recurring principles can be identified and generally applied to circular construction.

Focus on connectors
Detachable connectors are key to damage-free disassembly (see "Reversible Assembly and Connection Methods", p. 104ff.). Temporary buildings display an enormous range of such connectors – most of all, screws and bolts. Similar to a scaffold, system solutions are possible, as well as atypical connectors, such as tension belts (People's Pavilion) or cable ties. A standardisation of types is also possible in terms of the material context: Straps, cables and ropes are suitable for fabric and membranes. Clamps, screws and bolts work well for steel and timber.

Effective nodes and joints
The construction-based logic of temporary buildings is often rooted in nodes and joints that can be disconnected, are suitable for more than one application, or permit adaptation to different situations through rotation (scaffold) or different types of geometrical flexibility [5]. These nodes can achieve a remarkable degree of complexity [6] and are mostly combined with linear elements, such as tubes or other semi-finished products. They comprise a system in which complexity and simplicity can synergetically complement each other and, thus, achieve extreme efficiency with simultaneous flexibility.

Separation of functions on the level of buildings
Temporary buildings often display a clear structural separation of different functional domains. For instance, auxiliary functions are bundled and often comprise a different construction type than the other functional domains. Examples include the designs of Jean Prouvé or, simply put, temporary large-scale tents that feature annexes with toilets or kitchen areas [7].

Functional separation of construction components
On the level of construction components, functional separation often takes place in terms of layers. This can be observed in the separation between a load-bearing structure, a sealant layer and a thermal insulation layer (see "Layering as a Circular Principle", p. 118ff.). In this manner, the individual construction components become separable and less complex, while demonstrating lower weight and in many cases, higher flexibility. Examples include the buildings of Jean Prouvé (see p. 50f.) or also the Lustron houses (p. 49f.).

Interaction between specialisation and standardisation
The materials as well as the construction components of temporary buildings can often be distinguished according to two categories: standardised and specialised. Standardised materials and construction components display very good accessibility and high flexibility in terms of function. They

include, for instance, the cardboard tubes of the Paper Log House (see p. 51f.) or the prefabricated foundation piles of the People's Pavilion (see p. 52f.). Specialised materials and construction components are, however, often specifically developed for a project. The flexibility and performance of standardised construction components and materials can be increased by intelligently connecting related products.

Modularity

It hardly comes as a surprise that temporary structures display a high degree of modularity. As a consequence, they can be configured in different ways and, thus, can be adapted to different construction sites or to different user requirements.

Functionally neutral construction components

Many of the examples presented here demonstrate potential for reuse that arise when construction components are used differently as originally intended. Good examples of this are the repurposing the steel elements of the Crystal Palace for a skyscraper or the cardboard rolls of the Paper Log House as a load-bearing system. With building components that are constructed in a functionally neutral way from the very beginning, meaning by creating a type of modular system of interlocking elements, new opportunities arise for circularity (see "Contemporary Examples of Circular Construction", p. 56ff.).

In the case of temporary buildings, constraints exist in terms of flexibility, material efficiency, construction speed and economy – often triggered by crisis, such as described above (e.g. Lustron and Paper Log House). Such pressures resulted and continue to result in the creation of extremely innovative, surprising and sometimes highly specialised solutions that always represent the quest for a certain kind of simplicity. We can see that requirements placed on temporary buildings are often similar to those of circular construction. We can also see that damage-free disassembly is the lowest common denominator. Beyond that, temporary buildings reveal a high degree of innovation density. This is the result of the previously mentioned demands, as well as the fact that other specifications, such as building laws or code requirements for thermal insulation or soundproofing are dealt with more flexibly. Temporary buildings, thus, offer a treasure trove full of detail solutions, material usages or structural engineering concepts for circular design and construction.

Notes
[1] Schneider 2005
[2] Balzer 1973, p. 46–49
[3] Huber/Steinegger 1971; Vitra Design Museum 2006; Sulzer 2005
[4] Bureau SLA 2017; archdaily 2017
[5] mero-tsk.de
[6] see also the construction systems developed by Konrad Wachsmann
[7] neptunus.de

Contemporary Examples of Circular Construction

Katharina Blümke

In 1923 Marcel Duchamp provoked the art scene with his "Fountain", a urinal tilted by 90 degrees and signed "R. Mutt". Up to this point in time it was unthinkable that a simple everyday object could be considered art. Today, it is one of the most important works of art of the 20th century [1]. This has to do with the fact that it offered a new perspective on the valuation of an everyday object – aside from its shock effect and the question on what art is in the first place. Interestingly, Duchamp – by tilting the object and improperly mounting it, without connecting any pipes – deprived it of its original function. Hence, the form and the traces of use, as well as the story surrounding the object, were elevated to the high echelons of art. Simply by using an object that is industrially mass-produced to this day, in numbers ranging in the millions. Thus, the value chain that leads into the art world is not based on the act of artistic object-oriented creation, but instead, on the reinterpretation of a banal object within the museum context, with simultaneous visualisation of the history of the object. The inflicted traces and the psychological space that opens before the observer create a unique and unmistakeable experience, a new, one-of-a-kind object. This added value is accessible to artists and observers free of charge in the form of repurposing. Its importance for circular construction is discussed in the introduction to this publication (p. 12).

Wasteland exhibition

About 100 years after Duchamp, the Lendager Group opened an exhibition in 2017 under the name "Wasteland" [2]. With its focus on building materials and construction components intended for disposal, it once more elevated the topic of uniqueness to the status of art in a museum context. Similar to Duchamp, a discussion began on why we even consider used materials and objects as waste and discard them, instead of returning them into a cycle. The Danish architecture firm managed to reveal hidden added value and its opportunities based on built projects and posited that what we describe as waste today is not valueless. The exhibition expressed the need for extracting added value from so-called construction waste, discarded objects and their unique stories, in order to understand them as partial to an endless cycle of the built environment (Fig. 1a and 1b).

In the first exhibition space, visitors were confronted with facts and numbers on the

1
Wasteland exhibition, 2017, Lendager Group
a Reused windows form a double-skin facade.
b Different sorts of materials are mapped and exhibited as palpable objects.
c Waste heaps as a valuable resource.

1a

b

c

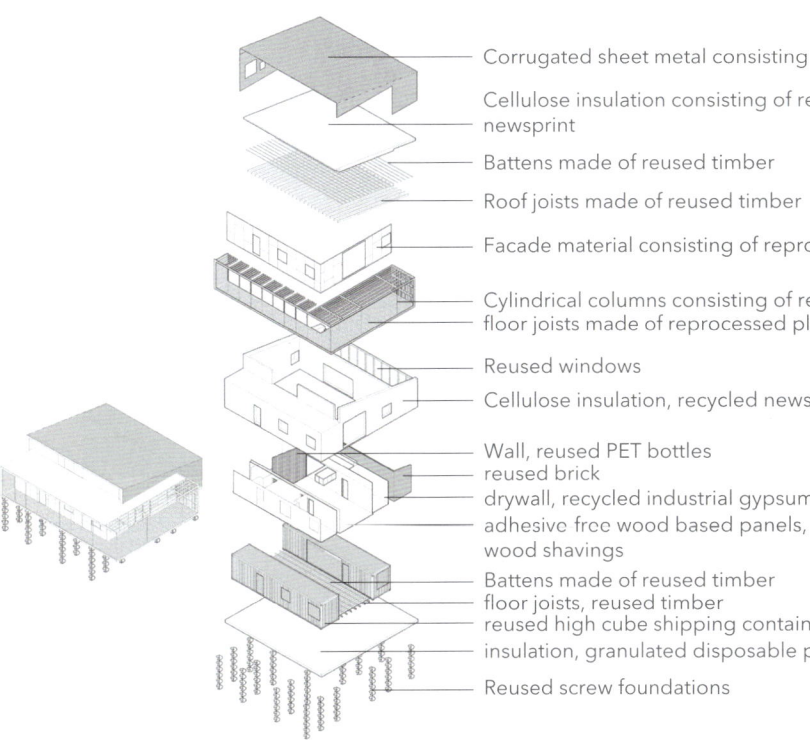

2
Layer composition of the Upcycle House, Nyborg, Denmark, 2013, Lendager Arkitekter

carbon footprint of the building industry. Then, they were led into an area in which heaps of discarded material lying on the floor were separated into different building material groups: Concrete, brick, plastic, metal, glass and wood (Fig. 1c). Each material category was correlated with a building project by the architecture office that demonstrated the quality of dealing with "waste" and, beyond this, proved that new aesthetic opportunities and even a new type of construction-based language can emerge from it [3].

Upcycle house

An example of such projects is the Upcycle House in Nyborg in Denmark, which is comprised almost entirely of recycled materials (Fig. 2 and 3). Two old shipping containers comprise the load-bearing structure. After completion of the building, they were no longer visible on the exterior. The house was insulated with paper wool created from processing old newspapers and equipped with roofing and facade cladding consisting of recycled aluminium beverage cans. When aluminium is recycled, 95% of the energy can be saved that would otherwise be used for the production of primary aluminium [4]. The dark facade panels feature recycled paper granulate that was pressed and heat-treated. All windows are reused items, for example from an old school in Copenhagen [5].

In the interiors, special OSB panels serve as wall panelling and flooring. In their manufacture, wood shavings as a byproduct of the wood industry were compressed without using glue [6]. Such byproducts of sawing, including sawdust, rinds or chips com-

3
Upcycle House, Nyborg, 2013
Lendager Arkitekter
a Two shipping containers serve as the load-bearing structure for the building, remaining partially visible in the interior.
b Champagne cork remains are used as flooring material in the kitchen. OSB panels were compressed mainly on a lignin basis.
c Recycled aluminium saves 95% of energy compared to primary raw material processing.

3a

b

c

Contemporary Examples of Circular Construction 57

prise a large share of secondary materials in Europe, due to the established cascading use of wood. The underlying linear concept needs to be designed in a more efficient manner or, in the long term, replaced by a circular approach. In 2005 in Germany 14.3 million m³ of residual wood sawing products were created, which corresponds to a share of 38.5% of cutting wood in its entirety [7]. This valuable resource unfortunately often ends up in synthetically bonded composite materials which do not meet the requirements of sorting by type nor are they circular. In 2011 in Germany alone 5.6 million m³ of particle board was produced, which means that Germany is the greatest particle board producer in Europe [8].

As the Upcycle House demonstrates: Only few primary materials are required for a home that meets all functional specifications for a family (living room connected directly to a kitchen, bedroom, three smaller rooms, bathroom, utility room, terrace, greenhouse). Reuse, repurposing and reprocessing conserve natural resources and also prevent causing further CO_2 emissions that would be created if these materials were newly provided or processed. Anders Lendager, the founder of the Lendager Group, states that, thanks to the uncompromising sourcing of used materials and construction components, greenhouse gas emissions for the Upcycle House were 86% lower than for a conventional house [9].

At the University of East Anglia in the UK, Buckminster Fuller asked a young architect, the designer of the Sainsbury Centre: "How much does your building weigh, Mr. Foster?" The question was aimed at the use of materials and the idea of conserving them based on qualified professional knowledge while promoting lightweight construction, as Buckminster Fuller demanded time and again and which he prominently demonstrated in his projects. Unfortunately, the answer to this provocative question was not recorded. However, nowadays the question should be put differently: "How much greenhouse gas did your building cause?" All planners and those bearing responsibility in the building industry need to be capable of answering this question, in order to cast new light on their work and face further questions on whether sufficient action has taken place in this area. In the late 20th century, we were still able to speak of a knowledge gap. In the early 21st century, we need to come up with an all-embracing paradigm change. This change needs to be both quantifiable and comprehensible, while providing future generations with the same possibilities and opportunities previous generations enjoyed. Anders Lendager's statement is, therefore, important and very timely.

In their projects, the Lendager Group deliberates on building materials and components and, thanks to reprocessing and recycling, provides pre-used materials with a new usage life. The Swiss office baubüro in

4
K.118, Winterthur, Switzerland, 2021, baubüro in situ
a The existing building was renovated and received a three-storey rooftop addition for office use.
b The reused facade elements were mounted reversibly.

4a

b

situ shows a path to reuse and repurposing and a corresponding means of construction that, in return, becomes unmistakeable.

baubüro in situ

For more than 25 years the Basel-based office of baubüro in situ has dedicated its work to the conversion and remodelling of existing buildings and the reuse of construction components. For this purpose, the collective has founded several companies over the years. It has also supported others that, for instance, deal with the trade and distribution of used construction components, or the coordination of their procurement. In 2020 baubüro in situ created Zirkular, a company that focuses on the professional planning, development and implementation of circular construction processes. In numerous built examples, the collective has demonstrated how buildings can be converted and how materials can be sourced from demolished structures and deployed again. An important component of this work is cataloguing, assessing and evaluating further fields of implementation of construction components as they become available. The planning process is reversed, since it always responds to the existing context of the materials and the building components: The design follows the available construction components [10].

Rooftop extension K.118

Project K.118 is a rooftop extension to a former three-storey machine factory located in the Sulzer industrial district in Winterthur (Fig. 4). The task was to add three additional storeys to the existing building in order to house studios and craft spaces. The client (Stiftung Abendrot) requested a markedly material-efficient architectural design, the preservation of the existing building including all required repairs, in addition to creating the rooftop extension with reused materials as far as possible [11]. The result is remarkable: According to their own statement, the new storeys consist of pre-used materials and building components to about 70 % [12]. The search for building components became the most extensive planning phase. The entire planning process repeatedly required readjustment according to the accessibility of construction components and materials. A reused steel frame of the former distribution centre of a supermarket chain on the Lysbüchel site in Basel now constitutes the primary load-bearing structure of the extension. The simple, modular system with its bolted connections enabled easy disassembly in Basel and reassembly in Winterthur. The stairwell on the exterior of the northwestern facade – amended by an elevator shaft – was sourced from a former office building in Zurich. Other reused items include construction components from the neighbourhood and the Sulzer site: Radiators, washbasins and the rooftop photovoltaic system, as well as the windows, doors and even the letterboxes. The new facade system consists of prefabri-

5a

b

c

cated timber elements into which reused windows and doors from two demolition sites in Zurich and Winterthur were set (Fig. 5b and c). In some cases, the collective created double box-type windows, in order to meet the necessary thermal insulation requirements. The facade elements were insulated with straw and feature loam render on the interiors, sourced from earthworks in the area. The facade envelope – the unmistakeable trademark of this pilot project – consists of orange-reddish corrugated sheet aluminium panels sourced from a former printing press in Winterthur, reused without having to adjust their size (Fig. 5a). Different types of sections required assembling the facade in an overlapping manner, which in return emphasises the unique character of the design and the reuse-based construction methods [13]. Doors and three-layer panels that previously belonged to stage sets were reused and placed into the wood stud structure of the interior walls. The solid timber flooring and the roof elements originated in a previous temporary timber edifice. In this case, the mineral materials required to meet soundproofing and fire safety specifications were applied in the form of loose fill that permits reclamation sorted by type at any time.

According to statements by baubüro in situ, circular construction resulting in the saving of 60% of greenhouse gas emissions and 500 t of primary materials compared to conventional building methods [14]. The remarkable thing is that initial calculations indicated that investment costs on the part of the client did not exceed those of conventional construction. However, a distinct shift took place: From costs for the acquisition of materials and components to planning, coordination, treatment and logistics. This indicates yet another new approach to sustainability: The production of local value chains in craftsmanship and logistics connected to the generation of technological knowledge that remains with the involved companies and can also be communicated to others.

RoofKIT

The RoofKIT project is the contribution of the Department of Architecture at the Karlsruhe Institute for Technology (KIT) to the international student competition Solar Decathlon Europe 2021/22, held for the first time in Germany in the city of Wuppertal (Fig. 6a). The city was also the object of planning – the participants were required to select one of three sites for an urban intervention and propose a related typological solution (urban infill construction, expansion or rooftop addition). Altogether 18 teams from different universities competed against each other. The task was to design a project that would combine sustainability aspects from sociology (integration into the surrounding neighbourhood, mobility concepts, participatory processes, building methods), ecology (questions on energy and resources), as well as economy (business efficiency). The RoofKIT project became the overall competition winner.

The unique aspect of the Solar Decathlon Europe 2021/22 was that teams had to present an overall design for a greater architectural project and create a typical building element as a demonstration unit at full

5
K.118, Winterthur, Switzerland, 2021, baubüro in situ
a The facade elements were sourced from the demolition of a former printing plant in Winterthur.
b Delayed planning process: In the beginning, the emphasis is on hunting and gathering, after which the design takes place according to the parts at hand, shown here: Windows.
c Reused windows form a new box-type window.

6
RoofKIT, Solar Decathlon Europe 2021/22, Department of Architecture, Karlsruhe Institute of Technology, KIT
a Temporary raised structure
b All construction components and materials are 100% circular

6 a

b

scale. These units were set up for a duration of three weeks on the site of the former Mirke railway station in Wuppertal, where they were tested and, eventually, awarded. The individual timber dwelling units of the RoofKIT overall design encircle a public atrium. Similar to the demonstration unit, they are arranged around a core unit that includes a kitchen, bathroom and all building services and storage spaces. This concept supports the efficient construction of prefabricated timber modules. Only one of the four built modular timber construction units on display in Wuppertal contains the larger-scale building services elements. Prefabrication guarantees a work process that is advantageous in social and safety-related terms for all actors involved in the realisation of the project. It also supports highly precise construction, which is an important precondition for creating buildings and connections sorted by type. Hence, approaches to mitigating tolerances typical to construction sites, such as the use of building foam, were avoided.

Designed as a prefabricated modular system according to building methods that are to 100% circular, the completed demonstration unit shows that currently available technology and design capacities are sufficient to meet the demands of the European Green Deal for circular economics in the building sector. No adhesives, no impregnations or colour coating, no foam or wet sealing were used (Fig. 6 b). Further, in the construction of the units, care was taken to exclusively use materials sorted by type. Neither composite materials nor material mixes were employed. Even the bathroom mirror consists of highly polished steel, which made it possible to avoid the metal/glass combinations typical to conventional mirrors.

Reuse
The idea of RoofKIT is not limited to showing what is possible in the future. Many construction components and materials were sourced from the so-called urban mine and used within the project in their second, third or even fourth cycle: Wood from old barns in the Black Forest, an entrance door from a 19th-century building, windows from a demolished structure in Basel and bathroom and kitchen fixtures from exhibition stand returns. This reuse strategy is certainly the most simple and direct pathway to establishing a circular way of building.

Recycling
The demonstration unit contains materials that were reprocessed or recycled. The team used 100% recycled copper as roofing material. The wall panelling in the kitchen and bathroom was created from old yoghurt pots (Fig. 7 b, p. 62). The toilet and the shower enclosure are clad in ceramic glass comprised of glass shards. The walkways of the outdoor spaces are paved with stones sourced from reclaimed demolition debris.

Bio-based and cultured materials
Aside from mineral, metal or synthetic circular materials, most of all natural biological materials found use: All walls are clad in loam construction panels in order to control humidity and, thus, the quality of indoor air in the unit. Natural felt covers the unit core and the ceilings. All insulation material con-

7a

b

7 RoofKIT
a The facade consists of reused wooden boards. The substructure is protected by living fungi cultures.
b The kitchen cabinetry features pre-used and reprocessed yoghurt pots.

sists of 100% dried seagrass without any further treatment or additives. Panel materials and lampshades were created from mycelium, the root system of mushrooms. Even living fungi were applied to the substructure of the exterior wall as weather protection. The timber used for the structure is free of adhesives and untreated.

Connections and joints sorted by type
The RoofKIT project features a timber frame structure with separate functional layers consisting of building elements sorted by type that were connected in a manner enabling reversibility: Instead of typical OSB boards, the timber frame was covered in diagonally arranged solid wood boards without any synthetic treatment. The frame structure was infilled with seagrass as insulation. PE film without adhesive bonding and sorted by type serves as a vapour barrier. The interior wall surfaces were finished in a layer of fine loam render. The wood sheathing was covered with loam construction panels fastened with screw connectors. The battens that hold the facade membrane in place were also connected to the exterior with screws. All joinery in the demonstration unit was realised in a manner that enables loss-free reversibility. Screws were arranged in a generally easily accessible manner. The reversible wood joints were created by use of digital production methods (CNC). The required thermal mass for the lightweight timber structure was provided in the form of air-dried, simply laid-out loam block as underfloor heating substrate. For the timber floors, untreated solid ash boards and reused floorboards from an old barn, connected by screws, were used (Fig. 8a). In many parts of the unit, the aspect of circularity defines the design without restricting functionality. Connections sorted by type can also be found in the bathroom. Here, the use of silicone was avoided completely and only dry sealing methods were applied. The shower tub consisting of canted stainless sheet steel is bordered by walls for which the team set ceramic glass panels consisting of recycled glass into a metal stud frame (Fig. 8b). Only a dry seal was placed between the panels and the studs. As a result, all materials can be separated from one another and built into a successive project, without creating any residue.

Combination of advantages and synergies
A circular system in the construction sector is only effective if it uses renewable energy sources. For the overall design of RoofKIT as a rooftop extension meeting high energy efficiency standards, the necessary power demand for the building (including appliances and e-mobility) is mostly covered by solar units mounted to the building envelope. In order to guarantee a pleasant indoor climate in the summer months, a passive cooling concept with minimal use of technology was developed that includes effective shading, thermal mass and

8
RoofKIT
a The flooring was sourced from an old farmhouse in the Black Forest and combined with new wood elements along module joints.
b The interior walls of the bathroom are clad in protective clamped ceramic glass panels made of recycled glass.

8a b

nighttime cooling for thermal mass heat discharge. The topic of sufficiency (avoiding complex technical infrastructure) is implemented in the demonstration unit in the form of decentral pendulum fans without ventilation ducts, built into the facade, as well as wireless light switches. Similar to a bicycle dynamo, by pressing a switch, electrical power is generated that turns a light fixture on or off by use of a radio signal. The entire lighting concept follows the same idea: Avoid unnecessary lamps wherever possible, use flexible and wireless portable elements in order to only illuminate areas where light is required.

Product as service
In Wuppertal, the RoofKIT unit was set on top of a scaffold, since it was principally designed for rooftop additions. Scaffolds are generally tried and tested systems created for temporary use with elements that can be reused again and again, in the same manner as intended and without loss of quality. This is also true for the conclusion of the competition: The scaffold, as well as all other construction components and materials that elevate the unit, were loaned for a specific period of time. On the competition site, traditional foundations weren't permitted. Thus, gabions were used to transmit horizontal loads into the ground (Fig. 7 a). Following the end of the competition, they were returned in their entirety to a local provider.

Material bank
At the end of its utilisation period, the RoofKIT project comprised a material bank for future projects: Instead of connecting elements and construction components in a manner that does not allow their separation, the unit featured reversible screw and bolt connections, clamps and fully latched systems, in order to be able to reclaim all used materials in a clean and sorted manner and successively return them to their specific material cycles without loss of quality.

Conclusion
The process of rethinking the way we build against the background of the unmistakable call of the European Union to comprehensively implement a circular economy by 2050 has already informed recent and pioneering built projects. They demonstrate the successful and circular use of materials and related construction methods. Questions on the valuation of materials and objects, as formulated by Marcel Duchamp 100 years ago, require renewed evaluation, discussion, implementation and, eventually, also education in the building sector, in order for architecture to retain its relevance and responsibility.

Notes
[1] Emke 2017
[2] Lendager Group, Wasteland Exhibition 2017
[3] Lendager Group, Wasteland Exhibition
[4] Bundesanstalt für Geowissenschaften und Rohstoffe 2020
[5] ArchDaily, 6.4.2017
[6] Lendager Group, Upcycle House
[7] Sörgel/Mantau/Weimar 2006
[8] Umweltbundesamt 2014
[9] ArchDaily: Upcycle House/Lendager Arkitekter
[10] baubüro in situ: K.118, 2021
[11] Feil 2021
[12] BauNetz, 25.11.2021
[13] ibid.
[14] see note 9

MATERIALS – CONNECTIONS – LAYERS

Materials Selection for Circular Construction

Elena Boerman, Dirk E. Hebel

The establishment of the new "Building Class E" is proactively aimed at reinforcing the responsibility of planners in terms of creating detailed solutions – detached from product-oriented DIN standards – and, as a result, contributing to sensible and context-based materials selections and related methods of connection and joinery. This can be considered a paradigm change in building culture in Germany (see "Circular Construction", p. 24ff.). At first glance, detaching practice from established, standards-oriented approaches seems astounding. One reason is that the self-conception of our discipline, to this day, is seen as based on decision-making in a self-evident, independent and free manner, on all levels. Despite this self-conception, a different kind of culture seems to have become dominant in recent decades. This culture is no longer oriented on the development of details according to contextual principles or sensible craftsmanship. Instead, increasingly predetermined and standardised system solutions find application. They resemble black boxes that are assembled without exact knowledge on which materials are used and how they are connected. The sole aim is to adhere to predetermined technical standards, specifications or norms, liberated from individual responsibility – which is assigned to the manufacturers. This is also a reason why architects find it difficult to employ new circular and sustainable materials in a sensible way, since they don't provide an all-inclusive, care-free package. It is necessary to develop new details and standards and to make mistakes while doing so, in order to gain experience that can lead to case-by-case approval. This is intrinsically connected to expenditure in terms of time and, as a result, money. In this regard, the call for once more assigning responsibility to architects is absolutely understandable. Very likely, it is also a promising way forward in order to uncompromisingly advance circular economy in the building sector. Such an understanding of construction will make it possible to use new building materials either from secondary streams or from biological sources on a large scale. From this perspective, calls for "building simply" (see "(Re)Building Simply", p. 98ff.), designing connections unmixed based on circular construction principles and the use of secondary raw materials are viewed as corresponding tasks and represented within this publication accordingly. New buildings that promise a sorted responsibility, yet abstain from using secondary resources, are merely lip service and viewed as inconsistent. As of today, architects need to develop a new understanding of these aspects. Also, generations of future planners need to feel a sense of excitement for reprocessed and recycled materials.

Building material demands and circular material banks

The term "more than a mine" used in this book is a call primarily addressed at maintaining existing buildings and changing the way we deal with them as resources. As such, it also encompasses the aspect of materials. As a result, the question is how architects can succeed in returning valuable raw materials from the anthropogenic stock into existing cycles (see "More than a Mine – Existing Buildings as Material and Cultural Resource ", p. 34ff.). The need to do so is overwhelming when considering the sheer mass of required building materials: In 2020, 30.1 million tonnes of cement were used in Germany [1]. Calculations show that this quantity, when applying a mixing ratio of 1:4 for concrete, corresponds to approximately 120 million tonnes of gravel aggregate. This equals more than one fifth of the entire amount of mineral raw resources consumed in Germany, with a per-annum figure of about 600 million tonnes. The entirety of consumed raw materials in Germany equals approximately 1.3 billion tonnes per year. This includes the previously mentioned non-metallic minerals (45%), as well as fossil fuels (29%) and biomass (21%) [2]. Extracting and producing such enormous amounts of primary resources seems all the more problematic, considering the approximately 230 million tonnes of building and demolition waste produced per year in Germany. The majority of this figure is subject to cascading use, landfill or incineration [3]. Combined with the fact that the average raw materials comsumption in Germany is 10% higher than the European average and even 100% above the global average figure, as reports show [4], it is urgently necessary to utilise the existing material stock in a more economic way and, thus, radically reduce the dependence on primary materials.

For this kind of economic utilisation, a sophisticated and simple strategy for circular and sorted reversibility, digital management of stock, as well as processing and logistics based on regenerative energy sources is tremendously relevant (see "Digitalisation in the Circular Economy", p. 92ff.). The political instruments required for establishing tax advantages for resource-efficient technologies, products and materials have been debated for years. Yet, they still await implementation. New products created from recycled materials are still niche products. They are used primarily in flagship projects and in minor quantities. In hardware stores, we tend to find them in the aisles reserved for decoration, but hardly ever in the major areas for common building materials. Unfortunately, they share this destiny with alternative materials of biological origin. Thermal insulation consisting of synthetic materials is offered in the form of system solutions in different variations and thicknesses, complete with adhesives, fleece, render and sealing tape. Pure natural insulation materials are, however, mostly unavailable for market purchase in large quantities. They haven't yet found sufficient acceptance among customers and, thus, are economically unaffordable.

At the same time, a tremendous benefit of circular construction is neither recognised, nor is it discussed in terms of its immense health benefits. In buildings consistently realised with circular and natural construction materials, instead of the still dominant synthetic substances (not to forget all the adhesives and composites), research shows that the pollutant load in the corresponding interiors is significantly lower than in rooms created with conventional construction methods and products. This is the result of intentionally avoiding composite materials and those not sorted by type and also adhesives, foams, impregnations or other types of synthetic bonding agents (see "Pollutants in the Cycle", p. 80ff.).

Valuation of building materials

Currently, the valuation of building materials is often derived from the characteristics of production and manufacturing processes: The more intensive or elaborate processing is, the more valuable the material becomes. Thus, differences are established between rough-sawn and planed wood products or variations of finishes that often lead to a product that is no longer circularly sorted by type. As of now, neither the value of being pure, nor the share of secondary raw materials, nor maintaining value in the case of potential reprocessing or recycling play any relevant roles. For the most part, they aren't even declared as material characteristics.

In the context of construction materials, however, other criteria are becoming increasingly important, such as life cycle assessments. They comprise methods for identifying the impact of construction activity (distinguished according to the separate phases of production, operation and demolition) on the natural environment. They can be applied to individual building materials, entire building components or even the building as such, including its use and the related consequences for operation. The most common factor is the so-called global warming potential (GWP). It is investigated according to the so-called CO_2 equivalent. The term equivalent is used, because greenhouse gases other than CO_2 exist that demonstrate a comparable and, hence, equivalent impact on the atmosphere: Methane (CH_4) and nitrous oxide (N_2O), as well as fluorinated greenhouse gases (F gases), hydrogenated fluorocarbons (HFC), perfluorinated hydrocarbons (PFC), or sulphur hexafluoride (SF6). Based on calculations, their impact is correlated to the effects of CO_2 on the climate: The CO_2 equivalent is, thus, a unit of measure intended to compare different climate impacts. For instance, methane displays a CO_2 equivalent of 25 to 33 by calculating its impact within a 100-year time frame. This means it is 25 to 33 times more effective and, hence, more harmful to the climate than CO_2. By selecting an observation horizon of only 20 years, it becomes even 70 to 80 times more effective. This is due to the fact that methane has a much shorter atmospheric lifetime than CO_2. The oxidation of methane in the atmosphere eventually produces CO_2 and water, in certain circumstances even ozone (O_3). Beyond the effects of pure gases, the unit of measure is also used to evaluate the impact of travel behaviour or objects such as buildings by investigating their GWP and comparing it. The method can also support establishing political guidelines in a targeted manner, for instance regarding whether the extent of particular climate impacts of a building should remain permissible in the future. The European Building Directive states that, by 2030, the building stock under observation within the EU is supposed to demonstrate no more than 55% of the GWP of the reference year 1990. This means that architects need to be educated in how to conduct these calculations – after all, the related parameters will become the essential legal data basis of a building.

Focus on materials and building components

For existing buildings, most emissions during operation are documented (how much energy is required for heating and cooling, how much electricity is consumed, how is power generated?). In the case of renovation, extension and new construction, the required construction components are increasingly becoming the focus of attention. A few years ago, the assumption was that about two thirds of climate impacts were related to operation and only one third to the construction phase. This ratio is reversed when using improved and increasingly climate-neutral technologies: The operation phase has a smaller impact while the production of materials and the construction of buildings have a greater impact. This is why databases that can reliably show the impact of materials and permit recording related changes will become increasingly important. Different factors – from raw material extraction, transport, processing, finishing and packaging – can continuously change as well. More and more manufacturers are in the process of analysing the so-called value chains or supply chains while attempting to optimise the individual related steps and measures. The climate impact of materials is becoming a central issue of architectural design. Yet, the origin of materials also needs to become the focus of attention. Here, the

European Union has formulated clear specifications. The aim is to use the instrument of the EU taxonomy to reduce the consumption of primary materials in buildings by 50% by 2030 and, at the same time, establish shares of 25% of secondary materials. The taxonomy is a system that is implemented across the EU with the aim of classifying sustainable economic activities. It comprises three levels of observation (activities that contribute significantly to climate protection; activities that contribute significantly to climate change adaptation; activities that contribute significantly to sustainable use and protection of water and marine habitats). It was enacted between 2021 and 2022. In 2023 circular economics are the primary topic of debate and enactment. The aim is to supply investors with guidance on the provision of capital for the green reconstruction of the existing economic system. The EU Commission is aware of the fact that the financial system plays a key role in the transition to a low-emission, resource-efficient economy. The taxonomy is intended as an instrument for financially strong and large corporations, allowing them to avoid so-called greenwashing. The European specifications state that the reuse and reutilisation of current and future built-in materials is decisive in order to meet ambitious climate and resource goals. Now and in the future, how can we realise buildings in order to use them as a material bank at a later point of time? Circular and sorted constructions are the answer, presented in this book by way of example on pages 130ff. Aspects related either to climate impacts or reusability and recyclability all need to be considered and evaluated in the design phase, equivalently to Vitruvian questions on beauty, utility, durability or permanence.

A new material library

It is necessary to meet the demands of material usage in the construction sector according to its multiple dimensions of relevance described here and to propose a new understanding of value aimed at truly modern, future-oriented, responsibly usable building materials. For this purpose, the three architecture departments of the Karlsruhe Institute of Technology (KIT), the University of Wuppertal (BUW) and the University of Applied Sciences Münster (MSA) founded an initiative for material libraries at German universities in 2022. The intention is to establish an association of material libraries at German universities. The aim is to create a new hierarchy of valuation of future building materials and develop data sheets on materials that feature lists of described characteristics for purposes of comparison. It is important to provide planners with instruments enabling them to make well-informed and responsible decisions that are relevant to society, in order to supply the built environment with resources from the greatest reservoir that humankind has access to: The existing and future building stock. Echoing Mitchell Joachim's 2013 statement, cities of the future will no longer distinguish between waste and resource [5]. For a related paradigm change to succeed, we need planners to be capable of designing our buildings in a manner that makes them suitable for such purposes. What is also needed are manufacturers who supply materials and products according to the principles of their sorted use. The materials presented on pages 70ff. are from the database of the above-mentioned initiative for material libraries at German universities and offer an impression of a new category of building materials that permit endless (material) circulation.

Notes
[1] Verein Deutscher Zementwerke e.V. 2021/2022
[2] Umweltbundesamt Pressemitteilung Nr. 39/2018
[3] Destatis: Abfallbilanz 2022
[4] see note 2
[5] Joachim 2013

Materials of the Circular Economy

Elena Boerman

Protecting

Roof and facade material: Recycled copper (unweathered and after multiple months under weathering)	Facade and flooring material: Reused spruce (lumber cleaned by brushing)

Material class: Metallic

Material class: Biological

Description

The global reach of copper ore deposits, according to current knowledge, is calculated at about 40 years (see "Resources, Reserves and Ratios", p. 17ff.) [1]. However, excavations are becoming increasingly extensive in order to extract these deposits. In return, this makes the production of primary materials less effective and more expensive. Extraction based on surface mining impacts large areas of land and requires significant amounts of power. Further, during production of primary copper, the environment is heavily impacted, among other things others due to poisonous emissions and the depositing of toxic solids. However, copper can be recycled very well and without loss in quality. The valuable raw resource can even be reclaimed from composites. The end-product recycling rate of the material in the construction sector has currently reached a figure of up to 95 % [2]. A significant advantage of the secondary material is the lower amount of energy required for repeated smelting and the significantly reduced environmental impact compared to the primary material. When attached to a building and exposed to weather impacts, the material develops a non-toxic patina with colours ranging from brown to green and providing high degrees of corrosion resistance, giving the material its characteristic appearance.

Description

European spruce is among the most important domestic wood species of the continent and, thus, one of the most frequently used wood species for building and construction purposes. In 2016 a figure of about 10 million tonnes of matured timber was recorded in Germany, of which about 40 % comprised waste from construction or demolition sites. Currently and for the most part, following its initial deployment, matured timber is subject to cascading use. This process concludes with incineration or landfill. The major share (more than 90 %) is used for thermal recovery right away. More sensible approaches to reuse or repurposing often fail due to adhesive bonding, impregnation, finishes and metallic impurities from connectors, all of which prevent composting. In order to reuse and repurpose reclaimed matured timber, the material may not comprise any synthetic bonding agents or finishes, impregnations, contaminants or metal impurities. In the course of renewed handling of the material, wood can be cleaned by use of a mechanical brush, after which it is cut and newly set for integration in a new structure. By cleaning the wood, different colourations can appear based on the use of harder or softer fibre bundles, resulting in a characteristic aesthetic appeal.

Circularity potential after dismantling:
reuse,
recycling

CO_2 emitting

Circularity potential after dismantling:
reuse,
reprocessing,
composting

CO_2 binding

Protecting

Translucent panel material:
Recycled glass ceramic
(reprocessed waste glass from window and glass container production)

Material class: Mineral

Description
For the production of glass-based items consisting of primary raw materials, high power demands are typical. Glass is perfectly suited for recycling without loss in quality. By collecting waste glass based on sorting by colour, the recycling rate for container glass in Germany exceeds 90% [4]. For plate glass, this figure is significantly lower. The reason is that plate glass from demolition projects features a high degree of contamination due to adhesive bonding, wet sealants or vapour treatment. The EU has defined a contamination threshold of 25 g of mineral and 2.5 g of metallic impurities per tonne of waste glass, above which melting and processing of the material into new plate-glass products is prohibited [5]. Currently, cleaning waste glass is considered economically unfeasible and is not practiced, even though aesthetically appealing and mechanically advanced reprocessed products exist. Translucent glass ceramic panels with thicknesses of 20 mm, for instance, consist of colour-sorted waste-glass shards. They permit melting at relatively low temperatures, while the individual shards remain recognisable. This makes visible the secondary and, hence, very individual character of the material.

Circularity potential after dismantling:
reuse,
recycling,
reprocessing

GWP

CO_2 emitting

Masonry:
Recycled brick made of mineral debris
(reprocessed construction and demolition waste)

Material class: Mineral

Description
The construction sector is responsible for an enormous degree of resource consumption. It produced 230 million tonnes of waste in 2020, equalling 55.4% of total waste production in Germany [6]. Of this waste, only about one third is currently reprocessed into building materials. In most cases, a loss in quality occurs and materials are used for tasks such as earthworks or road construction, where lower quality materials are sufficient. The large mass of mineral construction and demolition waste, however, could demonstrably serve to create new and aesthetically appealing building materials. Brick comprised of high-quality recycled stone, concrete and brick debris mixed with clay can reduce waste streams in the construction industry significantly. They also render obsolete the use of already scarce primary raw materials required for the manufacture of building components. Prior to production, the secondary raw materials sourced from the construction industry are sorted. The results are different types of new brick or block and related colour mixes with their own intrinsic value.

Circularity potential after dismantling:
reuse,
recycling,
reprocessing

GWP

CO_2 emitting

Insulating

Insulation panel:
Reed canes (100% compostable material of natural typha reed fibre)

Material class: Biological

Description
Typha reed fibre sourced from local sweet grass forms the naturally renewable raw material for compostable insulation panels. Reed and typha plants grow along the coasts of European inland lakes or ponds. Crops are harvested annually in winter. The harvested fibre is mixed with a binding agent consisting of corn starch, spread into a fleece and heated by warm air. The stability of the panels is achieved without the use of additives. The fibres fuse when heated and merge into a stable matrix during cooling. Reed as a natural insulation material can be applied to walls, ceilings and roofs. After its use period it can be reclaimed for reuse and reprocessing or returned to a biotic cycle through composting.

Circularity potential after dismantling:
reuse,
reprocessing,
composting

Insulation fill:
Seagrass (natural and untreated Neptune grass – Posidonia oceanica)

Material class: Biological

Description
The natural insulation material consists of dried seagrass fibres of the Posidonia oceanica plant, growing in large quantities in the world's oceans at depths of 3 to 40 m. Once the plants have died of natural causes, wave movements on the seafloor lead to the creation of spherical agglomerates that eventually wash ashore. These balls of seagrass are collected along the shore and then dried and pulled apart. This results in a natural, untreated, soft insulation material that can be used as breathable insulation for roofs, walls and ceiling constructions. The material is also suitable as impact soundproofing. Due to its naturally high silicate and salt content, seagrass is resistant to rotting and naturally non-combustible. It requires no additives in order to meet current fireproofing requirements. Further, it is impervious to infestation by mould, other microorganisms or pests. After its use period, the loosely infilled insulation material permits simple reversibility and reuse, reprocessing or, eventually, composting.

Circularity potential after dismantling:
reuse (manufacturer take-back)
reprocessing,
composting

Insulating

**Insulation granulate:
Foamed clay
(sintered material from expanded clay pellets)**

Material class: Mineral

Description

For the production of foamed clay granulate, a type of clay is extracted that forms air bubbles when subjected to high temperatures. In a so-called shaft mixer, clay is first formed into small pellets. These are then heated to temperatures of approximately 1200 °C in a rotary furnace, which incinerates embedded organic compounds [7]. The resulting gases contribute to the bloating of the pellets. The surface of the pellets is sintered, which significantly increases the compressive strength of the material. The current use of fossil fuels, however, means that the material releases high carbon emissions, due to the required process heat. Yet, this could be prevented by a corresponding shift to regenerative power sources. Foam clay granulate is resistant to rot, frost and fire. It also demonstrates high compressive strength and is breathable [8]. The insulation material is suited particularly well for soundproofing, due to its high density, or for thermal storage, in the form of fill or as a levelling layer. After its use period, foam clay insulation granulate is reusable or recyclable. It can also serve as an ecologically friendly material for gardening or landscaping in the form of fill or surface material.

Circularity potential after dismantling:
reuse,
recycling,
repurposing

GWP
CO_2 emitting

**Insulation panel:
Foam glass
(reprocessed waste glass)**

Material class: Mineral

Description

Foam glass consists of reprocessed waste glass. Crushed glass melts when subjected to heat. It is then processed, extruded, reduced in size and pulverised. While adding carbon, the material mix is heated again. Gas bubbles form that are enclosed within the mass as it begins to cool. Here too, it would be desirable to shift to regenerative forms of power generation in order to significantly reduce the currently large carbon footprint of the material. Foam glass finds use as a lightweight, frost-proof fill material, especially as insulation underneath flooring panels, where a capillary break or an additional levelling layer can be omitted. In the form of rigid panel material, it is often applied as exterior insulation layer in the area of basements or as walkable flat roof insulation layer. The material is durable and can be recycled completely without loss in quality.

Circularity potential after dismantling:
reuse,
recycling

GWP
CO_2 emitting

Materials of the Circular Economy

Sealing

Vapour barrier
PE recycling film
(recycled polyethylene)

Material class: Synthetic

Description
Pure PE recycling film consists of recycled polyethylene film that was reclaimed from the existing building stock, sorted, cleaned and finally shredded. The resulting small-scale granulate can then be melted and extruded into new film products. The construction material is translucent, remains flexible at low temperatures and is durable. It also displays a large degree of resistance to tearing force and water impacts. PE recycling film is, for instance, used as a vapour barrier in wall and roof constructions. Due to its material characteristics it is declared a highly circular construction material [9] and can be completely recycled after the end of the building life cycle.

Circularity potential after dismantling:
reuse,
recycling

Trickle proofing:
Kraft paper
(breathable separation layer
of 100 % cellulose)

Material class: Biological

Description
Coniferous wood waste chips and shavings resulting from industrial timber production are processed in a further industrial step into cellulose pulp that eventually serves as raw material for the manufacture of kraft paper. After cleaning and removal of possible contaminants and impurities, the pulp is formed into a thin paper sheet, dried and rolled. Kraft paper, due to its high impact and tear resistance, is well-suited for use in the building sector, for instance as trickle proofing in ceiling structures. Following the end of the building life cycle, the construction material, when properly and reversibly integrated in a building, allows simple reclamation. It can then be reprocessed or returned into biological cycles through composting.

Circularity potential after dismantling:
reuse,
reprocessing,
composting

Sealing

Windproofing:
PE film
(vapour-permeable sealant layer)

Material class: Synthetic

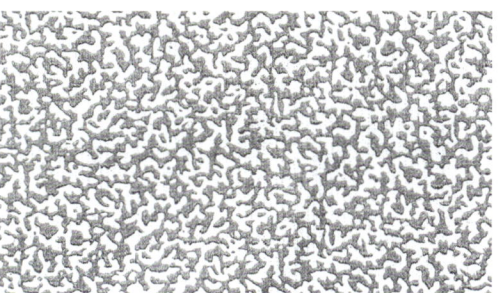

Description
Polyethylene (PE) is a pure plastic and very resistant to acid, brine, oil or grease. In order to create windproofing products, the basic material in granulate form is melted and processed into the desired shape. Special manufacturing methods provide sealant membranes with vapour permeability. Based on this characteristic, the material is used for roof and facade constructions. It both allows moisture to exit the building and prevents rainwater from entering. Windproofing is available in different colours and degrees of permeability. It is also impact- and tearing-force-resistant, weather- and temperature-proof. It is highly durable when protected from direct UV radiation. Following the end of the building life cycle, the polyethylene-based material – as long as it is brought in construction circularly – can be reclaimed and recycled endlessly in a technical cycle.

Circularity potential after dismantling:
reuse,
recycling

GWP
CO_2 emitting

Dry sealant:
Extruded sealing profiles
(elastic, moisture-proof TPE seals)

Material class: Synthetic

Description
Thermoplastic elastomer (TPE) is a type of synthetic plastic that is highly formable when heated. The production of dry seals sorted by type of plastic is generally based on plastic extrusion or injection moulding processes. Once the desired shape is achieved, the different types of dry seals can be further handled or cut with simple tools. In construction they are used for sealing windows and doors. Elastic dry seals are characterised by high resistance to changing weather and temperatures as well as chemical substances. After reclaiming extruded profile seals, given they have been integrated in a reversible manner, sorted building materials can be reused or completely recycled.

Circularity potential after dismantling:
reuse,
recycling

GWP
CO_2 emitting

Load-bearing

Construction material:
Spruce
(untreated coniferous wood)

Material class: Biological

Description
Spruce is one of the most widespread domestic timber species and the most commonly used construction grade timber in Europe. The very soft coniferous wood with its light to medium weight displays very good strength and elasticity characteristics with a medium raw density of 470 kg/m². By debarking, followed by technical drying, cutting and further processing into construction grade timber, spruce can be used either for load-bearing elements, as facade cladding or siding, or for interior finishes. Building with adhesive-free and untreated solid timber elements by use of reversible, circular connection and joinery methods can ensure simple dismantling of wood components in order to reuse them without loss in quality while maintaining their value. If the material can no longer be used in the same or a similar function, the aim should be to utilise it in its pure form, i.e. in the case of wood, mostly adhesive-free, cascading use in the form of biologically bonded wood-based materials that eventually permit returning the natural material into biotic cycles as a source of nutrients. Incineration and simultaneous release of captured CO_2 is not desirable and, in terms of sustainability, should be avoided.

Circularity potential after dismantling:
reuse,
reprocessing,
composting

CO_2 binding

Construction material:
Steel beam
(reused or repurposed construction grade steel)

Material class: Metallic

Description
Steel includes all iron-carbon alloys with a carbon content of less than 2%. Construction grade steel can contain 0.1 to 0.6% of carbon [10]. The material is sourced from iron ore and, increasingly, from reclaimed scrap metal. In the form of different construction elements, steel is contained in buildings either as reinforcement for concrete components, as sheet metal, screws and bolts, or door and window hardware. The degree to which construction-grade steel can be reprocessed or recycled today is very high, due to magnetic separation. The precondition is that the material has been integrated circularly and remained without finishes in order to avoid reducing its quality. If carbon steel is created exclusively from recycled material, 1.67 t of CO_2 emissions are prevented per tonne of steel, compared to primary material production [11]. Further, due to the natural durability of the material, given corresponding weather protection is in place, construction elements consisting of steel can be well-preserved and generally support reuse once they reach the end of the building life cycle. However, due to liability concerns, structural verification requires the documentation of material parameters from previous uses and of damage due to heat or force impact. Alternatively, following demolition and reclamation, experiments, examinations or test runs should serve to identify these parameters (see "Digitalisation of the Urban Mine", p. 94ff.).

Circularity potential after dismantling:
reuse,
recycling

CO_2 emitting

Load-bearing

Masonry:
Loam brick
(air-dried brick from natural clay)

Material class: Mineral

Description
Loam bricks comprise a mix of 100% naturally sourced minerals: sand, silt and clay, differing only in terms of grain size. When creating loam, this mix is either soil-moist or wet. In most cases, biological aggregate such as straw or grass is used as reinforcement against crack formation. By using formwork elements, brick or block shapes can be created as desired and dried by exposure to the sun or other heat sources. Then, the loam brick is used for load-bearing or non-load-bearing construction elements, such as masonry walls or vaulted ceilings. In addition and in parallel, machine-based production methods for soil-moist loam mixes have been established that allow the rapid creation of blocks or bricks by applying high pressure. Bricklaying with loam mortar is an appropriate monomaterial connection method for loam blocks or bricks. In order to change the physical characteristics of the building material, further natural substances such as biological fibre can be added to the natural loam mix, e.g. to improve insulating effects [12]. However, this can also alter the load-bearing characteristics of the material. In order to maintain circularity, loam mixes must include neither synthetic nor cementitious additives or aggregate, nor must such materials be used for bricklaying.

Circularity potential after dismantling:
reuse,
recycling,
return to natural deposits

GWP
CO_2 emitting

Masonry:
Sand lime brick
(thermally bonded artificial stone of sand and lime)

Material class: Mineral

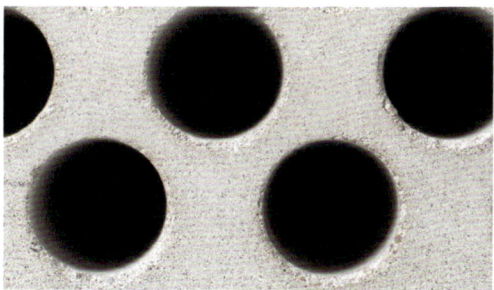

Description
In order to produce sand lime blocks or bricks, calcium oxide (burnt lime), sand and water are mixed. The resulting mix is formed and pressed into bricks. Moderate heat (200 °C) is sufficient to support hardening within a day [13]. The advantage of sand lime bricks compared to fired clay bricks is that the latter requires significantly higher temperatures for firing and, thus greater amounts of energy. However, the overall balance of the global warming potential (GWP) is still extremely high and comparable to concrete. The energy-intensive production of calcium oxide from limestone (calcium carbonate) emits very large amounts of CO_2 due to the required process heat and the resulting chemical reaction [14]. Due to their high weight and their appropriate insulating effect, sand lime blocks or bricks are suitable for use as masonry units, especially for multilayer exterior wall construction types. In the case of reversible assembly – which also calls for further in-depth research – the blocks or bricks can be reused after the end of the building life cycle, or recycled as new aggregate, which is why they are featured here in the context of circularity.

Circularity potential after dismantling:
reuse,
recycling

GWP
CO_2 emitting

Cladding

Wall and ceiling material:
Loam construction panel
(rendered construction panel of natural clay)

Material class: Mineral

Description
Loam construction panels (or loam panels) are used in drywall construction in order to clad interior stud frame structures. The panels support easy sawing and cutting and only require simple screw connections to wall or ceiling constructions. Loam panels consist of natural materials, including loam, biological fibre and starch. On their undersides, the panels feature planar reinforcement in the form of jute or other plant-based fabric [15]. They serve to stabilise the panels during transport, construction and use while also to preventing cracks. After joints are spackled, further natural fabric types are applied as reinforcement for render base layers or finish layers and also prevent crack formation. Loam surfaces provide buildings with an excellent indoor climate. They absorb moisture from indoor air and release it in a delayed manner while binding contaminants. Further, loam panels serve particularly well as thermal storage for lightweight (timber) construction. They absorb heat and release it in a delayed manner – a characteristic that is becoming increasingly important in the context of global climate change. Built into floors, loam panels or air-dried loam bricks can support the discharge of energy and heat in the summer months by use of water-driven systems. At the end of the building life cycle, loam panels with loam render finish can be recycled completely and without loss in quality, sourced for manufacturing new products, or returned to natural deposits.

Circularity potential after dismantling:
reuse,
recycling,
return to natural deposits

GWP
CO_2 emitting

Wall and ceiling material:
Wool felt
(fabric from untreated sheep wool)

Material class: Biological

Description
Wool felt is a textile created from high-quality sheep wool. Sheep wool is a byproduct of pasture farming. After it is collected, it is washed in order to remove excess wool grease and impurities. Moth or bug protection is provided by a biocide-free, electro-physical treatment method. The surface of the wool fibre is modified in a manner that enables permanent protection. Then, the material is processed into mats by applying moisture, heat, pressure and movement, without adding binding agents or reinforcing fibre. Used as wall cover for interiors, wool felt demonstrates good soundproofing and thermal insulation properties. By absorbing moisture and releasing it again, it contributes to improving indoor air quality. The natural and pure material can also be used as fabric wall cover. After its utilisation period it can be fully reused and recycled.

Circularity potential after dismantling:
reuse,
recycling,
composting

GWP
CO_2 emitting

Notes
[1] USGS 2020
[2] Kupferverband e.V.
[3] BMUV 2021
[4] Umweltbundesamt 2022
[5] ibid.
[6] Statistisches Bundesamt 2021
[7] Linden/Marquardt 2018, p. 97
[8] Hillebrandt et al. 2018, p. 90
[9] AMANN Die Dachmarke, 2016, amann-dach-marke.at
[10] see note 7, p. 568
[11] Fraunhofer IMWS 2019
[12] see note. 7, p. 361
[13] ibid., p. 321
[14] oekobaudat.de
[15] see note. 7, p. 364

Cladding

Panel material:
Recycled plastic panels
(recycled yoghurt pots)

Material class: Synthetic

Description
Panels consisting of recycled synthetic materials offer the potential for high-quality reprocessing of plastic items, such as yoghurt pots or packaging material, large quantities of which are contained in household waste. In addition, the basic product consisting of secondary materials can be used to conserve natural as well as fossil resources in a targeted way. After cleaning and sorting the different types of plastic and by applying heat and pressure, unique panel-based materials sorted by type can be created with a diverse range of patterns and colours. The material displays a high degree of resistance to corrosion and water. It is also resistant to acid, brine, oil, grease and scratches. It can be used in diverse ways as a surface materials for interiors, sanitary rooms or exteriors. Further, due to being sorted by type of plastic, it can be fully recycled without loss in quality. Possible impurities, such as those shown in the example in the form of intentional colour markings from fragments of aluminium yoghurt pot lids remaining visible in the material, can be removed in a problem-free manner following each life cycle.

Circularity potential after dismantling:
reuse,
recycling

GWP — CO_2 emitting

Flooring:
Recyclable carpet tile
(100 % circular materials)

Material class: Synthetic

Description
This type of carpet tile consists of completely circular recycled and recyclable materials. As an item of use, the specially designed carpet tile is taken back by the manufacturer free of charge after its period of use. The customer pays for the use of the item, not for the ownership of the item. The included raw materials are of such value to the manufacturer that they are returned into the in-house production process following the use phase. The visible carpet tufts consist of 100 % pure nylon, reprocessed, recycled and sourced from waste material. The underside of the carpet tile additionally contains calcium carbonate as a byproduct of drinking water treatment. It also permits sorting by type in the course of recycling. Further, the tile is not adhesively bonded to floor surfaces, but instead, loosely laid out and clamped. The recyclable flooring material combines innovation, functionality and simple application of circular economy principles within a new ecological model based on use and not on ownership.

Circularity potential after dismantling:
recycling,
manufacturer take-back

GWP — CO_2 emitting

Pollutants in the Cycle

Daniela Schneider

The selection and use of materials sorted by type with reduced pollutant content is a basic precondition for the circular construction of buildings. The composition of common and conventional building materials is varied. Their impact on humans, flora and fauna is very complex, due to the pollutants they contain and the emissions they release. When selecting materials for conventional buildings, legal specifications and frameworks need to be adhered to. A comprehensive declaration of materials, however, occurs only on a voluntary basis. For circular construction and "healthy" indoor spaces, materials need to be part of biological and technical cycles without discharging dangerous pollutants, contaminants or emissions in an uncontrolled manner. Per se, pollutants cannot be distinguished according to categories of "bad", "good" or "green chemistry". Evaluation and identification always take place on a case-by-case basis and in the overall context of all substances contained. A particular substance isn't necessarily toxic to humans or the environment. To quote Paracelsus: "All things are poison and nothing is without poison; only the dose makes a thing not a poison" [1]. Thus, at a lower dosage, a substance can be of use to the environment and to humans. A higher dose of the same substance can, however, be harmful. An evaluation of related substances in the construction sector is of particular importance, since their impact on the environment and on humans must be non-hazardous. By definition, pollutants can display characteristics that can potentially cause harm to the health of humans, wildlife and plants, or the sustainability of entire ecosystems [2].

Classification of pollutants

Potentially harmful substances and emissions are either of a biological, chemical or electrical nature [3]. Even "natural" biological substances, from a natural sciences perspective, can be toxic to humans. Toxic substances existing in nature often serve as a repellent. Natural products such as wood, loam, wax and types of oil can contain natural organic compounds such as terpenes, formaldehyde and other solvents [4]. These are harmful to humans under certain circumstances. Further, mould can release emissions that lead to allergic reactions among humans [5]. Chemically harmful substances comprise the contents and emissions of synthetic materials that are also hazardous to human health and the environment [6]. These include solvents, plasticisers, biocides, formaldehyde, organohalogens, heavy metals, flame-retardant substances, artificially produced mineral fibre and matter categorised as Substances of Very High Concern (SVHC) according to the REACH regulation of the EU. The REACH regulation (EC) 1907/2006 is part of European legislation on the registration, evaluation, authorisation and restriction of chemicals.

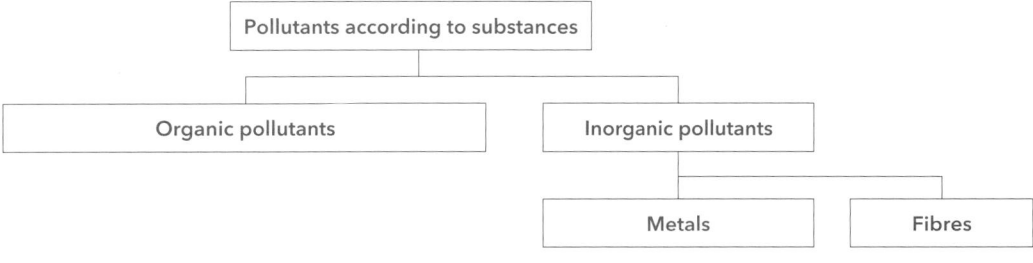

Hydrocarbon-based pollutants (selection of typical pollutants specific to buildings):

- pentachlorophenol (PCP)
- polychlorinated biphenyls (PCB)
- polycyclic aromatic hydrocarbons (PAH)
- volatile organic compounds (VOC)
- dioxins
- furans
- formaldehyde

Possible heavy metals and semi-metals in interiors:

- antimony
- arsenic
- lead
- cadmium
- cobalt
- copper
- nickel
- mercury
- thallium
- zinc
- tin

Inorganic fibre:

- asbestos fibre
- synthetic mineral fibre (SMF)

1

Aside from biological and chemical influences, electrical and magnetic effects are also relevant, including external and internal electrical, magnetic and electromagnetic fields inside and surrounding buildings [7]. This chapter's focus, however, is on natural and chemical impacts (Fig. 2).

Organic and inorganic pollutants

Organic and inorganic pollutants can be distinguished according to their active principle (Fig. 1). The group of inorganic pollutants includes select heavy metals, most of all cadmium, arsenic and lead, as well as copper, mercury and inorganic fibre [8]. Organic pollutants mostly consist of carbon. Such compounds, including potentially toxic ones, exist in nature in oil and gas deposits or coal tar [9]. Inorganic and organic pollutants are biologically metabolised at different speeds. They can be widely distributed and display a high degree of persistence [10]. Organic pollutants can have an acute or chronic effect and lead to reversible or irreversible harm to the environment or the human body.

We encounter organic and inorganic pollutants in different areas, such as agriculture, within food, in packaging, clothing, cosmetics and in buildings. To remove them from built structures constitutes a major challenge. In existing building stock, materials can be found that were legally approved in the 20th century and qualified as "problem-free" [11]. Among others, this includes asbestos and substances such as polychlorinated biphenyls, pentachlorophenol, or polycyclic aromatic hydrocarbons. At the time of product development, permit approval and building planning, related items were highly popular among planners, since they permitted multifunctional application. As cases of chronic illness among individuals who lived or worked in these buildings began to rise, further knowledge was gained and scien-

1
Classification of pollutants according to substances
2
Classification of pollutants according to hazard

Pollutants according to impact

Chemical impact
- substances of very high concern (SVHC)
- biocides
- volatile organic compounds
- formaldehyde
- solvents
- synthetic fibre

Biological impact
- bacteria
- mould
- virus
- endoparasites
- microbial volatile organic compounds (MVOC)

Electrical impact
- radioactivity
- ionising radiation
- non-ionising radiation
- alternating electric fields
- alternating magnetic fields
- electromagnetic waves

2

Pollutants in the Cycle

tific research was conducted on these substances. Yet they were legally banned only decades later. As of today, chemical substances are used with the aim of increasing the durability of building products. This includes preservatives, stabilisers, phthalates (plasticisers in plastic), biocides, aldehydes, flame retardants, as well as binding agents and solvents. The pollutant groups named above, however, comprise only a small selection of the existing range of such substances.

Threshold values and accumulation of pollutants with HBCD as an example
Flame retardants containing the brominated chemical hexabromocyclododecane (HBCD) were in use for decades in conventional building products, such as synthetic rigid foam insulation panels (e.g. expanded and extruded polystyrene). Its biological degradation is only possible under anaerobic conditions (without oxygen) and it remains detectable in the atmosphere, in water and in soil [12]. The REACH regulation categorises it as "highly toxic for aquatic organisms with long-term effect" [13]. It can enter the human body through high-fat foods. Its unrestricted use has been generally banned in the European Union since 2015. From 2016 onwards, products used and integrated in buildings containing more than 1000 mg/kg of HBCD must be discarded as hazardous waste [15]. As long as threshold values of 1000 mg/kg are not exceeded, however, products containing HBCD are categorised as non-hazardous and, in terms of their disposal, as mixed construction or demolition waste [16]. The result is that flame retardants that have already been known to be and qualified as toxic for a long time can, in low quantities and due to improper waste disposal, continue to pollute the environment and accumulate there.

Pollutant accumulation due to lacking boundary values
The example of HBCD shows that products containing pollutants also lead to impacts that defy definition below certain boundary values, by accumulating in the environment without any or only partial degradation within natural processes. In their entirety, they can enter biological or technological streams in an uncontrolled manner and negatively impact reuse or even render it impossible. Boundary values can, in principle, serve to regulate the extent of distribution, use and disposal of pollutants, not only in the case of building materials. However, not all boundary values of known pollutants are legally binding. In some cases, dealing with them is based on voluntary action, including environmental labels and certificates. Existing stock can contain materials with pollutants that should not be reused or recycled. Should the hazardous substance groups named above be distributed into biological or technical material streams in an uncontrolled manner, this would have unforeseeable consequences.

Legal regulations for hazardous substances in building products

New building products are being developed continuously across the world, for a global market. They can display a chemical effect in any kind of context. The Chemical Abstract Service (CAS) registers all known organic and inorganic substances in a database. Since records began in 1907, more than 183 million registered chemical compounds had been documented by 2021 [17]. Of these, more than 350,000 chemicals produced worldwide were also used in building products in 2022 [18]. This sheer number makes evaluation in terms of possible health and environmental hazards difficult, even for materials experts. The sustainable management of chemicals in the European Union and an increasing sensitivity among the population have contributed to making information on hazardous substances, also those contained in building products, available to the public [19]. Since 2007, a new EU directive on marketing new chemicals is in place. The previously mentioned REACH regulation (see p. 80f.) is intended to ensure a high degree of protection of the environment and humans [20] by registering hazardous chemicals. However, registering as such doesn't guarantee that substances are harmless. Currently, the appendix to regulation number 224 lists particularly alarming chemicals [21] that are banned from use within products based on proof of scientific compliance. However, such legally based exclusion of pollutant substances always entails a lengthy process and never constitutes a forward-looking precautionary measure. According to a study conducted by the European Environmental Bureau (EEB), it takes 13 years and 8 months on average in order to remove hazardous chemicals from market circulation [22]. In the meantime, the diversity of materials is growing demonstrably faster than a related hazard assessment can take place. Even for materials experts, an evaluation of individual chemicals that holds weight, in particular in terms of gathering information on their hazardousness, comprises a complex and lengthy affair, since it is dependent on the information provided by manufacturers. Detailed information with regards to contents can, for instance, be found in technical data sheets and safety data sheets, albeit mostly incomplete. They include advice on the use of products as well as safety-related information on substances and mixes thereof that can support the selection of building products. Often the included information features only parts of what should be required for circular planning and construction. It is therefore recommended to obtain a full declaration from manufacturers, including information on the contents of individual components, all the way to the level of suppliers (see "Full declaration", p. 85f.). The need for confidentiality is an obstacle to this.

Selecting low-pollutant building materials sorted by type

Ecologically non-hazardous, durable and universally applicable materials that retain their value are an important component of a strategy aimed at circularity. Products that display these characteristics can, in theory, be reused again and again or, at least, reclaimed while maintaining high levels of quality. This includes natural and synthetic chemical building materials. They consist of non-hazardous substances and, following their utilisation period, can be fully returned to natural and technical material streams as "nutrients". Thus, they display positive added value (Fig. 8, p. 25). Materials sorted by type and demonstrating minimal resource use should be preferred at all times and chosen for their effectiveness.

Exclusive use of known material substances

In principle, planners should avoid the use of materials with substances that are not fully known. In recent decades, the development of extensively modified and multi-functional products with time-saving, cost-efficient and practical application characteristics has increased. These building products are, to a major degree, considered safe and flexible, while enabling easy assembly. By using them, elaborate calculations and preparatory steps can be avoided and lacking craftsmanship skills or the use of particular tools can be compensated. Most of the time, these construction materials – often composite types – consist of synthetic materials that, however, result in problems for circular reuse.

No CMR substances

Substances contained by building products must neither include factors toxic to humans or the environment, nor substances that are sensitizing, carcinogenic, teratogenic or with mutagenic effect on germ cells. Such so-called CMR substances (carcinogenic, mutagenic and toxic to reproduction) according to the European GHS regulation (EC) no. 1272/2008, called CLP regulation (classification, labelling and packaging of substances and mixtures), thus, fall into two possible categories. Corresponding warning symbols and information are included on product labels and safety data sheets [23]. The related boundary values of the individual substances are, however, not explicitly regulated by legislation (see "Threshold values and accumulation of pollutants with HBCD as example", p. 82). Today, it is often the case that hazardous substances in building products that should be avoided when selecting materials sorted by type only need to be excluded on a voluntary basis. This calls for political intervention. In order to change the status quo, statutory regulations such as REACH could be amended by requiring a full declaration of the substances contained.

Avoiding "compound products"
Compound products consisting of natural and synthetic materials should be avoided in order to ensure circularity. For instance, cross-laminated timber adhesively bonded with artificial resin constitutes a mix of natural and technical components. Subsequent use is only possible in a technical material stream based on the cascading use of wood. Untreated solid timber without any synthetic impregnation etc. can, instead, be safely reused or returned into a biological cycle.

Building component joining sorted by type
A further relevant parameter for the implementation of a circular strategy and a related use of materials is to start thinking about building components that are reversibly assembled and layered already in the design phase. Reducing the complexity of the individual building layers and functions helps to avoid composite solutions (see "(Re)Building Simply", p. 98ff. and "Layering as a Circular Principle", p. 118ff.). Even the most diligent selection of building products cannot prevent having to choose materials that are not completely free of pollutants. Reasons for this are the need to adhere to norms (e.g. related to fire safety), the implementation of technical regulations, as well as market availability and/or lacking healthy product alternatives. In principle, synthetic, liquid, paste-like or insoluble coatings, render, spackle, finishes etc. requiring curing or hardening should be avoided. They prevent reversibility sorted by type and, when they comprise layers near to the surface, often contain pollutants and easily release emissions.

Full declaration
In order to be comprehensively aware of the contents of building products, manufacturers should ideally provide a full declaration upon request. Knowing the chemical composition of materials, their contents and emissions is key to their potential and safe subsequent use [24]. The C2C Banned List of Chemicals [25] can serve as a basis for a full declaration. It defines and excludes hazardous chemicals that accumulate in the biosphere and technosphere, lead to irreversible harm to humans, or are hazardous to the environment in terms of manufacturing, use and disposal. The list is not simply intended as a checklist aimed at eliminating hazardous chemicals. Instead, it is supposed to set an example and provide an incentive to avoid substance mixes with similar structure and effect as a replacement. The list distinguishes between technical and biological nutrients. It, thus, recognises two separate cycles. For a full declaration, the list can be supplied to manufacturers in order to adhere to and confirm the selection of listed materials.

Content and emissions certificates

Aside from a desirable full declaration, products generally find use that feature a recognised quality seal or environmental label (ecolabel) and related certification. According to the online platform "Label-Online", currently 169 different ecolabels for building products are available (as of October 2022) [26]. They can contribute to the transparent communication of substances, contents and emissions of building products. However, comparing them is difficult. Each represents a different aim, focus and approach to evaluating and ensuring quality standards:

- Testing content and emissions: When testing building products, some environmental labels focus on select and individual contents and emissions (e.g. Blue Angel) [27].
- Testing emissions: There are ecolabels that exclusively evaluate specific emissions of building products. The related tests and evaluations take place, in the case of Emicode, after 3 and 28 days respectively in a test chamber and based on Building Product Health Evaluation Board decisions on the lowest concentration of interest (AgBB-NIK) [29]. In the course of the test process, individual substances contained are neither analysed nor evaluated.
- Testing contents, emissions and odours: There are certifications that test contents, emissions and odours according to an institutional testing methodology and evaluate results accordingly (e.g. GUT carpet label) [30].

Labels and certificates are provided according to a voluntary exclusion list defined by the particular institution. Within this process, it is determined which substances a product or its emissions may contain at all or according to certain boundary values. This kind of definition is a common and simple form of regulation and evaluation of individual substances. Depending on the particular specification, it corresponds to a boundary value (see "Threshold values and accumulation of pollutants with HBCD as example", p. 82). However, there is a risk that problematic substances contained are omitted from the observation and, thus, retain their toxic effect. This is due to new substances and compounds continuously entering the market. The exclusion list – depending on institution – is amended only after a delay. Thus, when selecting building products featuring a particular label, it is recommended to precisely enquire on the following parameters:

- number and completeness of reviewed contents
- number and completeness of reviewed pollutants
- reviewed emissions and volatilisation
- determined boundary values

In addition, consideration should be given to the fact that the criteria employed by a certifying body can change over time. Review of related data on recent test processes is, thus, recommended. For planning to be circular, it is further recommended to exclusively select and use building products of which all contents are known, which equals a full declaration. Currently, the range of products available on the market is insufficient for the construction

3
ABC-X declaration list for the evaluation of building materials: Each individual content of a product is determined and its risk evaluated. Substances that receive the grade "A" are optimised. Substances with the grade "X" describe CMR substances that are not permissible within a product in the long term and need to be replaced with improved substances.

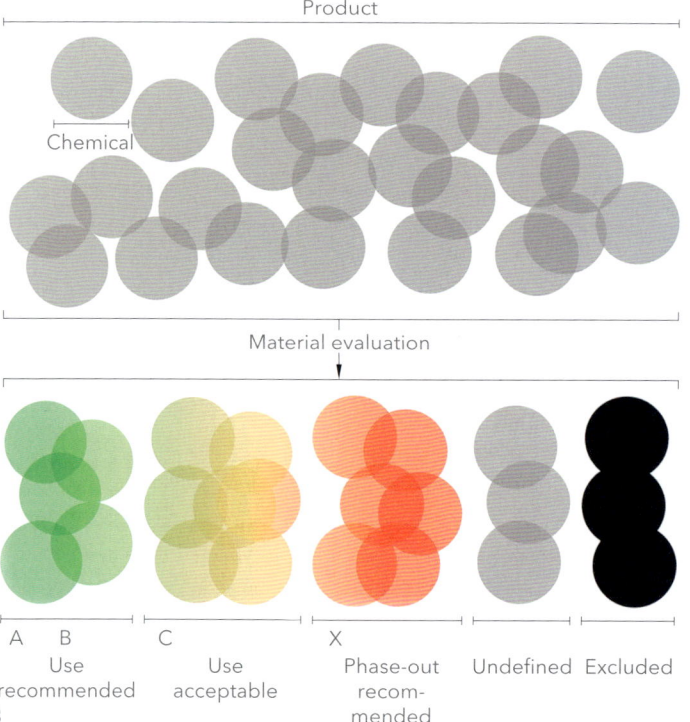

of a circular building based on a full declaration. It is necessary to request and collect individual declarations.

Example: Cradle to Cradle certificate

The Cradle to Cradle certificate aims at a complete review of all contents. The evaluation programme doesn't follow pre-defined material groups. Instead, it aims for a transparent inventory of all contents including supply chains. The precise identification of contents conducted by materials experts within the certification process excludes all substances toxic to the environment. If the certification process reveals any of these, the opportunity exists for the manufacturer to replace them with non-hazardous substances [31]. The basis for the evaluation of substances is the Cradle to Cradle Certified Restricted Substances List (RSL) [32]. This label, with its transparent award and control process, places high demands on circularity. Five categories serve to review whether materials are healthy, where they originate from, whether they are reusable, whether they were produced with renewable energy, which greenhouse gases were emitted, whether water management was applied and how they score in terms of social responsibility. Building products certified according to the Cradle to Cradle principle include flooring, wall paint, furniture, as well as entire systems for interior finishes and building services. These products are designed to positively impact the environment and humans, they are considered safe, transparent and actionable [33]. Product certification for the C2C banned substances list includes the classification and evaluation of materials according to the ABC-X system. Each individual substance contained in a product is graded according to the risks it poses. Substances that receive an A grading are already optimised. Category X products are CMR substances that are unacceptable as product content in the long term and need to be replaced with improved substances (Fig. 3). Further, building products are recommended that include the ecolabel Natureplus and the Eco-Institut-Label.

Misleading product labels and specifications

Aside from building products with eco-labels and certificates, there are building products on the market featuring seals created by the manufacturers themselves, for instance with confirmation that certain standards were met. By doing so, manufacturers save costs for an independent testing process and use their own labels and logos for advertisement purposes [34]. Statements made by the companies on labels resembling certificates often provide consumers with misleading information lacking transparency regarding the substances used. The European Society for Healthy Building and Indoor Air Quality (Europäische Gesellschaft für Gesundes Bauen und Innenraumhygiene, EGGBI) points out that "comprehensive criteria catalogues and test reports, for the most part, cannot be obtained from these manufacturers or trade organisations" [35]. In principle, manufacturer information of this kind used in advertising and product descriptions calls for critical enquiry. When encountering such ecolabels, it is recommended to research the testing methodology, the analytical approach, the scope and legal testing basis, as well the individual results. A further important quality attribute is to name an accredited institution typically mentioned in a test report [36]. A misleading description would be to advertise a product as a "formaldehyde-free wood product". Instead, it would be correct to say: "formaldehyde-free adhesive, timber contains natural formaldehyde" [37]. The description gives no information on further pollutants that are potentially to be expected when using adhesives, coatings or finishes. What can be observed is that manufacturers are becoming increasingly bold in their use of so-called greenwashing statements without providing well-founded evidence [38]. Different definitions of key terminology and calculated sums are featured with different degrees of significance. The time has come to formulate clear political regulations and, additionally, introduce a certification system for circularity. This would obligate manufacturers to establish compre-

Acronym	Description	Boiling point
VVOC	very volatile organic compound	< 0 to 50–100 °C
VOC	volatile organic compound	50–100 to 240–260 °C
SVOC	semi-volatile organic compound	240–260 to 380–400 °C
POM	organic compound associated with particulate matter, or particulate organic matter	380 °C
TVOC	total volatile organic compounds, the term describes the concentration of all individual compounds	

4
Overview of all volatile organic compounds (VOC) according to boiling point

hensive positive declarations for building products by presenting test results in a transparent manner in support of achieving the goals of the European Green Deal by 2050.

Building product emissions
Building products, interior finishes (e.g. flooring, furniture, carpets), yet also the particular function, indoor maintenance or users present can all release different kinds of emissions impacting indoor air. Indoor emissions can also be described as volatile organic compounds (VOC). This term encompasses numerous solvents and other chemical-organic substances that exist in the indoor air in gaseous form. According to the WHO, these compounds can be distinguished according to their boiling point and/or the resulting volatility (Fig. 4). They are predominantly inhaled and, thus, play the most important role in the evaluation of indoor air quality. The degree to which indoor air is considered "healthy" and whether it features a low pollutant content also depends on the quality of the outdoor air supply, as well as pollutants and contaminants inside the building. Odourless emissions can also lead to health hazards by triggering allergic reactions. In contrast to substances contained in building products that are precisely defined and describable without ambiguity, indoor air pollution generally features complex mixes of compounds demonstrating a most diverse range of emissions characteristics.

Quality control in the planning and implementation process
The selection of "healthy" building products with low pollutant content requires a necessary concept and testing process for adhering to quality standards. For this purpose, an extensive process for circular planning and construction is recommended that is organised into six steps with standardised service phases and stages (Fig. 5).

Planning team expertise
In addition to process sequences and service phases in place to date related to the German Fee Scale for Architects and Engineers (deutsche Honorarordnung für Architekten und Ingenieure, HOAI), the creation of "materials competence" in the planning team is important. For instance, a qualified person from the field of building ecology, or a person with expertise in materials health, or an architect with corresponding training can offer related input during the planning and implementation process. In an ideal case, at the beginning of a project, a team will develop a materials concept. This includes selecting materials, obtaining necessary declarations prior to the construction phase, testing materials used on the construction site for their compliance with standards, as well as following requirements placed on indoor air quality.

Step 1: Concept development for a preliminary design materials concept
At the beginning of this process, the preliminary design phase entails developing a

project-specific materials concept. Among others, it contains the general requirements placed on the origin of the materials, on the conservation of resources, on the preparation of the site, as well as the description of how to deal with construction waste. Further, material types relevant to the design, the production processes and the environmental impacts are clearly defined and selected accordingly. This constitutes an important aspect of the overall process of quality control. Individual material types are selected according to the following list:

- biogenic material (meaning renewable and of biological origin), not from fossil sources
- technological material from a fossil source, yet designed in a manner supporting reversibility and reuse while maintaining the highest possible level of quality
- reusable biogenic and technical components
- hybrid systems solutions that feature biogenic and technological materials supporting reuse
- possible reversibility and available manufacturer or supplier take-back programme
- repurposing and reprocessing [39]

In general, materials should already be tested in an early planning phase with regards to their environmental impact, the embodied energy, the CO_2 emissions released within the production process, the use of secondary raw materials, as well as the use of renewable energy sources for the manufacturing of products and materials [40].

The responsible planning team needs to define the individual reclamation pathways for related material types, as well as connection and joinery methods. It is recommended to structure a comprehensive, project-specific materials concept according to an already familiar system corresponding to particular trades. The materials concept should feature an effective reusability strategy according to the materials selection and specification.

Step 2: Design
The design should determine the positive contents and substances for individual material types according to possible boundary values defined by regulations. The established materials concept is amended by further detailed information on contents and emissions that are either excluded or demonstrate a positive impact on humans and the environment. For every product, clear individual quality criteria are determined that address circular planning and building. This includes definitions on requirements for:

5
Quality management process for the selection of materials sorted by type

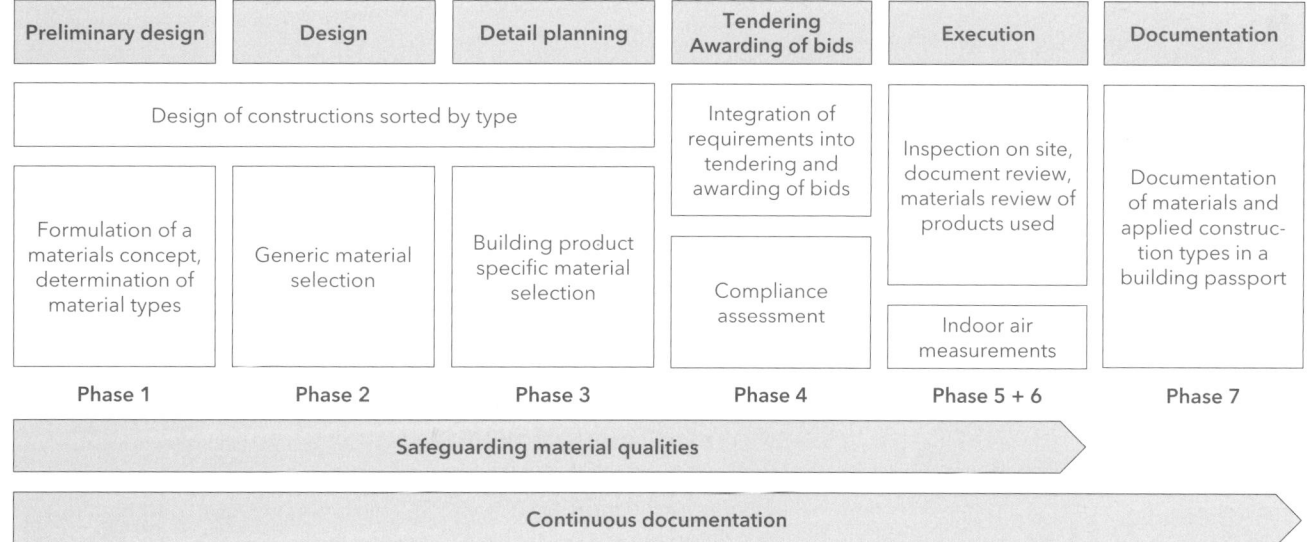

5

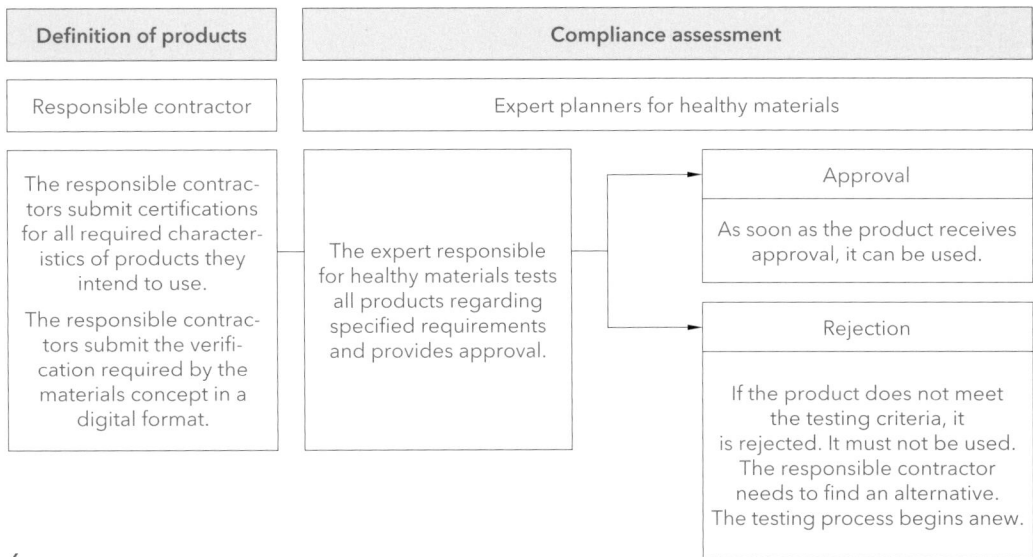

6
Material testing sequence

- contents
- positive formulations
- full declarations
- boundary values for hazardous chemicals, pollutants and byproducts
- quality labels and ecolabels

During this phase of sampling, generally applicable building materials are selected and the defined leading products are tested by experts according to specified quality criteria. The concept also outlines the verification process for every product. Such processes can be documented by technical data sheets, safety data sheets, manufacturer declarations as well as further substance groups given consideration in quality labels and ecolabels, or the highly important full declaration.

Step 3: Detail planning
The more design and planning progresses, the more detailed information on substances and emissions needs to be, evaluated by qualified experts. The selection of declarations and materials comprises a continuous process of deliberation on and identification of suitable building products.

Step 4: Tendering
The requirements placed on materials are integrated into the service specifications down to the level of individual items. In order to adhere to and realise the required material qualities, a contractual amendment to the service specifications is recommended. Contractual agreement on specific material qualities provides legal entitlement with regards to responsible contractors and/or tradespeople actually delivering stated levels of quality.

Step 5: Material declaration prior to realisation
In the course of a material declaration, the compliance of materials with requirements stated in the materials concept is subject to testing. When awarding contracts, the determined building products and materials need to be reviewed according to the defined quality criteria. In the context of a full declaration, processes for the verification and evaluation of materials require obtaining product information from the manufacturer, as well as technical data sheets, safety data sheets, sustainability data sheets and specific manufacturer declarations. According to the information received, possible substances can be identified and evaluated in relation to their circularity. Material declarations should be obtained prior to building execution. Only building materials and products approved by expert planners are permitted for use on the construction site.

Notes
[1] Schäfer 2021
[2] Kaub 2021, p. 120
[3] Bavarian Chamber of Architects 2018, p. 146
[4] BAM 2007, p. 8
[5] see note 3
[6] ibid.
[7] REACH
[8] Gesamtverband Schadstoffsanierung 2010, p. 170 and p. 179
[9] ibid., p. 66
[10] Glücklich 2005, p. 121
[11] Zwiener/Lange 2015, p. 5
[12] Linden/Marquardt 2018, p. 247
[13] ibid., p. 248
[14] Umweltbundesamt 2017, p. 8-12
[15] see note 12
[16] see note 14
[17] American Chemical Society: CAS Content
[18] Scinexx 2020
[19] see note 12, p. 522
[20] ECHA: REACH
[21] ECHA: List
[22] BUND 28.7.2022
[23] BfGA: CMR
[24] Schneider. In: Heisel/Hebel 2021, p. 127
[25] Cradle to Cradle: Banned List of Chemicals
[26] Label-Online
[27] Blue Angel
[28] Emicode
[29] Umweltbundesamt 4.11.2022
[30] GUT-Produktprüfung
[31] C2C Products Innovation Institute
[32] Cradle to Cradle Certified
[33] EPEA
[34] Spritzendorfer 2022
[35] ibid.
[36] ibid.
[37] EGGBI Holz
[38] EGGBI Greenwashing
[39] EPEA
[40] ibid.
[41] Material Building Scout
[42] DGNB – ENV1.2

For the implementation of material and compliance tests, as well as consistent documentation of approved and rejected materials, the use of an online service platform such as Building Material Scout is recommended. It features consistent documentation of all materials defined and tested for planning and realisation, which allows simplifying the selection and related communication between all actors involved in planning and construction.

The material-testing process shown in Fig. 6 has been established since the beginning of conducting building certifications in the construction sector. The template for this testing process was determined, among others, by the German Sustainable Building Council (DGNB) criterion ENV 1.2 "Risks for the local environment" [42]. Due to the aspects relevant to the evaluation of environmentally friendly materials and building products, a material-testing process was explored and established that companies, manufacturers and tradespersons should be familiar with. As a companion to the implementation of the process, a control function remains necessary – in this case, conducted by a qualified expert – for securing material qualities in all service phases (see "Planning team expertise", p. 88f.). For this purpose, political decision-making is required for establishing a material testing process for circular planning and building.

Step 6: Indoor air measurements
In order to safeguard the quality of indoor air, it is necessary to select and apply building materials with low emission behaviour. Known, expected and avoidable emissions should already be considered and documented in the materials concept. Qualified experts can, on the basis of manufacturer data, analyse and evaluate emissions early on. The risk of unforeseen reactions of emissions in the indoor air (secondary emissions) can hardly be predicted by a material declaration, because it depends on the moment of integration into the building, the ventilation of interiors, indoor temperatures and solar radiation. Replacing materials on the construction site is typically only possible with significant effort in terms of construction and related high costs. Further, indoor air measurements should be partial to the construction process. In an ideal case, they are conducted by an accredited testing institution according to specifications stated in DIN 16000 prior to completion of the building.

Step 7: Documentation of materials used
In order to repurpose and reprocess materials after their projected period of use, select information on the materials used and products declared is required. In the course of individual service phases, they should be documented and updated. It is also recommended to refer to an online platform, such as the Building Material Scout, in order to compile all information. Such documentation is complex because of the need to manage the verifications of the individual products. Materials containing contaminants that cannot be avoided due to technical requirements or legal preconditions (e.g. in the case of fire safety specifications) also need to be documented. In an ideal case, a building material passport serves this purpose (see "Digitalisation of the planned and built environment", p. 96f.). This allows compiling all information on the building elements realised that are relevant to the subsequent use of materials within a single document.

Digitalisation in the Circular Economy

Hanna Hoss

Since the 1980s the digitalisation of the building sector has been advancing rapidly. In the beginning, computers were used as a digital drafting tool. Eventually, planners began to apply the immense technological capabilities to the design of increasingly complex tasks and forms. They also employed digital means to fundamentally reorganise and increase the efficiency of administration, logistics and management of processes in the building sector. Often, approaches to adaptation from other industry branches served as inspiration. This applies to the use of robots in the production of buildings. In the automotive industry, robots had already been more commonly in use at an earlier point in time. However, the building sector features a lesser degree of repetitive tasks. New pathways and approaches became necessary on how to sensibly employ such technologies for construction. In general, accelerated processes, increased precision and reduced staff costs are considered advantageous. The interactions between humans and machines can also be advanced on the construction site. This also includes large-scale 3D printing by use of mineral building materials, which is becoming increasingly relevant. Further, small-scale drones will play an important role in the future, mostly for construction supervision tasks, documenting construction progress, measuring and surveying, or for the maintenance of buildings. The digital documentation and communication of information in real time also suggests employing artificial intelligence. This includes recognising certain patterns and sequences, proposing resulting actions by use of applied technology and detecting damages for controlled building maintenance. Many opportunities exist that are increasingly oriented on issues of sustainability and circularity. Planners work with digital tools in order to optimise the flow of forces within structural systems and, thus, drastically reduce the amount of materials required for construction.

Digital design tools

In his Block Research Group, Philippe Block, professor at the ETH Zurich, uses parametric design principles in a targeted manner in order to develop geometrically complex load-bearing systems with a focus on topics of sufficiency and effectiveness. Inspired by the know-how of master builders of the past, who used materials in an

1
Armadillo Vault, Block Research Group, 2016
Venice Biennale

2 a
2 a–b
Robotic prefabrication, roof structure of the Arch_Tec_Lab, ETH Campus Hönggerberg, Zurich, 2016, Chair of Architecture and Digital Fabrication, directors: Fabio Gramazio, Matthias Kohler

3
Material savings through geometrical optimisation by use of digital design tools, prototype of a roof structure for the HiLo project, Zurich 2021, Block Research Group with EMPA and NEST

optimal way according to their intrinsic characteristics, thereby spanning great widths with as little material usage as possible, the Block Research Group is particularly oriented on compression-only shell structures. For this purpose, they developed the plug-in RhinoVAULT, which identifies ideal load flow patterns. Based on its optimised form, the compression-only design of the Armadillo Vault [1] for the 2016 Venice Biennale required no reinforcement whatsoever (Fig. 1). The team advanced principles that were initially applied to a floor slab by the High Performance – Low Emissions (HiLo) Unit of the Swiss Federal Laboratories for Materials Science and Technology (Eidgenössische Materialprüfungs- und Forschungsanstalt, EMPA) in Zurich in 2021. The floor slab consists of a thin, doubly curved shell with vertical bracing. The required formwork was created by use of a 3D printer. Here as well, the use of materials follows the flow of forces. As a result, more than 70 % of concrete and 90 % of reinforcement steel [3] were saved, compared to a reinforced concrete slab subject to bending forces (Fig. 3). Hence, "weak materials" [4] become suitable for construction according to their specific characteristics, without requiring additional capabilities resulting from connections or bonding to other materials, which typically does not allow sorting by type.

An historic example for load-bearing structures based on the optimised use of materials is the Zollinger roof: Created after World War I and motivated by the housing and material shortage of the 1920s, the roof design impressively demonstrates how Friedrich Zollinger succeeded in creating a timber roof based on a lightweight, serialised and, most of all, cost-efficient construction method. The roof type and its vaulted tectonic form can serve to enclose large spaces. Longer beams or columns aren't required. The self-supporting construction system consists of prefabricated, relatively short pieces of timber with minor cross sections. They are assembled into a diamond lattice bar net structure, resulting in material savings of 40 % compared to a purlin roof [5]. The bolted connections of the construction permit problem-free disassembly. At the time, approximation methods were necessary for elaborate structural stability verifications, which made structural testing complex. As recent applications of state-of-the-art production technologies indicate, digitalisation plays a decisive role for such tasks. Without a doubt, Fabio Gramazio and Matthias Kohler, the architects of the roof construction of the Arch_Tec_Lab on the ETH Hönggerberg campus (Fig. 2), were familiar with the Zollinger roof. The living lab for robot-based production methods demonstrates in the design of its own roof how to arrange nearly 50,000 timber slats into a self-supporting roof structure with curved trusses spanning more than 15 m [6]. The timber roof structure became possible by use of a digital, interdisciplinary planning process. The resulting building model was eventually fed into a gantry robot for production. Unlike conventional architectural construction with material usages of 400 kg/m³, only 240 kg/m³ were required for building the Arch_Tec_Lab [7].

In the building sector, most resources are used for shell construction [8], the typical 50-year life cycle of which is twice that of facades [9]. If an alternative or continued use of an existing building is impossible, enormous potential exists for reducing related emissions and embodied energy. By contributing to the creation of actual buildings, the research projects mentioned above demonstrate that optimising load-bearing characteristics – combined with innovative digital production methods – supports the development of a type of architecture that is functional, aesthetically appealing and sustainably qualified, while offering a glimpse of the future of digital construction.

Digitalisation of the urban mine
In terms of the administration and provision of big data and information, building information modelling (BIM) plays a central role for digitalisation processes in the construction field. Not only does it offer new opportunities for visualisation. It also supports managing complex construction processes, documenting them and recording changes by use of a so-called digital twin. This digital image of an edifice is becoming increasingly important for long-term building maintenance. Its aim is to include data on where, when and at which quality levels which building materials will be available again for future use. The digitalisation of buildings will also contribute to the intelligent management of future resources and the provision of information on their continued use in data form. Currently and unfortunately, such processes – related to materials sourced from the so-called urban mine – present a very different case: Information on reclaimed materials or construction components and, thus, their specific material characteristics or capabilities, are hardly available. In 2018 the team of the Professorship of Sustainable Construction of the Karlsruhe Institute of Technology (KIT), in cooperation with students, began planning the so-called Added.VALUE.Pavilion (Mehr.WERT.Pavillon) for the German national garden show (Bundesgartenschau, BUGA) in Heilbronn. The task was to design and realise a temporary structure with the intention of demonstrating that exhibition architecture following the principles of circularity in an uncompromising manner is possible today. Following the end of the exhibition, the aim was to dismantle the building sorted by type and return its materials into the corresponding cycles (Fig. 4). All materials and building components used were actually reused or recycled and, thus, had already experienced one or more life cycles. One challenge was to develop the load-bearing structure based on pre-used steel sections from an old and recently demolished coal power plant near Cologne. With a current share of 88 % the use of recycled scrap steel is a common practice, simply due to economic reasons. Reuse, however, is still scarce at 11 % [10], despite its potential for saving costs and emissions. In the case of the Added.VALUE.Pavilion various steel elements were successfully reclaimed sorted by type and without destroying them. However, data on the sections, the steel quality, or the mechanical properties weren't available. The project shows that the reuse of steel currently still requires significant effort in terms of planning, time, as well as economic expenditure. Which amount of suitable steel can be reclaimed from the power plant? How are the construction components or their cross sections dimensioned? Are there material locked connections, such as welding seams, that make reclamation difficult? What kind of steel was used, which are its characteristics? Can a manufacturer certify its quality?

4
100 % reuse and recycling, Added.VALUE.Pavilion, German national garden show in Heilbronn, 2019, Professorship of Sustainable Construction, Karlsruhe Institute of Technology (KIT)

How was the steel handled during operation? Most of these questions remained unanswered. The only way to use the sections and evaluate them from the perspective of structural engineering was an elaborate and, thus, costly technical examination. This testing took place at KIT in Karlsruhe. A diverse selection of samples was extracted from the hollow sections and tested for their steel quality. These tests eventually resulted in case-by-case approvals that, however, are only valid for the particular construction project. Construction-grade steel is only defined according to its mechanical properties. Steel of the same type can, depending on manufacturer or batch, display different chemical compositions. After disassembly, the pavilion was intended to serve as an example for a future material stock sorted by type. This also means that a renewed, elaborate certification process would have to be avoided for the next cycle of using the steel, in order to actually ensure that "buildings become the material stocks of tomorrow" [11]. In order to meet this demand, the availability, quantity and quality of these material stocks need to be digitalised, also in the existing urban mine. The precondition is, thus, the creation of a digital twin of every building that is fed into a material database featuring information on materials and construction components. This process, however, is very elaborate and costly.

Yet, new principles of digitalisation are becoming relevant here as well: Across the globe, different initiatives are researching the generation of data on and the management of existing materials, elements and components integrated in existing buildings. Algorithms permit conducting building surveys in real time with common tablets or smartphones. Digital devices are used for image recognition of individual components or elements of an existing building. They identify the potential of materials and elements while giving advice on resale values or reclamation strategies. By employing these tools, a digital marketplace for used building materials and components is emerging at an increasing rates.

Since 2014 a number of firms has specialised in this field: RotorDC in Brussels, Zirkular in Switzerland and Restado in Germany [12]. The latter has, in the meantime, become the largest digital marketplace in Europe for the reuse of construction materials. New building materials that, for instance, are created due to planning errors and, as a result, are normally disposed of or landfill, can be traded on the digital marketplace and procured by private or professional customers. At the same time, materials and products that have reached the end of their first cycle of use are offered for sale, reclaimed and processed. The company advertises item prices that are up to 70% lower than for comparable new products [13]. The corresponding search filter typical to commonly used digital marketplaces can limit the range of potential customers, resulting in a new regional significance. Thus, different actors can participate in a regional, circular value chain. This includes local demolition and reprocessing firms, test laboratories, planning and architecture offices, as well as so-called building material saviours, traders specialised in secondary or historic building materials, transport firms or craftspeople who eventually build the reclaimed material into another structure, until this phase of use concludes and the cycle begins anew.

Aside from saving resources and preventing production-related emissions, this allows avoiding globalised, high-emission supply chains that result from the place of origin and production of raw materials and semi-finished items in countries with often questionable labour conditions. In 2020 the founders of Restado created Concular, a Software-as-a-Service (SaaS) platform. Its aim was to digitalise the corresponding material streams before and after entering the marketplace and to make the data available and transparent, thus providing them with a digital identity [14]. The result is the creation of a "digital ecosystem for circular construction" [15]. Viewed as a process, it permits economic and ecological evaluation, from demolition to quality testing to logistics to reintegration into a building.

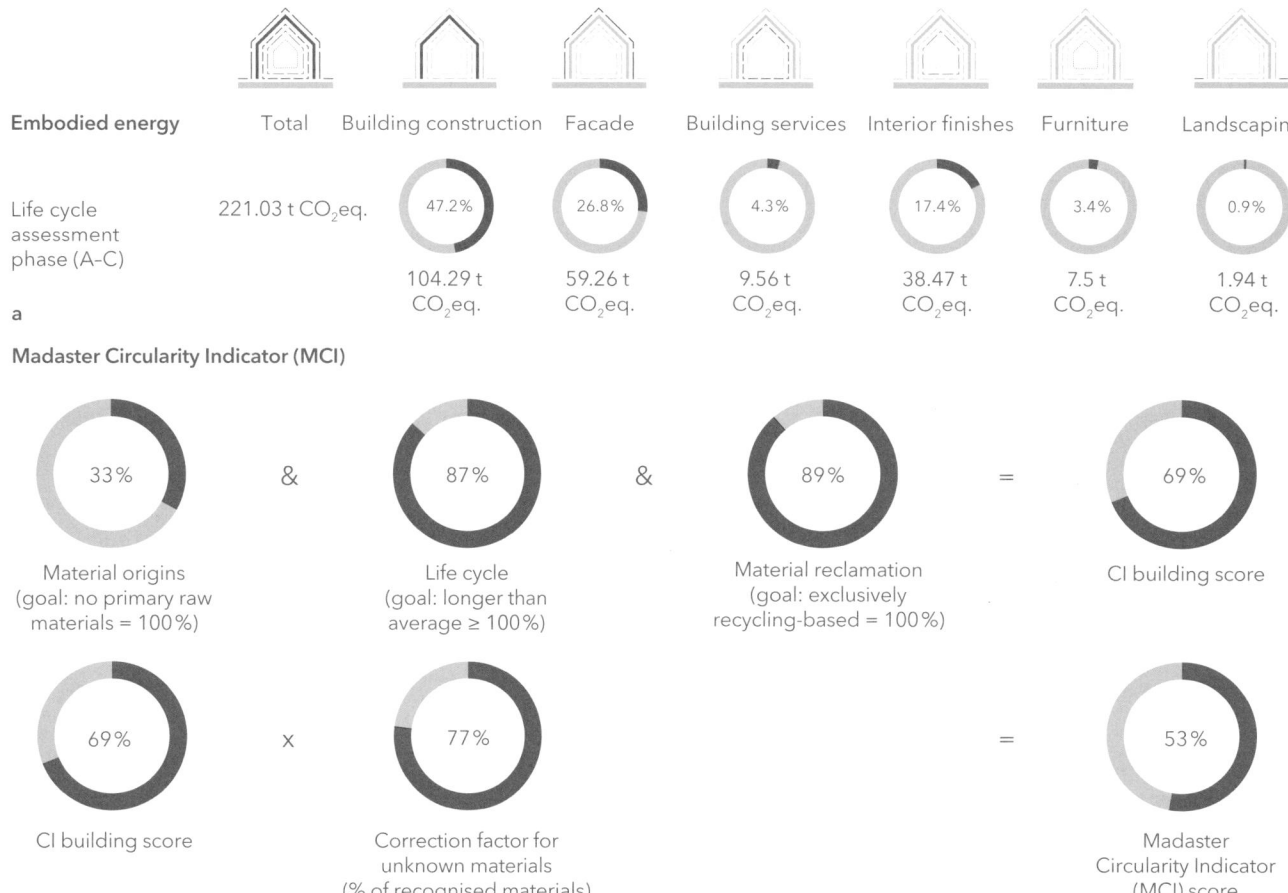

5 b

Digitalisation of the planned and built environment

"The buildings of today are the material stocks of tomorrow" [16] – to do justice to this premise, it is necessary to generate future data sets in order to illustrate and comprehend complex material flows. This results in the need for a material passport. It serves as a digital "travel inventory" for every newly erected building, containing all of its material components. Such material passports will become necessary on the (local) level of buildings and on the level of specialised platforms that compile corresponding data in the international context. An example for such a material passport is the Madaster platform (Fig. 5). In order to evaluate circularity, inspired by the 2015 proposal of the British Ellen MacArthur Foundation (Fig. 8, p. 25), the Madaster Foundation developed the so-called circularity indicator (CI). It evaluates the degree of buildings' circularity by use of information fed into databases on materials and connection types of each building component [17]. This categorisation allows displaying the circularity of buildings according to percentages. The result is a tool that permits comparison and enables future review of the material stock economy and, thus, shifts the view from investment costs to life cycle costs. For this purpose, three specific partial indicators are surveyed: The construction phase (origin of materials), the use phase (life cycle) and the demolition/reversibility phase (material reclamation). The Urban Mining and Recycling (UMAR) testing unit [18] is part of the research platform Next Evolution in Sustainable Technology (NEST) at the Swiss Federal Laboratories for Materials Science and Technology (EMPA). In 2018 it was evaluated as a case study by the Madaster platform. The digital twin of the UMAR unit consists of 32 material data sheets and 90 product data sheets. Prod-

5
Illustration according to the Madaster platform
a The material passport of the platform displays material quantities and positions (classified according to layers as per Brand, see p. 119ff.) for an example building and disaggregated according to life cycle assessment phases A–C (simplified illustration).
b The CI circularity indicator comprises the building use phases origin, life cycle, reclamation as partial indicators and displays the circularity of a project.

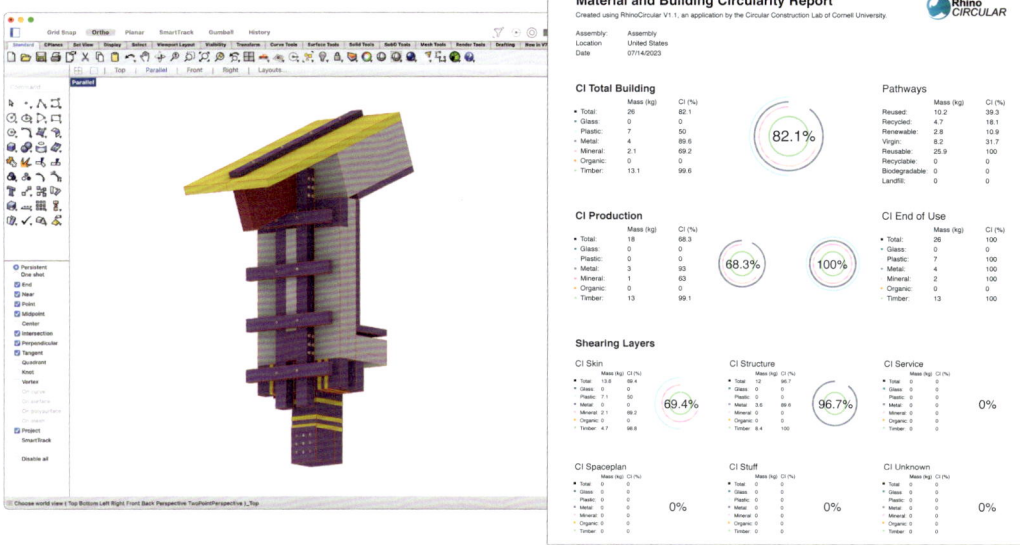

6
The screenshot shows RhinoCircular, a plug-in for the 3D modelling software Rhinoceros. It helps planners make design decisions in early design phases with a view to their impact on circularity.

Notes
[1] www.block.arch. ethz.ch
[2] www.empa.ch
[3] Block 2017, p. 33
[4] ibid. p. 17ff.
[5] Winter/Rug. In: Bautechnik 4/1992, p. 193
[6] Troxler. In: NZZ, 22.9.2016
[7] ibid.
[8] Hettinger/DGNB 2020
[9] Hillebrandt et al. 2018, p. 65 and 72
[10] Heisel/Hebel 2021, p. 99
[11] ibid. p. 157
[12] Presskit Concular 2021
[13] restado.de
[14] Campanella/ DGNB Sustainability Award 2020
[15] ibid.
[16] see note 10, p. 158
[17] ibid.
[18] Sobek with Hebel/Heisel
[19] Heisel/Rau-Oberhuber 2020
[20] Heisel/McGranahan 2023

ucts can refer to different material data sheets in relation to their composition. The study also distinguishes between material families and, thus, the separate evaluation of, for example, the timber structure (renewable raw materials), the use of glass ceramic (reprocessed raw materials), or the metal door handles by RotorDC (reused construction elements) [19]. The location of the individual product within the building is specified within a building element list. The building layer model developed by Stewart Brand (see "Layering as a Circular Principle", p. 118ff.) serves to estimate the duration of exchange or maintenance intervals of a particular product within the building. This also means that the digital twin must be more than a static data set. Instead, it is managed dynamically and is updated every time changes to a building take place, for instance in the case of renovations or if damages occur that can influence the material stock. The results of the evaluation are displayed as percentages and disaggregated according to partial indicators (construction, utilisation, reclamation). In the case of the UMAR unit, these calculations were established after completion of the building. In the future, it would be ideal to include these evaluation parameters already in the design phase – in the form of a novel design tool. The Circular Construction Lab (CCL) at Cornell University under the directorship of Felix Heisel is working on the development of such tools. Plug-ins are under development that permit CAD software integration (Fig. 6) and, thus, support design decision-making in terms of its effects on circularity during early design phases [20].

The use of digital tools and the compilation and management of large data sets will become the foundation of a functioning circular economy. Teaching and research need to increasingly address the related questions in order to facilitate the easy and widespread use of tools, ranging from the design process to the responsibilities of building authorities. In the future, this will require including certifications that permit evaluating and comparing circularity when submitting permit applications. Such instruments allow dynamically adapting and tightening boundary values, in order to actually achieve the declared goal of the European Union of comprehensively establishing a circular economy by the year 2050.

(Re)Building Simply

Thomas Auer, Andreas Hild

Focus on existing buildings

The German Federal Ministry for the Economy published its "Green Book Energy Efficiency" in 2016. The proposed theses and guiding questions on topics of energy efficiency were intended for discussion within a consultation process including experts as well as the broad public. The guiding principle of "efficiency first" is supposed to emphasise that a significant reduction of greenhouse gas emissions is only possible if the existing potential for energy savings in the building and construction sector is fully utilised [1]. Even if all future buildings are realised and operated in a climate-friendly manner, the compensation effect for emissions that result from the existing building stock not subject to renovation will be insufficient in terms of the envisioned goals. This doesn't even include compensation for emissions that have already been released in the building sector. Starting from the assumption that the approximately 1 % annual share of new construction in Germany will remain constant in the coming years, this means that 70 % of the building stock existing in 2050 already exists today. This is why it is urgently necessary to focus on the existing stock and to redistribute earmarked grant resources in order to reach stated climate policy goals.

Numerous studies [2] indicate that the residential and non-residential building sectors often fail to achieve ambitious energy targets, both for new construction as well as for renovations. The actual degree of consumption is often significantly higher than calculated needs. This is commonly known as the so-called performance gap. The difference between the projection and the actual preferences of users, in particular in housing construction, is an important element of the performance gap [3]. It is easy to conclude that numerous new and rather ambitiously designed buildings indeed consume a considerably greater amount of energy than actually necessary.

The complexity of buildings has continuously increased in the past 20 years. Still, these buildings fail to meet expectations on intended effects. A further outcome is the increase in costs for construction. This raises the question of whether currently prevailing practices in the construction and use phases (operation, maintenance, renovation), leading to this complexity, as well as the demolition phase are actually justified or even truly practicable.

The research project "Building Simply" [4] demonstrates how the concept of robustness [5] can counteract complexity in terms of design, construction and technology. The concept, on the one hand, focuses on the targeted omission of layers and, on the other hand, the additional integration of installation layers sorted by type in the manner of a circular economy, corresponding with the layer model according to Stewart Brand [6] and covering all phases of construction, utilisation and demolition. This way, complexity can be reduced and, at the same

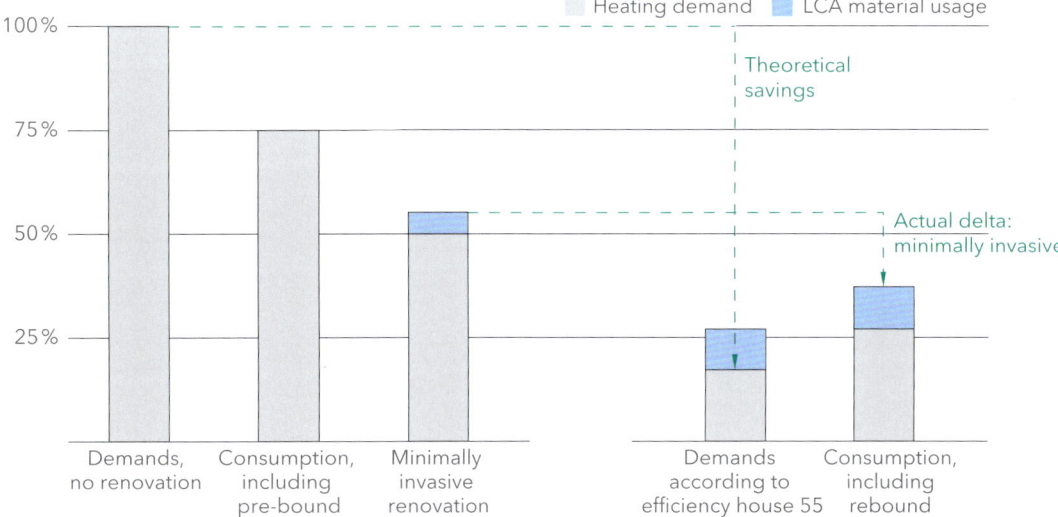

1 Hypothesis on what can be achieved by a minimally invasive approach to renovation compared to renovation according to the efficiency house 55 standard under consideration of pre- and rebound effects as well as material usage. 100% describes the theoretical, calculated heating demand according to the Buildings Energy Act prior to energy renovation.

time, circularity [7] can be optimised (see "Layering as a Circular Principle", p. 118ff.). Life cycle assessments show that the carbon footprint of the research project and its test buildings in Bad Aibling is comparable to the near-net zero energy building standard throughout the reference period. The performance gap mentioned above is absent in this observation. As measurements show, the inhabited houses don't display the differences expected between the calculated model and the actually metered demand [8]. Thus, the hypothesis that robustness within operation is possible was confirmed.

Transferring the approach of building simply to the existing context – most of all, non-residential structures – is highly relevant. Often, installing state-of-the-art building services systems can hardly be realised at all, or only with excessive effort. In general, neither existing shafts nor central building services facilities possess the necessary dimensions. Further, the often limited ceiling height complicates the horizontal layout of mechanical services. When attempting an integration of state-of-the-art building services into an existing structure, the results often entail highly specific adaptations of installations that make demolition difficult or impossible (most of all, air conditioning is critical due to the significant space requirements). Given these circumstances, the aim of circularity of making resources accessible and separable cannot be achieved.

Fig. 1 illustrates, in qualitative terms, the hypothetical, comprehensive life cycle assessment of carbon savings achieved through minimally invasive renovation. The pre- and rebound effects (performance gap) as well as the required material usage were taken into account. Hypothetically, carbon savings can be achieved by an energy renovation according to the efficiency house standard. The comprehensive observation shows these savings are actually much lower than suggested by calculations according to DIN 18599 (German Buildings Energy Act). Similar to the "Building Simply" projects, this hypothesis requires verification within living labs. An example for this approach is the Power Station (Maschinenhaus) in Schwabing.

(Re)Building Simply: Power Station Schwabing

An example of a renovation following the principle of "(Re)Building Simply", an extension of the "Building Simply" principle with a focus on existing stock, is the Power Station in the Munich borough of Schwabing. The historically listed boiler house at Schwabing Hospital was recently converted into an office space by Hild und K (Fig. 2, p. 100).

Generally and primarily, modern office buildings follow the principles of flexibility and economy. The latter, most of all, finds its expression in the assumed maximisation of efficiency. This principally refers to spatial efficiency, meaning the creation of maximum usable area in relation to gross floor area and plot size. Flexibility can be translated into different functional office scenarios, such as cellular (double, triple or quadruple width), combination or open-plan office types. Typically, office workspaces are arranged in rows along the facades. This inevitably leads to standardised facade designs and often grid-based and correspondingly monotonous building elevations. By correlating requirements for standardisation with those for quality of stay (lighting, air, acoustic and thermal quality), this increasingly leads to the creation of spaces devoid of any appeal. User requirements such as individual control of thermal comfort in winter and in summer, or a high level of air conditioning, are demanded as standards. This leads to a high degree of technical installations that can only be created with enormous effort within a renovation project. Often, targets for both energy efficiency and user satisfaction are missed.

Historic preservation and subsequent use of hall structures
The Power Station in Schwabing is a historic, listed building. This results in certain constraints, due to which "normal" office use and typical floor plans become impossible. Retrofitting building services according to common standards can only be achieved with enormous effort and disproportionate interventions. At the same time, historic preservation liberates designers from adhering to the Buildings Energy Act. In such cases, as is commonly known, historic preservation is considered more relevant than thermal protection. This opportunity for evaluation is also needed beyond the purview of historic preservation. It is self-evident that there are many buildings requiring renovation that are not historically listed, yet still constitute existing stock worthy of maintaining. Precisely this existing stock – still lacking a proper definition – is what fosters the identity of our cities. The Power Station, beyond historic preservation concerns, serves as a model for renovation in the manner of "(Re)Building Simply". Insulating the facade and exchanging the windows is – due to historic preservation regulations – principally not required. Regardless of possible historic preservation constraints, if we intend to take seriously the renovation of the immense stock of existing buildings from the perspective of climate and resources, many existing warehouses or halls in Germany and Europe should be maintained. Only in very few cases can subsequent use be ensured for such hall structures and mostly only cultural functions appear sensible. This, however, often leads to underutilised spaces, temporary uses or even the demolition of real estate. This is why it is a logical step to systematically research office functions according to their suitability for implementation within hall structures.

Power Station spatial and climate concept
Due to the fact that typical office layouts parallel to grid-based facade structures can hardly be realised in this particular

2
Historically listed Power Station in the borough of Schwabing in Munich (DE), 2024, Hild und K, energy concept: Transsolar, project development: Ehret und Klein

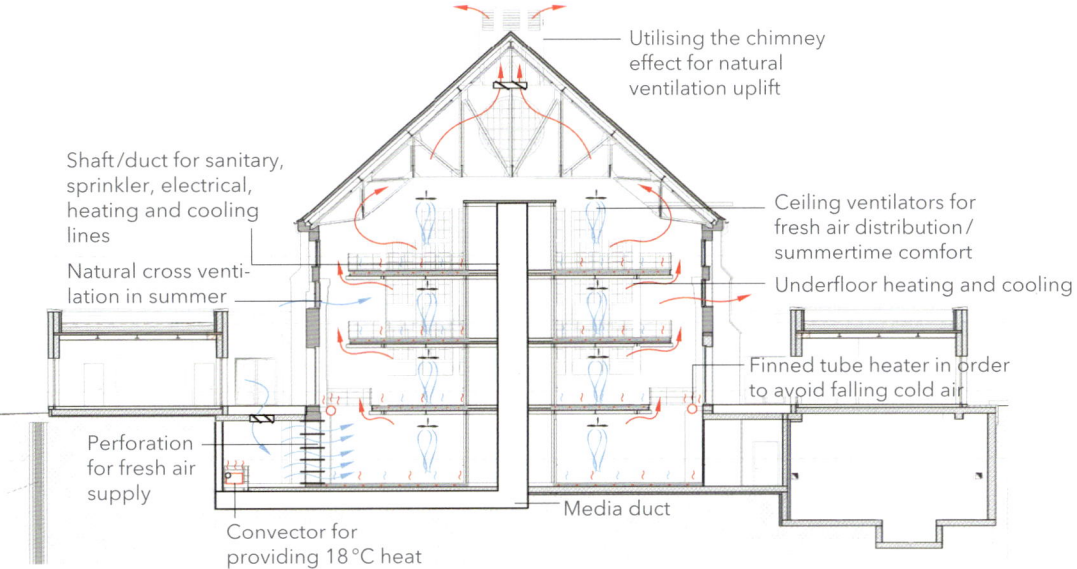

3
Climate concept, Power Station, Munich 2024, Hild und K, energy concept: Transsolar

existing building, the architectural solution inevitably leads to an open-plan office. The new storeys are placed within the tall existing space similar to tables set on top of each other, establishing an open and interconnected spatial continuum. The built-in elements are detached from the walls and, thus, the historic building as such. This leads to creating a buffer zone between the new uses and the existing building envelope. The absence of central building services and shafts leads to a consistently low-tech approach (Fig. 3). The office areas are heated and cooled by activation of the newly created floor slabs. Fresh air is naturally drawn into the basement level – due to the stack effect of the open hall and its existing ventilation towers – while exhaust air rises upward to the roof. The distribution of air in the building takes place by use of ceiling fans. As a result, air distribution through shafts or ducts is not required. Despite the minimal degree of building services technology, a high degree of comfort can be achieved throughout the year. This is due to the large, uninterrupted interior space and the thermal mass of the existing structure – combined with thermally activated slabs and natural ventilation (Fig. 4).

The heating and cooling demands (usable heating and cooling) can be significantly reduced through a limited set of measures (Fig. 5, p. 102). At less than 5 kWh per m^2 per year, cooling demands (thermal cooling demand) appear insignificant. Compared to

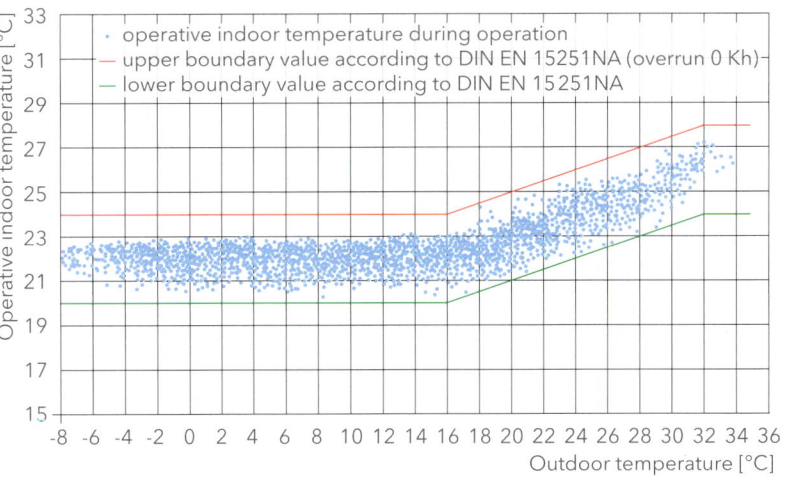

4
Operative indoor temperatures on the second floor higher than outdoor temperatures. Every dot represents an hour within annual use. Red and green lines describe the comfort range according to DIN EN 16798.

(Re)Building Simply

modern, extensively glazed buildings, they are even lower by a factor of 10. Interior insulation was omitted, since it would have reduced the effect of thermal mass and therefore compromised thermal comfort in summer. Calculated heating demands of above 100 kWh per m² per year exceed figures of modern office buildings by a factor of 2 to 3. The assumption was that the building is fully occupied five days a week from morning through evening, making a correspondingly high air change rate necessary. Increased heating demands need to be offset against the electrical power required for operating an air conditioning system with heat recovery. Since air exchange is based on CO_2 sensor data, while full occupation throughout a working day is unrealistic, real heating demands will be significantly lower. This refers to the so-called pre-bound effect, meaning that the actual energy demands remain below the calculated energy demands in the case of existing buildings. Results of field tests show that the primary energy demands of naturally ventilated office buildings are, on average, lower – in comparison to mechanical ventilation – despite absent heat recovery from exhaust air [9].

By applying the hypothesis illustrated in Fig. 1 (p. 99) to office buildings, the assumption is that the carbon emissions by floor area caused by construction and operation in the case of the renovated Power Station are significantly lower than the emissions of a new office building observed throughout its life cycle. On the one hand, the emissions caused by renovation as compared to new construction are clearly lower. On the other hand, it is a realistic assumption that energy demands during operation will also be lower than in the case of modern office buildings. However, only monitoring after completion – anticipated for 2025 – can verify this.

Open-plan office within an open spatial structure

The open structure, based on creating one large space in the Power Station, is a solution that utilises the potential of the existing building in an ideal manner and supports a consistently low-tech approach. It appears logical at first that this kind of new interpretation of the open-plan office by use of table structures set on top of each other can only be successful if the renovated space is no longer governed by the imperative of efficiency. The office space primarily becomes a space of encounter. Individual workplaces are secondary and, correspondingly, are no longer realised in the form of cellular offices. However, the result isn't a simple open-plan office. Without a doubt, encounters within a space require a corresponding architectural treatment with regards to quality of stay and, thus, parameters such as lighting, air, acoustics and thermal comfort. In the meantime, occasional remote work has become increasingly common and has led to changes to requirements for office buildings. Personal encounters on office working days are also becoming increasingly relevant. In contrast, the individual

5
Simulations show the reduction of heating and cooling demands based on individual measures. Renovation, Power Station, Munich 2024, Hild und K
a heating demand
b cooling demand

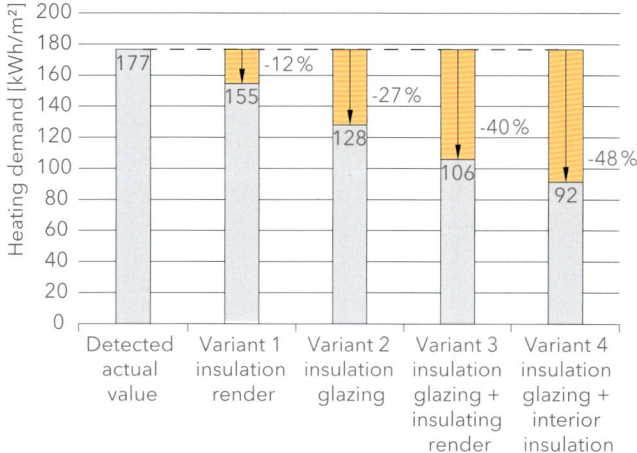
5a

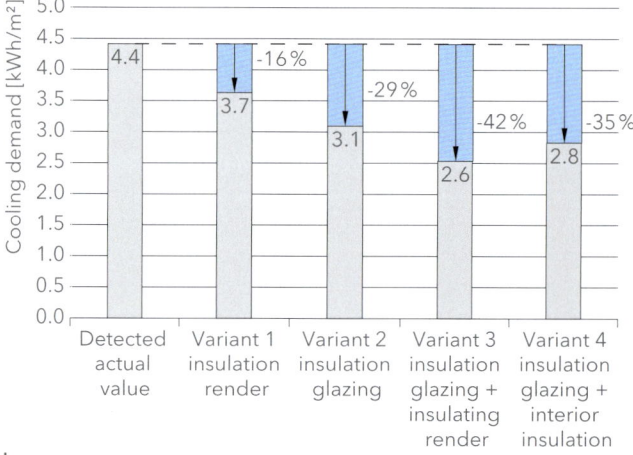
b

6
Visualization, interior situation, renovation, Power Station, Munich 2024, Hild und K

workplace typology is becoming obsolete. The solution developed for the Power Station in Schwabing, viewed from a purely market-oriented perspective, can be considered unusual. It is rather unlikely that the model will be transferred on a larger scale for new office uses in existing buildings. Still, such approaches are necessary and useful, since they expand the range of options for renovations of similar hall structures. The very limited scope of building services technology requires a corresponding architectural approach that addresses issues of interior architecture, such as the arrangement of common areas. In a prospering metropolis, it will be easy to find users for such a building. Still, prior to the COVID-19 pandemic, the target group for such renovation projects was smaller than for more conventional solutions. Scientific evaluation will show whether the concept of the Power Station will be transferrable to future conversion projects for the creation of office space.

Intelligent omission

The example of the Power Station demonstrates the potentials of "(Re)Building Simply" with regards to flexibility of use, circularity, quality of stay and energy efficiency. Each individual aspect serves as a model for the existing building stock and, to a certain degree, for new construction. The characteristics of the hall structure contributed to an approach aimed at consistent omission – to a degree that architects would not dare in the case of a new building design. The example of the ceiling of the Power Station demonstrates this: Due to fire safety and structural safety concerns, preserving the wire plaster ceiling of the original construction on the level of the steel beam bottom chord became impossible. Conversion into an office space, from an historic preservation perspective, therefore seemed impractical at first. Maintaining the ceiling would have made smoke extraction impossible with limited use of the ground floor. This shows that the conversion of a building worthy of preservation mostly means that deficits need to be taken into account, that building components of such historic buildings or relevant parts thereof need to be removed. The argument that the original building fabric must be preserved is understandable. However, the resulting limitation of functions as the outcome of an assessment process is difficult to communicate. Remodelling often demands solutions to problems that don't exist for new construction. From this point of view, deliberating on remodelling and renovation projects has the potential of becoming a driver of innovation in the construction sector. This also requires that solutions as such can be realised in a manner that permits generalisation. The related approaches demand simple construction methods and low-tech approaches so that they can be applied to both new construction as well as renovations.

The hypothesis illustrated in Fig. 1 (p. 99), thus, awaits confirmation after completion and within monitoring of operation data and user satisfaction. Nevertheless, the simulations conducted show that simple measures based on sensible key figures in terms of energy demands permit establishing very good levels of comfort. At the same time, a space is created that provides for a diverse range of experiences beyond common models of efficiency. Do we actually intend to succeed in achieving the building turnaround mentioned previously by systematically underutilising existing buildings? Across the globe, societies and the building sector need to change. Intelligent omission as described here does not have to be viewed as a deficit. The aim is to balance it with good ideas and architectural solutions.

Reversible Connection and Joinery Methods

Daniela Schneider

Beyond the selection of suitable materials, connection and joinery methods and their construction-based classification, variability and flexibility decisively influence circular construction sorted by type. Based on connection methods suitable for specific materials, material layers and components can enable disassembly for later reuse and repurposing. The aim is to already think of materials as future secondary resources during the design phase, while maintaining their potential and making them available to future generations by employing adequate construction methods [1]. As a result, for construction sorted by type, connection and joinery methods should be taken into consideration that correspond to the characteristics, functions and maintenance of assembly parts within all material cycles and phases. On the level of planning, a diverse range of connection methods exists, offering design-based and functional opportunities for the development of construction methods sorted by type. In terms of the design and production of assembly parts and regarding their quality and sustainability criteria, separating or detaching them without destroying them is unfortunately not yet a priority. The linear model of production, use and disposal still dominates (see "Take, make, throw", p. 16f). The option of repairing components and the resulting functional longevity are, more often than not, unavailable, due to the employed construction type. The simple removal and replacement of products based on reversible assembly methods is only planned for to a limited extent. Absent transparency and lacking knowledge, in particular on the level of products, can be reasons why architects prioritise and implement the planning and detailing of reversible connections only to a limited degree.

In addition, planning and building sorted by type as such still has not become an aspect of how architects understand their professional services. It would be highly important to integrate all construction principles, materials, detail planning, layer compositions, technical equipment and structural engineering concepts and designs within a shared vision of circularity that can bring together all professional planning experts and consultants – a joint task for all actors involved in the building sector. The development of circular constructions sorted by type in recent years indeed demonstrates budding innovation power and a new spirit of inventiveness. Suitable connection and joinery methods based on reversibility and demonstrating flexibility have been increasingly adapted from other industry sectors and employed for the implementation of architectural details. This refers to the following connectors from other fields:
- velcro and zippers used in textile production
- clips and interlocking fasteners from the automobile and aviation industries
- timber joinery used by traditional craftspeople, reinterpreted anew with digital methods

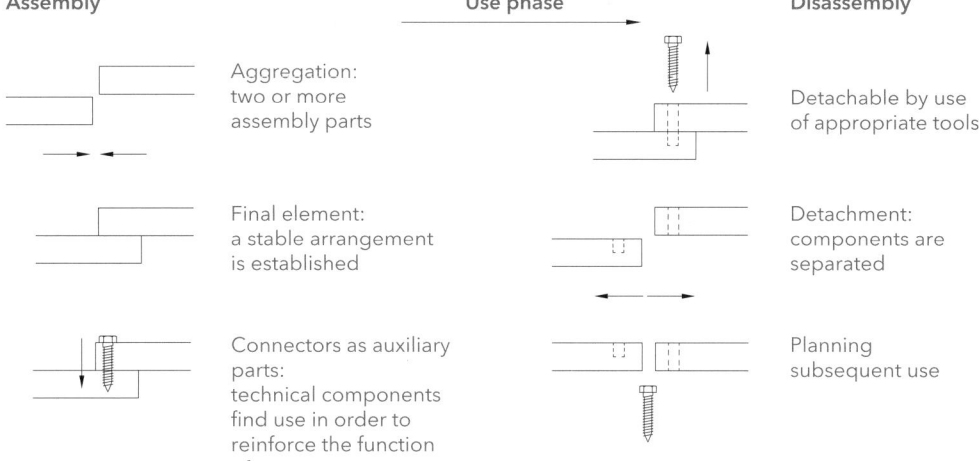

1 Overlapping, with screw connections applied during assembly and disassembly following the use phase

- magnetic connectors are experiencing a renaissance and being introduced to the building sector based on new research insights and high-performance products

So-called lighthouse projects of construction sorted by type increasingly demonstrate innovative principles of building. A pioneering project can inspire architects. Unfortunately, related examples are represented in professional publications only to a limited degree of detail. This calls for communication strategies that contribute to making these innovations accessible to the public and pave their way to standardised applicability. When selecting and implementing connection and joinery methods, the aspects of security and longevity – mostly due to lacking alternatives – still tend to contradict the intended reversibility of connections. Adhesives in use today for flooring products often have a significantly longer lifespan than, for instance, parquet flooring. The reversibility of flooring without destroying it, thus, becomes impossible. Further, insoluble impurities remain on the wood parquet and the substrate, which in return require landfill or, in the worst of cases, incineration. Durable adhesive connections can, in fact, support functional longevity. However, they also prevent the simple replacement of less durable layers sorted by type. This is also why it is necessary to refine connectors in order to correspond to the different functional life cycles of different assembly parts. Circular construction sorted by type should, thus, focus on the planning and implementation of reversible connection methods in particular that enable the disassembly of buildings and their components.

Basic assembly principles

Connecting and joining materials and parts is supposed to enable the transfer of loads, moments and movements. The aim is to create durable connections or other methods of joining two or more assembly parts of a geometrically defined and solid form [2]. The applied terminology encompasses processes of different shapes and forms in terms of the assembly sequence and related movement, as well as cohesion and separability. The components used can be, for instance, connected with screws. For circular construction sorted by type, the connection process is categorised according to the separability of each connection [3], as Fig. 1 shows.

Joining and connection methods

For the design and development of joining and connection methods, designers should inquire precisely how a specific form of assembly can contribute to achieving reversibility sorted by type. The aim is to preserve the function of components for subsequent applications. Aspects of joinery and flexibility need to be considered early on in the design phase and

	Form-fit connections	Force-fit connections	Material-locked connections
Description	Connections created by form-giving and interlocking components	Connection between two parts based on friction or force	Connection based on employed materials [4]
Physical action principle	AP1 AP2	AP1 AP2	AP = assembly part AP1 AP2
Number of assembly parts	at least two components	at least two components	two or more components
Examples	• tongue-and-groove-connections • zipper • dovetail joints • gear coupling (gearwheel) • pin connections • clinch connections • seamed connections • bolted connections • snap-on connections [5]	• clamped connection • shrink-fit connections • nails • elastic connections: screws and rivets • wedge, bolted, pinned and rivet connections • magnetic or gravity field connection • friction-based compression joints [6]	• compression/ fusion-welded joint • soldered joint • adhesive joint • riveted joint • coating [7] • painted finish [8]

2
Overview, different connection types: Reversible connection and joinery methods are defined by the capacity for damage-free separation of assembly parts.

demonstrate their potential for adaptation during different use phases. Aside from efficiency, security and consumer-friendliness in application, the integration of different factors relating to flexibility comprises a significant challenge in planning. Based on their physically active principles, connections are distinguished according to form-fit, force-fit and material-locking types (Fig. 2).

Form-fit connections
Form-fit connections are defined by their form and the way parts interlock. This simple and cost-efficient connection type is also one of the oldest used in the construction field [9]. The active surface transfers loads physically. Form-fit assembly methods serve to create fixed as well as unfixed connections. Typically, they are neither water- nor airtight. For such purposes, additional materials are used that provide functional support along the active surface. Form-fit connections are predominantly reversible.

Force-fit connections
Force-fit assembly methods combine parts based on forces, such as by friction or compression. Pressing components together leads to creating a friction surface. The active forces stabilise the connection. Compared to form-fit types, force-fit connections possess the advantage that they can bear greater dynamic loads. During use, they also permit simple retensioning or readjustment [10]. In the building sector, predominantly combinations of form-fit and force-fit connections are applied. In such cases, assembly parts are fixed with an additional connector (e.g. screws, bolts or nails). The threaded form of screws or bolts produces force-fit connections.

Material-locked connections
A material-locked connection achieves the assembly of parts by atomic or molecular forces. Typically, this is accomplished by bonding, welding or soldering. Cohesiveness results from using matching or different additives that, however, do not allow simple reversibility. The degree of

damage and destruction of components that occurs during separation is related to the specific materials and the connection surface.

Connectors

Connectors provide additional stability in the assembly of two or more parts. They are selected in relation to the raw material of the assembly part. They also need to meet specific requirements, such as for fire safety or certain building physics characteristics. In timber construction, predominantly mechanical connectors such as nails, clamps, bolts, etc. are used. Employing different materials for assembly can complicate the reversibility and the return of connected parts to specific cycles at a later date. Aside from the materials selection, the quantity of connectors is relevant to the disassembly and the preservation of components: Small-scale connectors applied in large quantities, such as clamps or nails, can lead to extensive damage to component surfaces. Separating them during demolition requires significant labour, which tends to be unproductive and complex.

Hybrid connection types

Combinations of force-fit and form-fit connections are possible in relation to how they are used and which function they perform. Related connections are distinguished according to their mechanical purpose, their functional requirements, force flows and structural characteristics. An example for a hybrid assembly method is a form-fit connection featuring screw connectors. The combination of multiple connection types can achieve a greater mode of action and, thus, provide additional security. However, it is detrimental to the dismantling process and the disassembly of parts. Case-by-case evaluations can identify whether hybrid types are required for construction at all. Adhesive bonding of form-fit connections or adhesive sealing of screws in order to create air-tightness or increase positive locking are examples that should be avoided.

Adhesives and sealants

In recent years, the use of highly specialised adhesives and sealants has increased in the construction industry. In order to save costs and time, elastic adhesives and sealants are preferably employed for the creation of large-scale surfaces, yet without any consideration of circular reversibility. Based on their characteristics, they limit or even prevent the need for subsequent reworking. Adhesives and sealants are applied in buildings by nearly all trades active in shell construction, facade construction, mechanical engineering, as well as interior finishes. The undesired outcome is reversibility not sorted by type that mostly leads to the destruction of materials or building components [12]. "Dis-connecting" adhesively bonded materials integrated in buildings is tremendously laborious and time-consuming. and should therefore be avoided. Alternatively, dry sealants permitting problem-free disassembly are recommended that enable simple "disconnection" sorted by type. Further, the development of new soluble adhesives should be intensified in order to meet circular economy demands.

Planning reversible connections
In order to plan circular buildings, it is important to critically enquire and test whether connection and joinery methods for a particular construction type can also serve purposes of reversibility. This functionality can have an expiration date or be completely independent of time. Therefore, planners should consider the following:
- determine the life cycle of the assembly parts
- research the durability of connectors
- define the functions of connections in relation to individual phases and cycles
- investigate the potential for damages in the case of connector disassembly
- plan the demolition, separation and reuse of assembly parts and connectors

The ISO 15 686 standard on "Architectural Construction and Buildings – Planning Life cycles" is a seminal reference. Already during planning, factors that may potentially impact the use phase need to be evaluated, for example, the life cycle of connectors, how damage-prone they are, the influence of building physics characteristics (e.g. temperature fluctuations), as well as repair needs and intervals. This also includes the quality and precision of assembly, the assembly sequence, the accessibility, the tools used, the influence of trade-related tolerances and quality management on the construction site. The proposed joinery and connections need to be defined by detail planning and tendering. The goal is to prevent the application of unwanted, non-reversible connection methods during assembly. Connections that enable separation require precise specifications from construction managers and reliable handling by craftspeople. This refers to, among others, the implementation of form-fit and force-fit connections that are complex to install and require screws, clamps, snap locks or zipper fasteners. In comparison to material-locked assembly, form-fit and force-fit connections mostly demand additional time, planning effort, as well as craftspeople's expertise and experience [13].

Disassembly
The process of disassembly includes separating building components along their point of connection. Whether this can be successfully undertaken depends on the characteristics of the materials used and

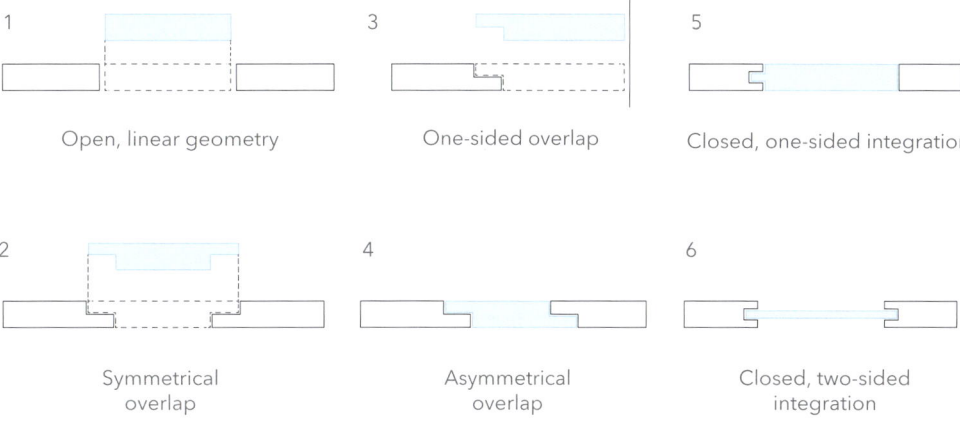

3 Types of geometries of connection surfaces that influence assembly and disassembly processes [16]. Creating connections and systematically arranging geometrically formed components influence the disassembly process. Depending on the sequence of disassembly, the geometry of components requires a corresponding design.

the selected construction type. Assembly components with form-fit connections are particularly decisive in achieving destruction-free disassembly. According to Durmisevic [14], the geometry of connection surfaces significantly influences assembly and disassembly processes (Fig. 3). In principle, three types can be distinguished: open, overlapping and closed geometries. Open geometry allows simple disassembly. Overlapping geometry limits disassembly to one direction. Closed geometry prohibits the disassembly of an individual part [15].

It is beneficial to consider factors that influence disassembly already during planning. They can involve accessibility, the force and energy required for disassembly, the time and resulting economic expenditure, consideration of additional tools and machinery, thermal treatment of connection surfaces as potentially required, subsequent reworking or processing of assembly parts, or the storage of components on the site of disassembly. Designers are advised to precisely consider and simulate the sequence of assembly and disassembly of planned layers and parts of a building component. In certain cases, disassembly does not equal a reverse assembly process. Based on schematic illustrations, potential weak spots that can impair disassembly and hamper the separation of connections can be identified and changed in early planning phases. The assembly method, thus, demands particular attention in relation to questions of circularity and sorting by type in building construction. Precise planning and detailed tendering can ensure reusability or recycling in a sustainable way and, thus, address responsibilities relevant to future generations.

Various assembly methods exist that are based on the physical active principles described on p. 106. A corresponding selection is displayed on p. 110ff. However, material-locked connection methods were omitted. Components assembled in that manner complicate or disallow the repurposing and reuse of materials.

Notes
[1] Schneider 2021
[2] DIN 8580:2003-08, p. 3
[3] Bender/Gericke 2021, p. 729f.
[4] Hering 2009, p. 120–130
[5] Mayer 2020, p. 35
[6] ibid.
[7] Rieg/Weidermann 2018. p. 156–253
[8] DIN 8580:2003-08, p. 12
[9] Steinhilper/Röper 2000, p. 1f.
[10] ibid. p. 36
[11] Doobe 2016
[12] see note 1
[13] ibid.
[14] Durmisevic 2006, p. 178–179
[15] ibid.

Assembly Principles

This chapter will look at the most important assembly principles for construction sorted by type. The proposed structure of the chapter is derived from terminology and definitions used in the DIN 8593:2003-09 standard on "Assembly as Production Method", which provides a comprehensive overview of industrial connection methods. Beyond the formulated and defined detachable assembly methods permitting application in the field of building construction, select additional craftsmanship-related assembly methods are illustrated that are not precisely described by the standard. The different assembly methods are distinguished into different categories according to type of cohesion and related detachability. The cohesion type is described in relation to physically active principles of form-fit, force-fit and material-locked connections (see p. 106) or combinations thereof.

Assembling
Assembling involves connecting multiple parts that fit together based on their form. The cohesion of two assembly parts is achieved by form-fit connections, gravity or a combination thereof.

Cohesion type: Form-fit connection, gravity
Detachability: In general, assembly parts can be detached without damage or destruction

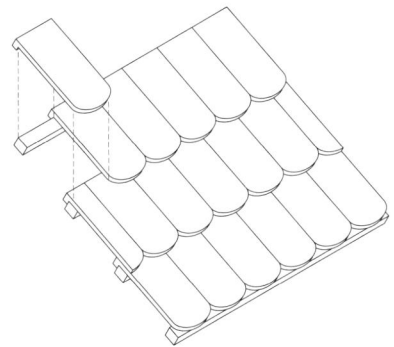

Laying roof tiles

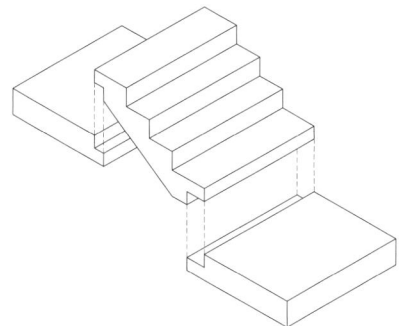

Installing a prefabricated staircase

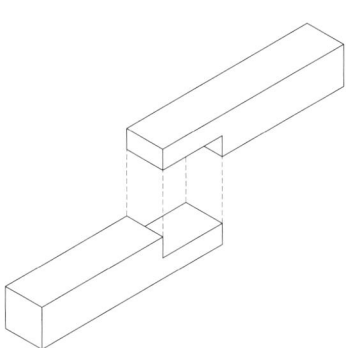

Creating a lap joint

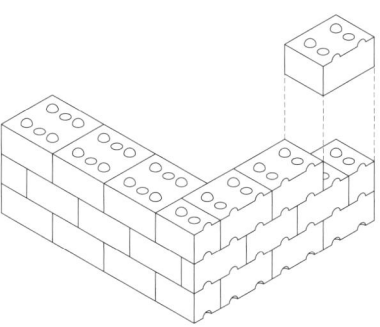

Laying a dry masonry wall

Laying, setting, layering
In the case of laying, setting or layering, parts that fit together are assembled by use of gravity. Individual components are placed on top of each other in a stable manner. The connecting surfaces can also be geometrically matched to create form-fit connections, thereby preventing parts from rotating or shifting.

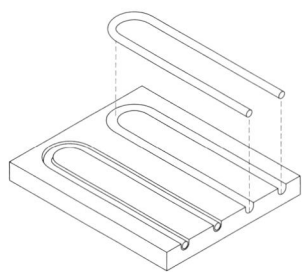

Setting underfloor heating into a moulded component

Setting, inserting
A building part is set or inserted into a counterpart of matching geometrical form. The connection is form-fit.

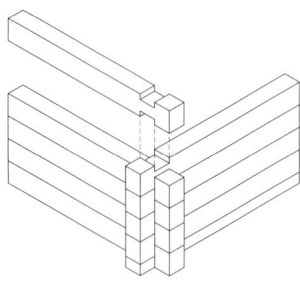

Interlocking corner connection in block or log constructions

Interlocking
Interlocking describes connecting two assembly parts placed on top of each other in an intersecting manner by use of notches.

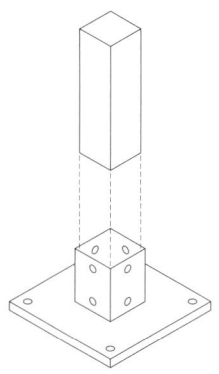

Pushing a column into a column cap

Pushing/sliding
This assembly principle describes pushing or sliding one construction component onto or into another component.

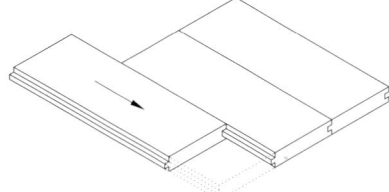

Sliding tongue and groove connections together

The principle can be distinguished according to whether one part is pushed/slid onto another part, or whether one part is pushed/slid into another part.

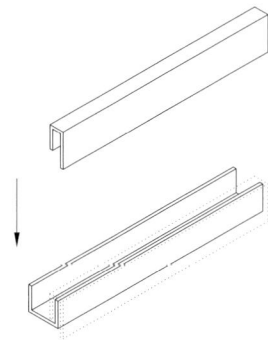

Hanging facade elements into a frame structure

Hanging
This assembly method is based on hanging one component into another. Typically, gravity ensures stability, mostly combined with form-fit connections.

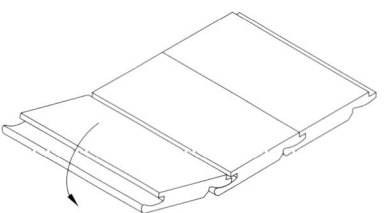

Installing parquet with click connections

Click-locking
Click-locking comprises a combination of pushing and turning to establish a connection between two parts. The connection is based on the resulting compressive forces.

Assembly Principles

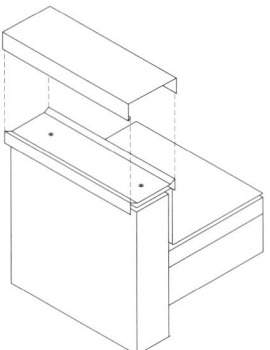

Snap-on sheet metal parapet coping

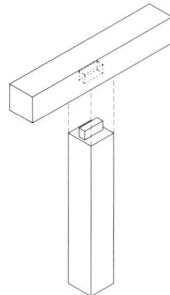

Mortise and tenon joints in timber construction

Snapping in place
An assembly part is set into a rigid counterpart or pushed onto it through elastic deformation. The assembly part subject to elastic deformation assumes its original form after insertion. A form-fit connection is established with the rigid part.

Creating mortise and tenon joints
Mortise and tenon joints are a traditional method of timber joinery, where the mortise is set into the tenon. The form-fit connection prevents sliding or turning of assembly parts.

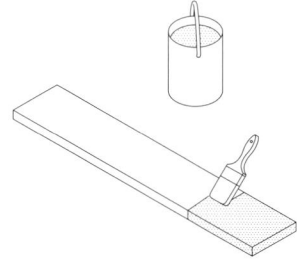

Oiling timber elements

Filling
Filling describes loosely introducing gaseous, liquid, pasty or other materials that permit filling into cavities or porous interstitial spaces.

Type of cohesion: Embedding
Detachability: Without damage to assembled parts

Saturating and impregnating
Saturating and impregnating with biodegradable substances can increase the resistance of suitable construction parts to external influences. Saturation describes deeply immersing and completely coating a construction part with a liquid.

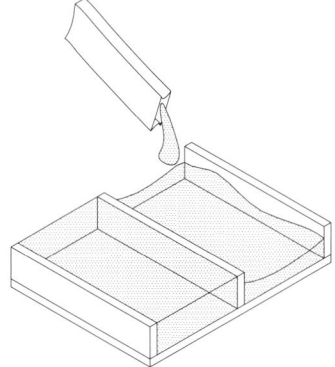

Infilling insulation

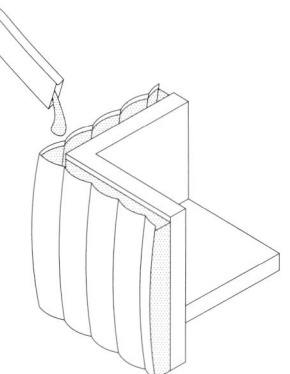

Infilling cavity insulation

Infilling
Introducing solid, liquid, pasty or vaporous materials into a cavity or other predetermined structures is termed as infilling.

Pressing and press fitting
Creating a connection through pressing and press fitting is based on the elastic deformation of the assembly parts and/or auxiliary assembly parts, such as screws or nails. Unintended detachment is prevented by the established force-fit connection.

Cohesion type: Force-fit
Detachability: Mostly without damage to or destruction of assembly parts

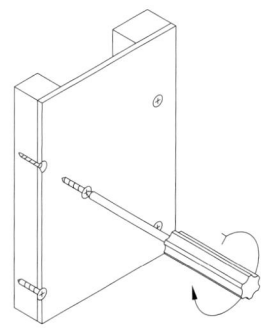

Screw connections for sheathing

Screw connections
Screws are connectors consisting of a cylinder with a threaded surface. They find use when connecting two or more materials.

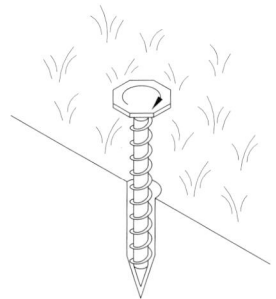

Installing a screw pile

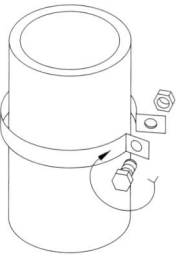

Bolt and nut connections

For this purpose, screws are introduced by penetrating one material and terminating in another material. The threading grips the assembled materials and establishes a stable connection between them. Further possible applications of screws or bolts include driving the connector through one or more assembly components. For this purpose, the threaded part is connected to a counterpart, the nut, on its opposite side. This establishes a force-fit connection based on applying compression.

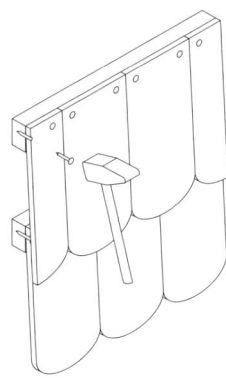

Nailing shingles onto battens

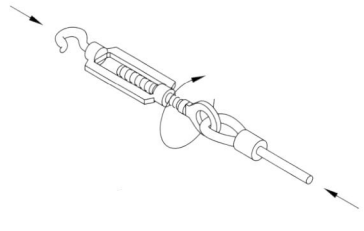

Steel wire – tension rope

Nailing
Nailing describes connecting two parts by driving a nail into them. The connection of assembly parts is based on pressing them together. The nail itself can be the assembly part. In that case, the term hammering is used.

Tensioning
Tensioning describes connecting parts by pulling them together tightly and fixing them in place by use of screws or nuts and bolts through the application of compressive force. By tightening the screws or bolts, this force can be increased.

Assembly Principles

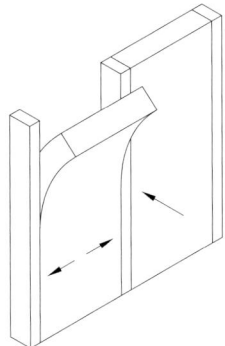

Fitting flexible insulation material

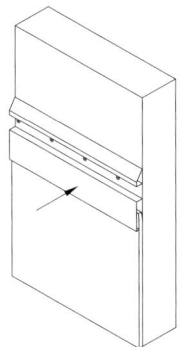

Sealing by use of cover bars

Fitting
Fitting describes the process of connecting two or more materials by using a rigid auxiliary part in order to tightly hold or press them together. Cohesiveness is based on the plastic or elastic deformation of the assembly part.

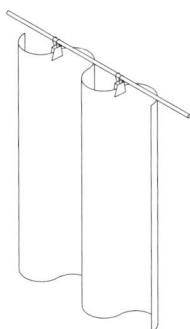

Clamping textile

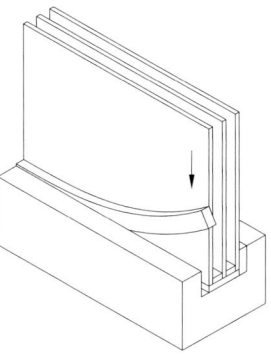

Applying a press-in gasket

Clamping
Clamps are flexible spring connectors that are used in order to hold together two or more materials or objects, or to fix them in place.

Pressing
Pressing involves connecting an interior part with an exterior part. A difference in size exists between the two. Elastic deformation of the assembly part results in a force-fit connection.

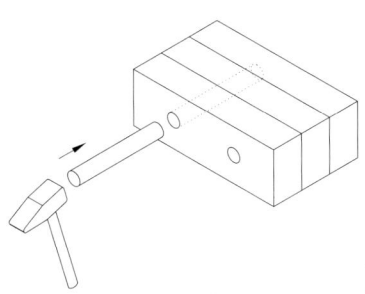

Pin connections with wood dowels in a stacked timber ceiling

Wood dowel expansion in a stacked timber ceiling

Dowelling
Dowelling entails pressing pins or bolts into holes or drillings in order to connect parts. A difference in size exists between interior and exterior part. A stable, durable connection requires high degrees of precision.

Expansion
The assembly of parts by expansion is achieved by loosely inserting an interior part with reduced volume (through cooling or drying) into an exterior part. Expansion (heating / moistening) of the interior part results in a force-fit connection between the two components.

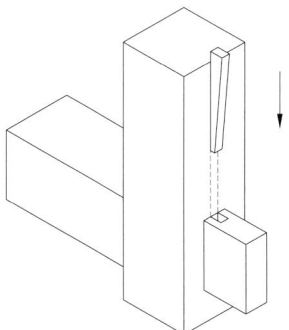

Wedging wood parts

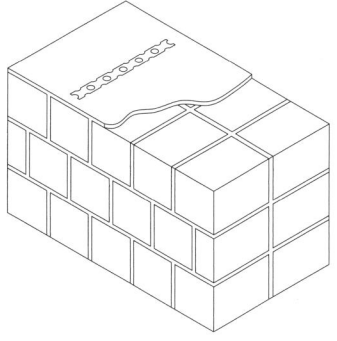

Masonry anchor

Wedging
Wedging means to drive a wedge-shaped part into the gap between two matching workpieces in order to connect them.

Anchoring
Anchoring comprises a tensile connection between two parts by use of an anchor as auxiliary assembly part. Inserting or introducing an anchor leads to a force-fit connection between two parts.

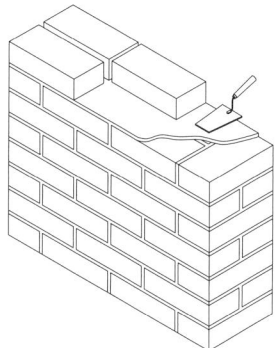

Laying loam brick with loam mortar joints

Shaping
Connecting parts through shaping leads to a connection between one part and a formless, liquid, pasty, granular, powdery or gaseous additional part. This refers either to connections between a construction component and a form-fit, formless counterpart or a connection between multiple construction components by introducing formless material.

Cohesion type: Form-fitting, resulting from original forms
Detachability: In the case of materials selection sorted by type, without damages or destruction

Using mortar
By using mortar, a soft and pasty substance is distributed onto a surface in order to create a connection. After drying, mortar must be reversible sorted by type.

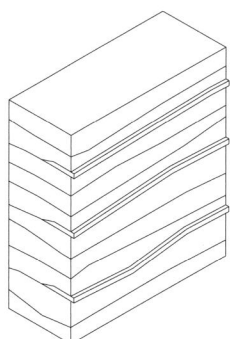

Embedding erosion barriers in rammed earth

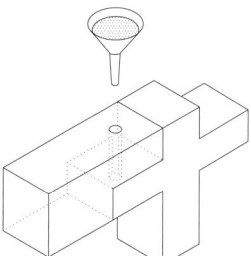

Grouting prefabricated concrete element joints with lime mortar

Embedding
Embedding is a process that is based on setting an object or material into a larger solid part in order to stabilise or protect it.

Grouting
Grouting involves pouring a liquid material into formwork or a cavity, in order to create a stable structure from within.

Assembly Principles

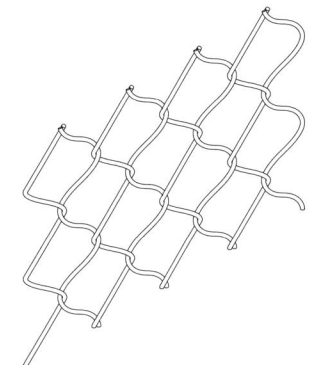

Forming
Forming is a type of mechanical assembly. It results in a connection between two workpieces based on plastic deformation. It is distinguished according to whether the workpiece itself or an auxiliary part, such as a rivet, are subject to forming.

Cohesion type: Form-fitting, based on forming
Detachability: In general only by damaging or destroying the assembly parts

Stainless steel mesh

Weaving wire
By weaving wire, the material is woven to produce particular patterns or braided into a flat element.

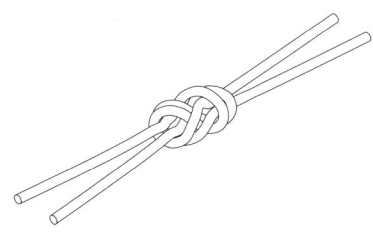

Knotting ropes together

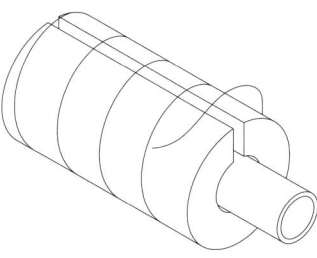

Wrapping wire around insulation

Knotting
Knotting means to twist and create knots along a rope or a string in oder to establish a stable and durable connection between two ends.

Wrapping wire
This assembly method consists of wrapping a thin wire around an object to cover it or hold it in place.

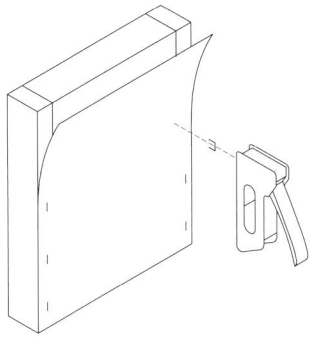

Stapling foil to a substructure

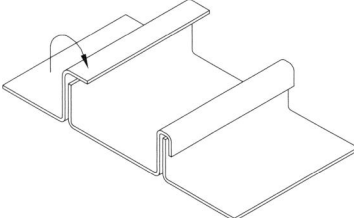

Sheet metal roofing with standing seams

Stapling/tacking
Stapling or tacking describes an assembly method based on connecting two materials by use of bent clamps, pins or needles in order to temporarily fix them in place. Then, they can be fixed permanently or sewn together.

Canting
Canting specifically refers to bending, flanging or folding a workpiece consisting of sheet metal. Folding two workpieces together results in a form-fit connection between both parts. For a precise, homogeneous edge, the material is typically handled with tools or machinery.

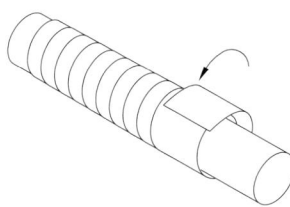

Wrapping textile around a handrail

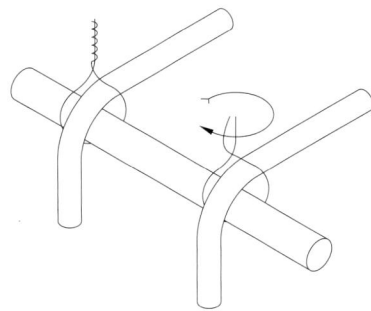

Braiding of steel reinforcement

Wrapping
Wrapping means to wrap strips of a material such as textile, leather or metal around an object or a surface in order to envelop or reinforce it.

Twisting together
The twisting together of parts comprises an assembly method for twining together two or more ends of wires or cables.

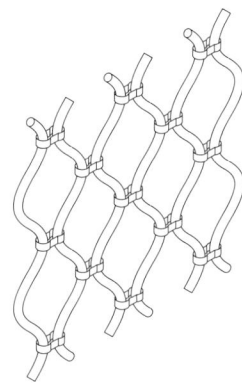

Crimp connection with metal mesh

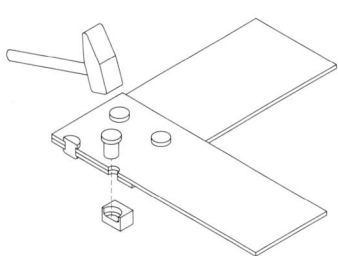

Riveted metal construction

Crimping
Crimping involves an assembly method in which two parts are connected by pressing them together, mostly by use of an auxiliary part.

Riveting
Riveting means to drive a rivet into a drilled or punched hole and form it through compression or hammering in order to create a form-fit connection.

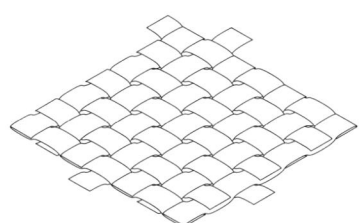

Woven facade cladding

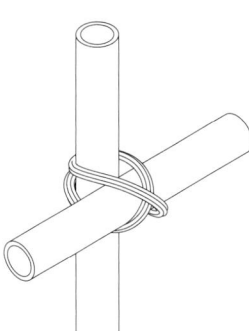

Connecting bamboo rods with rope

Weaving
Weaving is a method based on weaving together two parts according to a pattern that results in a flat or tubular textile object. It is similar to weaving fabric, yet other kinds of material find use.

Rope binding
Rope binding describes using rope as an auxiliary part in order to fix in place other construction parts. The method is used often in the building field, in order to connect construction parts of cylindrical form and increase the stability of the connection.

Assembly Principles

Layering as a Circular Principle

Daniel Lenz

"All buildings are predictions. All predictions are wrong." Stewart Brand [1]

Longevity and adequacy
Each building project requires energy and material resouces. In return, this requirement is related to the life cycle of the erected building. The longer a building exists and is in operation, the more beneficial the share of embodied energy in construction becomes. One conclusion is that a structure needs to be built for maximum longevity and robustness in order to minimise its ecological impact. However, this perspective remains incomplete. It does not account for the fact that only those buildings are truly durable that permit alteration and adaptation. In his book "How Buildings Learn: What Happens After They're Built", published in 1994, the US author Stewart Brand impressively describes the impact of the factor of time on the function and form of buildings [2]. Trends, changing requirements and user behaviour or natural events can be reasons for changes to buildings. In addition to these factors, changing standards and laws play a role (e.g., fire safety, soundproofing). Every small or big change and adaptation measure leads to renewed expenditure in terms of energy, material and, of course, money. Ideally, such measures also lead to extending the life cycle of a building as a functioning system. If we aim at keeping the total energy and material consumption across time (ideally, as long as possible) at a minimum, while also avoiding side effects harmful to humans, the environment and the climate (for instance, through toxic substances or greenhouse gases), two seemingly opposing ideals need to be balanced: The first ideal focuses on the use of materials intended for robustness and durability, which allows achieving a beneficial balance in terms of consumed energy and material resources across the entire use phase. The second ideal aims at a maximum degree of recyclability of the selected materials. The task at hand is to use all raw materials in such a way that the duration of their deployment is precisely predetermined, in order to keep them quickly and easily reclaimable in an energy-saving manner and ready for use as a resource in a follow-up cycle. In practice, most cases oscillate between the two ideals or partially meet the demands of both – for instance, by enabling the load-bearing structure to be as robust as possible, while optimising the facade for quick renewability.

Cycles, time as factor, separability
For each and every building, due to anticipated adaptations and changes, the replacement of material resources is to be expected in certain intervals. The most extreme case is to replace the entire building with a new construction. If a building is supposed to become a bank for materials and building components [3], with built-in elements that can be freely introduced into a new cycle, the inevitable question is which coordinated methods of planning and construction are suited to achieve this goal. It is neces-

sary to understand how the different components of a building are interrelated in technical and functional terms, how they behave across time during the use phase, and how they are assembled and connected in a manner that allows extracting them as high-quality raw materials at a later date.

A careful consideration of the various components of a building is important. All of them display extremely specific functions functions with greatly varying life cycles. One approach to addressing this circumstance is "Design for Disassembly". It involves identifying sensible interfaces between different building parts and, depending on their purpose and their envisioned maximum life cycle, designing them with separability in mind [4]. Each of the deployed building parts or materials is then to be realised with an optimum degree of reusability or, at least, recyclability. As a result, the factor of time becomes a influence on the design of the construction.

If we want to seriously consider a building as a materials bank, integrating all materials sorted by type and mapping their location (e.g. in a material passport) becomes essential in order to ensure optimum reusability or recyclability [5]. The connections among materials and between materials and parts need to be designed in a manner that ensures simple and practicable reversibility. Envisioning the reuse of an entire building part also means being able to separate it from the overall construction without causing damage. The simpler the logic of construction and the design of technology required for assembly and disassembly, the more likely achieving a closed cycle becomes. Beyond identifying intersections and interfaces, the aim is also to reduce them, as intended, for instance, by the "Building Simply" [6] approach (see "(Re) Building Simply", p. 98ff.). The underlying principle itself reflects simplicity: Plan for a sensibly limited amount of interfaces, or those that permit adequate separability. Which interfaces are required depends on the individual case. The preconditions can be very different, depending on the construction method and the materials used. In order to understand the general rules that underpin the definition of these interfaces, two models will be discussed in the following: First, the system of "shearing layers of change" formulated by Stewart Brand as a set of quite universally applicable observations and, second, a system that specifically focuses on building construction as an organising principle of the discussion on construction sorted by type presented in this book.

Stewart Brand's layer model

In "How Buildings Learn", Stewart Brand presents a model consisting of six "shearing layers of change". His perspective is inspired by Francis Duffy's research and experience. The British architect and designer of office spaces pointed out that the expenses for changes to the interior and exterior of a building exceed the costs for erection many times over throughout a 50-year time period [7]. These costs correspond to the consumption of resources and energy. In Brand's book, Duffy is introduced as the originator of a systematic method of analysing buildings according to different layers (of use). He defines buildings as a set of layers comprising elements of different durability and concludes that the factor of time is the decisive challenge of design [8].

The six shearing layers derived from this model, which Brand ultimately proposes as universally valid, are site (place), structure (load-bearing system), skin (facade and building envelope), services (building services and mechanical engineering), space plan (spatial configuration and interior finishes) and stuff (interior design elements and furniture). The corresponding diagram was originally created by Donald Ryan. Publications on the topic of sustainable or circular building have since become unthinkable without it (Fig. 1, p. 120).

Brand proposes life cycles for the individual layers as follows:
- Site (place): Lasts forever
  Site describes the place of construction

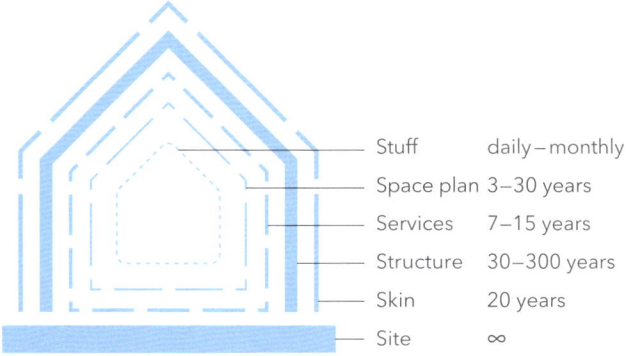

	Stuff	daily–monthly
	Space plan	3–30 years
	Services	7–15 years
	Structure	30–300 years
	Skin	20 years
	Site	∞

1

and, thus, the geographical location on earth, independent of human-made, external influences.

- Structure (load-bearing system): 30–300 years
  Structure comprises the foundations and the load-bearing system of a building. Brand refers to the circumstance that this layer is the most complex and difficult to change. The load-bearing structure typically also contains the major share of embodied energy (up to 75% of embodied energy and greywater of a building) [9].
- Skin (facade and building envelope): 20 years
  The skin layer includes the facade as exterior protective layer and medium of expression of a building. It can also function as the thermal building envelope.
- Services (building services and mechanical engineering): 7–15 years
  Services contain all technical equipment (electrical, water, sewage, heating, cooling, etc.) including the required wiring, plumbing and installations, in addition to elevators or escalators.
- Space plan (spatial configuration and interior finishes): 3–30 years
  Brand describes the space plan as the floor layout of a building, including its interior (partition) walls, (hung) ceilings and doors. The broad range of life cycles results from the different uses. Brand views retail shops as examples for a very short-lived space plan and considers dwellings occupied for longer periods of time as their opposite.
- Stuff (interior design elements and furniture): Daily–monthly.

The layer termed as stuff includes all movable furniture and interior design elements.

Brand observes that every building consists of slow layers (more durable and difficult to change) and rapid layers (short-lived and easy to change). The slow layers dominate the rapid ones and enable (or prevent) an incremental integration of desired changes facilitated by the more rapid layers. The slower layers of a building include the site, the structure and, to a limited degree, the spatial configuration and the building envelope. They typically permit reliable use for the longest periods of time. Services, space plan and stuff are correspondingly rapid layers. As Stewart Brand states, "because of the different rates of change of its components, a building is always tearing itself apart" [10]. Thus, it is sensible to consider the interfaces he proposes already in the conceptualisation phase of planning a building.

The most important insight that Brand's observations yields is that changes to a building take place continuously and without interruption. This occurs at different speeds and frequencies for different building parts. For circular construction, this means that intersections between slow and rapid layers in particular require specifically diligent consideration.

Building construction layer model

Against the background of the "shearing layers" model, primarily the aspect of time becomes important for the determination of interfaces that are as sensible as possible for the components of a building. In the

1
Model, six shearing layers according to Steward Brand, illustration based on the diagram by Donald Ryan: Six universally valid layers, including site (place), structure (load-bearing system), skin (facade and building envelope), services (building services and mechanical engineering), space plan (spatial configuration and interior finishes) and stuff (interior design elements and furniture) are associated with cycles of change according to Brand's assumptions.

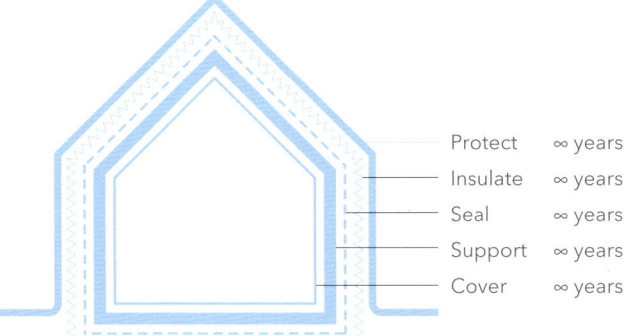

2
Model of the five technical and functional layers of building construction. Integration sorted by type and reversible connections are the preconditions for an optimal life cycle and potentially infinite circularity of the deployed materials.

following, building construction will be the focus of deliberation. From this perspective, Brand's model remains vague in certain areas. It is unclear, for instance, where he locates the thermal building envelope or where the dividing line is between structure and space plan. Still, the realisation remains that each and every building layer should be utilised in the most efficient possible manner in terms of function and life cycle. Further, the design of the individual layers is supposed to enable separability. This allows altering, exchanging or adapting them separately without reducing the life-cycle of the entire system.

The components in the focus of the building-construction-oriented layer model are designed to meet essential requirements of thermally sealed buildings in moderate latitudes. Compared to Brand's model, this enquiry is oriented on the layers of the skin, the structure and (as related to building elements) the space plan. The separation between building construction and building services, as well as the capacity for adaptation of the basic structure of the architectural design are considered essential. In the evaluation of a contemporary new structure in terms of how it meets requirements for comfort and building physics, viewed from the exterior to the interior, the following technical and functional demands must be met: Protect, insulate, seal, support, cover. Within this model, the technical and functional specifications of the respective layer define the interface. This model can also take different "speeds" into account.

- protect (requirement: Protection from weather, rain, wind): Exterior building envelope, facade cladding, roofing, typically supported by framing or substructure, with or without back ventilation
- insulate (requirement: Minimise thermal losses, reduce (heating) energy consumption): Insulation mounted to the exterior, infilled between the load-bearing structure, mounted to the interior
- seal (requirement: Protection from moisture from the exterior and vapour from the interior, decoupling of building parts as required): Film or membranes as vapour barrier, sarking membrane, wind barrier, etc.
- support (requirement: Transfer of building loads and live loads, stiffening, stability, e.g. also earthquake proofing): Different construction types (filigree or solid construction)
- cover (requirement: Proper finish of the building interior, surface subject to touch and wear, indoor acoustics, concealing installations, etc.): Wall coverings, hung ceilings, installation layers, surfaces, flooring materials

The proposed model involves a paradigm change in the way in which architects consider materials and constructions (Fig. 2). While the shearing layers model is focused on the life cycle-related interfaces of a building and remains applicable even in the context of a linear economic model (see "The linear system", p. 15ff.), the building construction model aims at the construction-related and technical interfaces. Its purpose is to establish reversible connections sorted by type and provide each deployed material with an optimum life cycle while enabling an optimum afterlife in the manner of a potentially infinite circularity.

Notes
[1] Brand 1997, p. 178
[2] ibid., in particular p. 12–23 and p. 178–221
[3] Heisel / Hebel 2021
[4] ZHAW / Stricker 2021, p. 129–142
[5] Heisel / Rau-Oberhuber. In: Heisel / Hebel 2021, p. 157–167
[6] Einfach Bauen. Forschungsvorhaben. TU Munich 2018
[7] Duffy. In: Brand 1997, p. 12f.
[8] ibid.
[9] Stahel. In: Heisel / Hebel 2021, p. 39f.
[10] Brand 1997, p. 13

Layer Compositions

The construction of a building is defined to a high degree by the structural system, the materials used for creating load-bearing elements, or the arrangement of the individual building layers in the overall system. This section will look at examples of wall and ceiling constructions sorted by type. They are categorised according to filigree or solid construction types, the arrangement of the insulation layer in relation to the load-bearing components, or the predominant building materials, including wood, steel, concrete, masonry and loam. In particular, the different building layers will be illustrated according to functions – protect, insulate, seal, support, cover (see "Building construction layer model", p. 120f.) – and their respective connections (see "Assembly Principles", p. 110ff.).

All isometric illustrations scale 1:25

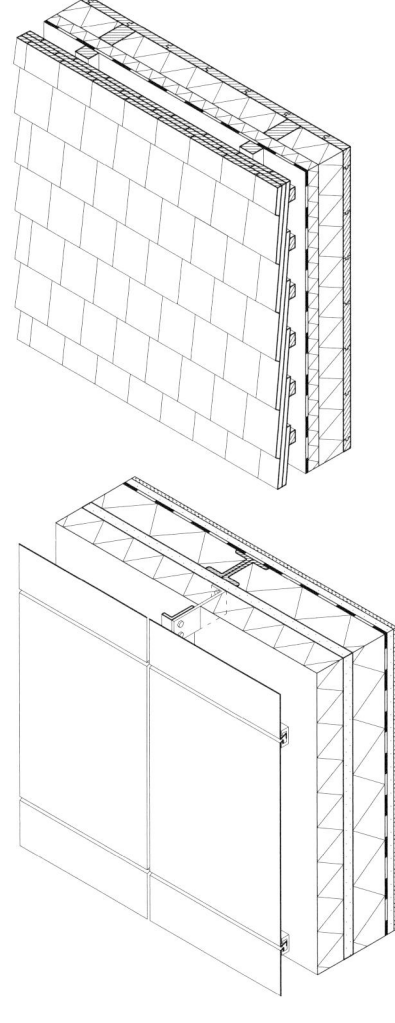

Filigree construction | cavity insulation | timber

Function	Composition
Protect	shingles, nailed
	battens, screw connections to counterbattens
	counterbattens for back ventilation, screw connections
Seal	sarking layer, breathable, overlapping installation, stapled
Insulate	rigid insulation panel, screw connections with battens to wood studs
	insulation, fit between wood studs
Support	solid wood studs and posts
Cover	diagonal siding, impermeable, screw connections to wood studs

Filigree construction | cavity insulation | steel

Function	Composition
Protect	sheet steel, concealed screw connections to framing
	steel section framing, set into bracket, screw connections to drywall panel, penetration of insulation layer
Insulate	insulation, anchored to drywall panel with insulation dowel
	drywall panel, screw connections to stud frame
	insulation fit between studs
Support	steel frame
Seal	vapour barrier, overlapping installation, clamped
Cover	drywall panel, screw connections to stud frame
	render primer and finish

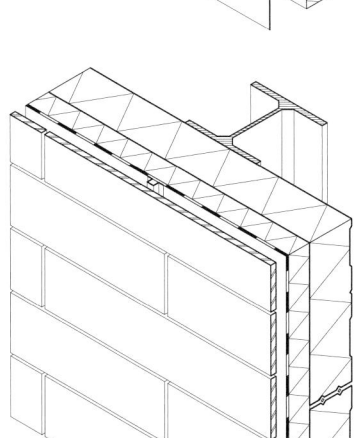

Filigree construction | exterior insulation | steel

Function	Composition
Protect	facade panels hung into brackets
	framing, steel channels, galvanised, screw connections to system coffer, penetration of insulation
Seal	sarking membrane, breathable, overlapping, clamped
Insulate	insulation, anchored to system coffer with insulation dowels
	sheet metal system coffer, inlaid insulation, galvanised, canted, screw connections to steel structure
Support	structural steel section, galvanised

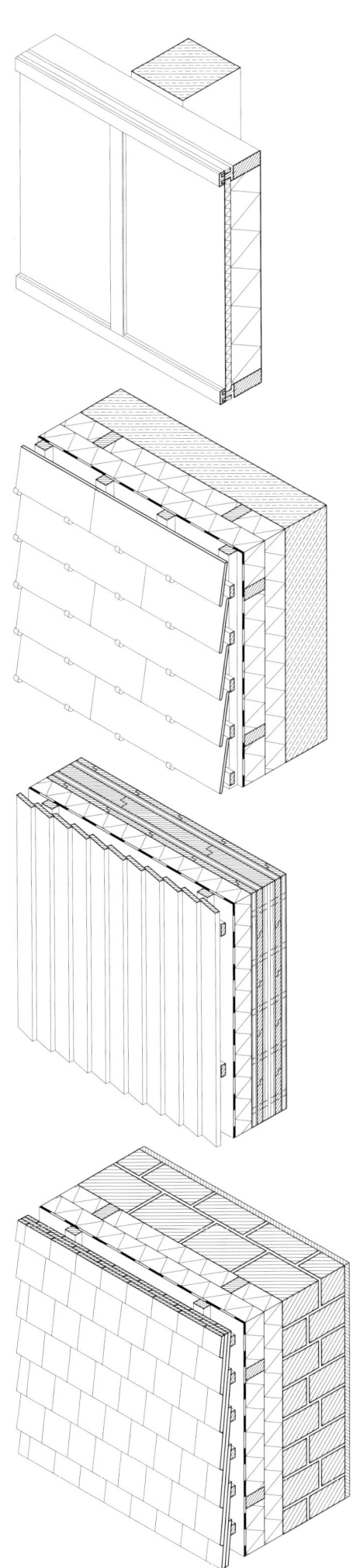

Filigree construction | exterior insulation | concrete

Function	Composition
Protect & insulate	mullion-transom coffer sections, sheet steel galvanised, fit between transoms insulation fit into steel coffer section
Support	prefabricated reinforced concrete column

Solid construction | exterior insulation | concrete

Function	Composition
Protect	facade panel hung into batten structure battens, screw connections to counterbattens counterbattens as back ventilation, screw connections
Seal	sarking membrane, breathable, welded
Insulate	structural battens, screw connections to counterbattens counterbattens, screw connections to prefabricated reinforced concrete element inlaid insulation panels, 2-ply, fit into structure
Support	prefabricated reinforced concrete element

Solid construction | exterior insulation | timber

Function	Composition
Protect	facade siding, visible butt joints, screw connections battens, screw connections to counterbattens counterbattens for back ventilation, screw connections sarking membrane, breathable, welded
Insulate	rigid insulation panel, anchored to solid timber wall with insulation dowels and counterbattens
Support	solid timber wall, wood dowel connections

Solid construction | exterior insulation | masonry wall

Function	Composition
Protect	shingles, nailed battens, screw connections to counterbattens counterbattens for back ventilation, screw connections
Seal	sarking membrane, breathable, welded
Insulate	battens, screw connections to counterbattens counterbattens, screw connections to masonry wall inlaid insulation, 2-ply, fit into structure
Support	loam brick masonry wall, mortar joints
Cover	render primer and finish

Layer Compositions

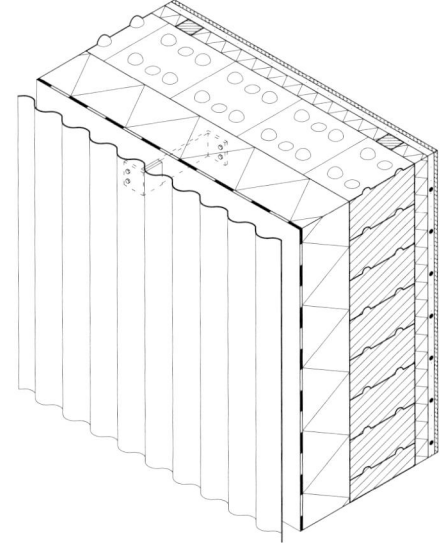

Solid construction | exterior insulation | dry masonry wall

Function	Composition
Protect	corrugated sheet metal, galvanised, screw connections
	framing for back ventilation, steel bracket
	galvanised, screw connections to masonry wall
Seal	sarking membrane, breathable, welded
Insulate	insulation, anchored to masonry wall with insulation dowels
Support	modular wall block, stacked
Cover	battens, screw connections to modular block wall
	insulation, fit between battens
	drywall panel with integrated heating pipes
	screw connections to battens
	render finish

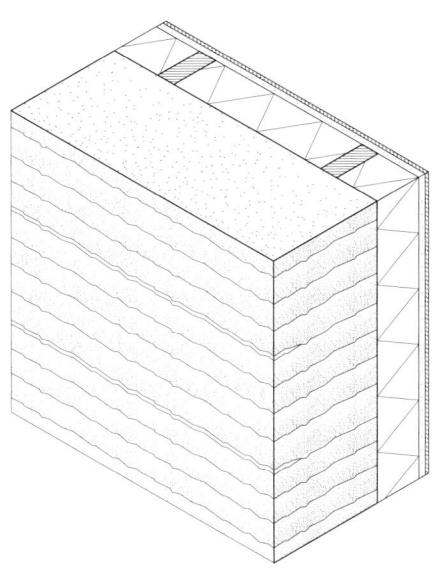

Solid construction | interior insulation | loam

Function	Composition
Support	rammed earth with embedded erosion proofing
Insulate	battens
	insulation, fit between battens
Cover	drywall panel, screw connections to battens
	render primer and finish

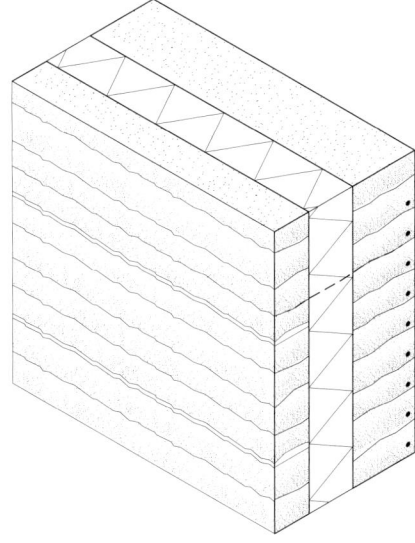

Solid construction | core insulation | loam

Function	Composition
Protect	prefabricated rammed-earth element, self-supporting, anchored to load-bearing wall plate with geogrid
Insulate	insulation panels, integrated in the workshop
Support	prefabricated rammed-earth element, self-supporting, with integrated wall heating

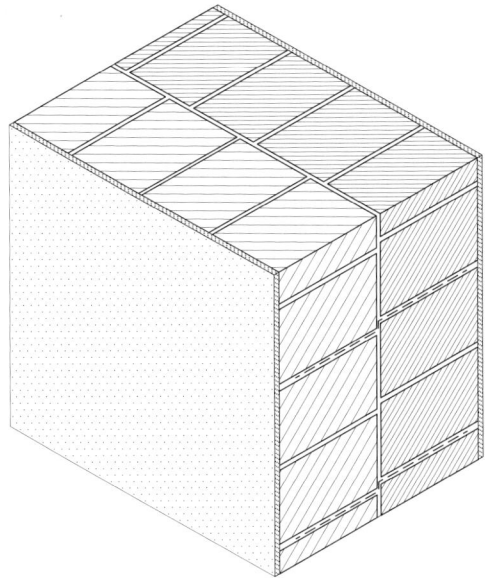

Solid construction | masonry wall

Function	Composition
Protect	render primer and finish
Insulate	vertical perforated brick, insulating, mortar joints, connected to load-bearing shell with masonry anchors
Support	flat steel anchor, embedded into mortar layer vertical perforated brick, load-bearing, mortar joints
Cover	render primer and finish

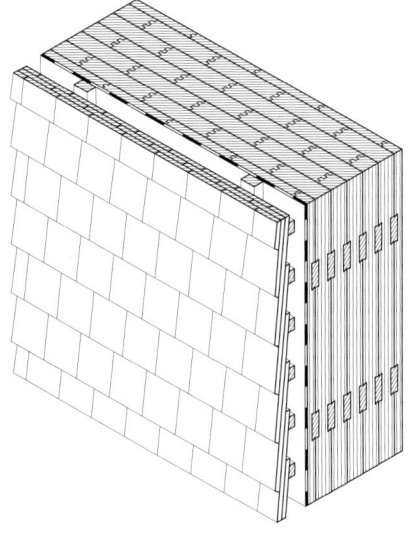

Solid construction | timber

Function	Composition
Protect	shingles, nailed battens, screw connections to counterbattens counterbattens for back ventilation, screw connections
Seal	sarking membrane, breathable, welded
Support & insulate	solid wood wall, connected with hardwood dowels or ridge strips

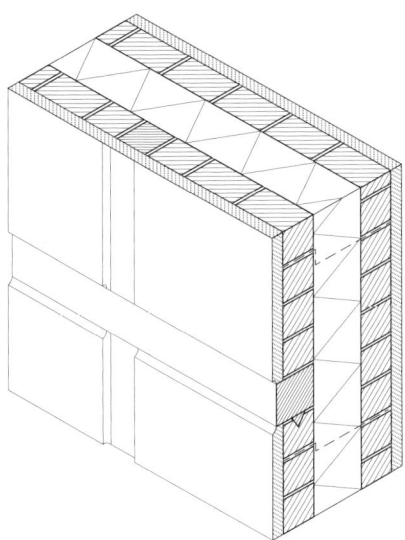

Timber frame construction | core insulation | loam brick

Function	Composition
Protect	render primer and finish
Support	timber frame construction, mortise and tenon, screw connections, infilled with loam brick
Insulate	insulation, anchored to timber frame with insulation dowels
Cover	loam brick, non-load-bearing, mortar joints, with masonry anchor render primer and finish

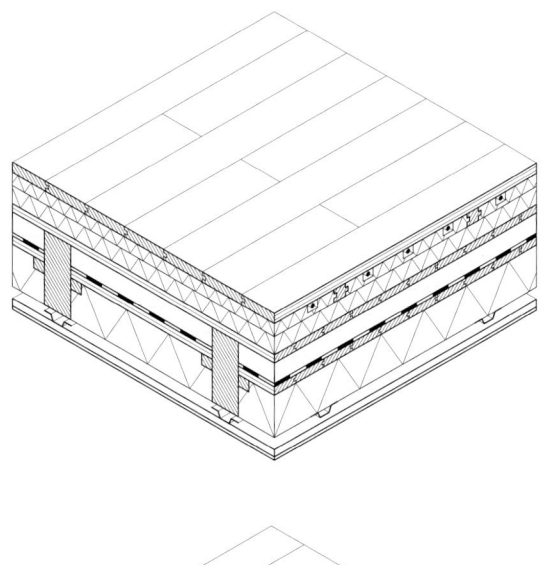

Filigree construction | cavity insulation | timber

Function	Composition
Cover	floorboards, tongue and groove, concealed screw connections to battens
	heating pipes on heat transfer plate embedded in floor panel
	floor panel with millings for heating pipes
	floating installation between battens
Insulate	impact soundproofing, floating installation
Support	sheathing, tongue and groove, screw connections to ceiling beam
	solid timber ceiling beam
Insulate	subfloor with heavy fill for soundproofing, on trickle protection, loose layout, supported by battens, screw connections to ceiling beams
	insulation fit between beams
Cover	hung ceiling system, steel spring rail, screw connections
	drywall panel, screw connections to framing, spackled

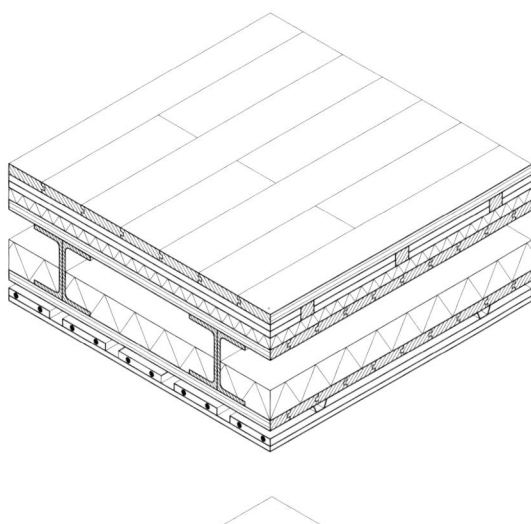

Filigree construction | cavity insulation | steel

Function	Composition
Cover	floorboards, tongue and groove, concealed screw connections to battens
	drywall panel, 2-ply, floating installation between battens, soundproofing as installation layer, floating installation
	sheathing, tongue and groove, screw connections to ceiling beams with support clamps
Support	ceiling beams, double steel section
Insulate	insulation, fit between ceiling beams
Cover	siding, tongue and groove, screw connections to ceiling beams with support clamps
	framing, steel section, screw connections to siding
	heating and cooling ceiling with embedded heating pipes, screw connections to framing

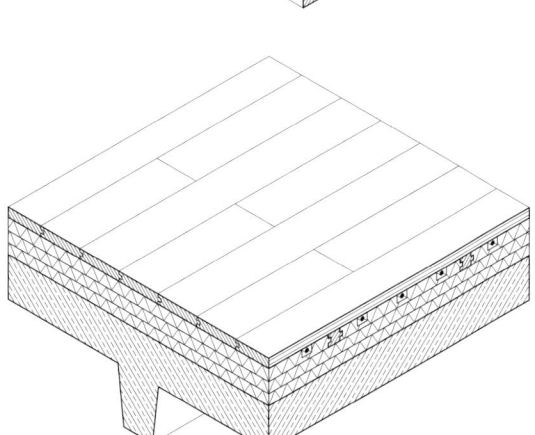

Filigree construction | timber

Function	Composition
Cover	rammed earth
	heating pipes embedded in loam block
	loam block with recesses for heating pipes, floating installation
	drywall panel, 2-ply, floating installation
Insulate	impact soundproofing, floating installation
	levelling fill, loose, for soundproofing
	trickle protection, loose installation
Support	sheathing, tongue and groove, screw connections to ceiling beam
	solid timber ceiling beam

Filigree construction | concrete

Function	Composition
Cover	floorboards, tongue and groove, concealed screw connections to battens
	heating pipes on heat transfer plates, embedded in installation panel
	installation panel with millings for heating pipes, floating installation between battens
Insulate	impact soundproofing, multilayer, floating installation
Support	prefabricated reinforced-concrete ribbed ceiling

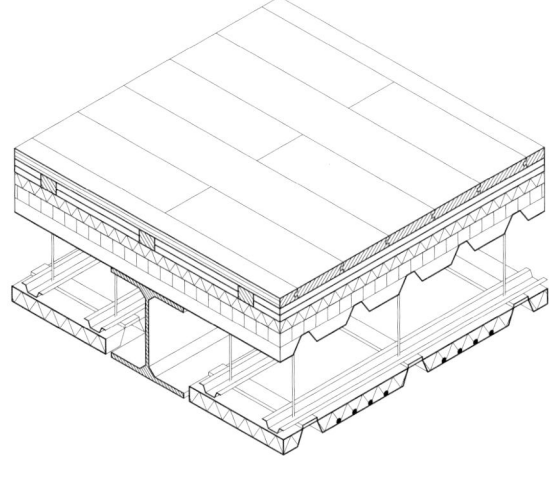

Filigree construction | steel

Function	Composition
Cover	floorboards, tongue and groove, concealed screw connections to battens
	drywall panel, 2-ply, floating installation between battens
Insulate	impact soundproofing as installation layer
Support	mass fill, loose, in cardboard honeycomb
	corrugated sheet metal with mass fill, bolted connections to steel beams
	structural steel section, galvanized
Cover	hung ceiling system, steel spring rail, screw connections
	heating and cooling ceiling, steel panel with heating pipes
	loose infill insulation, screw connections to framing

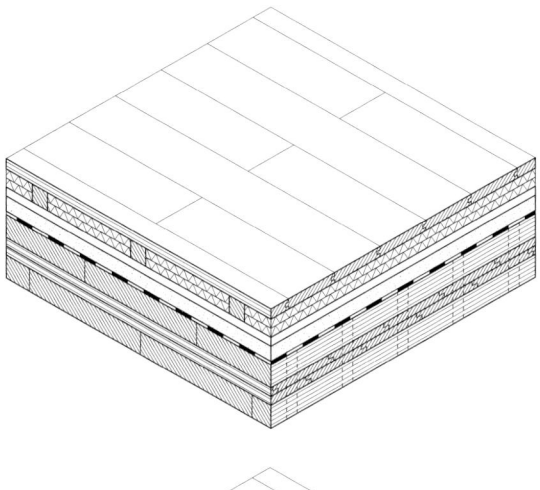

Solid construction | wood

Function	Composition
Cover	floorboards, tongue and groove, concealed screw connections to battens
Insulation	impact soundproofing, 2-ply, floating installation
	dry screed panel, floating installation
	loose levelling fill for soundproofing
	trickle protection, loose installation
Support	solid timber ceiling, wood dowel connections

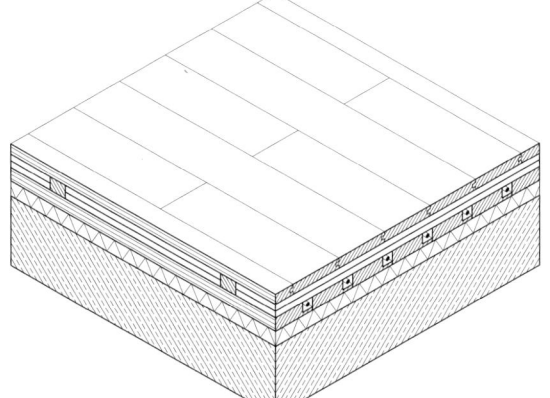

Solid construction | concrete

Function	Composition
Cover	floorboards, tongue and groove, concealed screw connections to battens
	battens, drywall panels fit in-between, 2-ply, screw connections,
	heating pipes on heat transfer plate, embedded in installation panel
	installation panel, millings for heating pipes
Insulate	impact soundproofing as installation layer, floating installation
Support	ceiling, prefabricated reinforced-concrete element

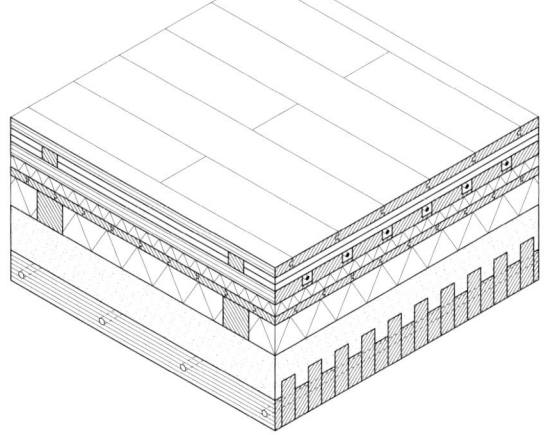

Hybrid | timber and loam

Function	Composition
Cover	floorboards, tongue and groove, concealed screw connections to battens
	battens, dry screed panel fit in-between, 2-ply, screw connections
	sheathing with millings for underfloor heating pipes
Insulate	impact soundproofing as installation layer
	diagonal siding, screw connections to battens
	insulation, loosely fit between battens
Support	timber-loam composite ceiling, rammed earth on stacked timber ceiling, dowel laminated ceiling with wood dowel connectors

Layer Compositions

DETAIL CATALOGUE

Focus on Wood

Wood is one of the oldest and most widely used building materials. It is sufficiently available, allows easy processing and displays high performance. As a result, characteristic timber buildings have been created in nearly every region settled by humans throughout history, ever since the onset of civilisation. For centuries, half-timbered structures, Scandinavian timber houses, as well as timber bridges and other engineering constructions have defined the culture of building in Europe. The weak spot of timber construction was, time and again, its susceptibility to fire. In 19th-century Paris, catastrophic fires raged repeatedly and, among others, destroyed numerous theatre buildings. In response, the French government advertised an award for fireproof means of construction. This primarily led to the advancement of building with concrete [1]. From this point in time, timber construction was on the retreat. In their fascination for new construction materials such as concrete, steel and glass, modern architects no longer used much wood at all. The bitter experience of both World Wars and the destructive fires in their wake further contributed to a tendency to avoid timber as a construction material for rebuilding in an urban context as far as possible [2]. Timber construction remained limited to rural regions and, beyond that, served for specific tasks, such as creating roof structures.

In the course of the industrialisation of the building sector, chemical wood preservatives were soon preferred, instead of traditional craftsman-based preservation methods. The application of such substances contributed to health risks and environmen-

1
Ways of sawing a tree trunk

tal pollution. As a result, the reputation of timber construction took a hard blow. This negative trend was slowly reversed from the late 1970s onward. The change was driven by a growing awareness of ecological issues that led to wood as a building material once more becoming attractive. The architectural avantgarde's approach to timber construction as an expression of a self-confident and contemporary form of building culture also contributed to this. Progress in building technology in timber construction further supported these developments and resulted in the widespread distribution that persists to this day. As of now, large-scale multi-storey timber buildings, such as high-rise buildings, can be found even in an urban context.

Availability as raw material

Wood, according to its most basic definition, is the lignified tissue of trees and shrubs. Lignification is based on the deposition of lignin in plant cells, which produces qualities that are indispensable to construction tasks [3]. Construction-grade timber as a renewable resource originates in trees. In sawmills, logs are typically cut into linear elements such as slats, boards, square timber products and beams (Fig. 1). In Central Europe, timber is regionally available in almost every place where humans have created settlements. In the course of history and in some regions, such as the Black Forest, excessive demand for wood led to deforestation beyond repair. This was caused by regional and super-regional demand for wood in the 17th century [4]. In 2021 in Germany the share of timber buildings approved for residential construction reached 21.3 % [5]. Sufficient timber is available in Germany to significantly increase this share [6]. However, there are natural limits to the use of the raw material. According to a study by the WWF, planetary limits of sustainable timber use have already been exceeded by the current levels of consumption. "If all of us in Germany used timber for construction, hardly any domestic wood would be available to other sectors" [7].

The linear character of the construction elements produced from this raw material has, for a long time, defined building with wood. This becomes apparent in, for instance, half-timbered construction, the patterns of formwork, as well as panelling for interior finishes. Then and now, the aim is to assemble smaller linear elements into a larger structure. This is comparable to textile production – at least in terms of appearance and logic. Here as well, many smaller elements are assembled or woven into a greater structure, leading to the creation of a pattern or rhythm. The linear character loses meaning once timber is processed into panels, boards and sheets by use of binding agents or form-fit connections. Such products feature significantly greater dimensions than sawn wood, which is limited by natural growth. Panel materials are available on the market in various forms, among others as plywood, laminated wood, particle board, stacked timber elements or cross-laminated timber products. The trend from linear to planar elements has led to a fundamental change in contemporary timber construction and made it more efficient.

Characteristics
Timber displays specific material characteristics that support its diverse use as a building material. It is of relatively low weight and can be processed and transported easily. It is highly flexible and can bear strong mechanical loads, as well as tensile, compressive and bending forces. Thus, it can be assembled into high-performance ceiling slab systems or load-bearing structures. Different species of wood display strong differences in terms of their characteristics and visual appearance (colour, grain). On molecular levels, wood consists of the following substances: Cellulose, hemicellulose and lignin. Its structure is defined by cells as the smallest basic component. Cells, mostly longitudinal in shape and following the direction of the stem or trunk, are also described as fibre. Cell walls consist of layers of lignin that are surrounded by cellulose and hemicellulose fibrils. While the latter can bear tensile forces, the lignin layers are only capable of absorbing compressive forces [8]. The directional arrangement of this structure allows wood to behave differently in longitudinal and transverse directions (anisotropy).

Timber is capable of absorbing moisture and releasing it again, in the course of which its volume and/or its length changes. This process is also known as swelling and shrinkage. In particular, moisture intrusion through the open-pored ends where the capacity to absorb water is exceptionally high should be prevented. As a result, this can lead to high degrees of deformation and to an acceleration of natural decomposition processes within the timber component.

Its low density qualifies wood as a relatively good insulation material. This simplifies construction in comparison to steel and concrete, since thermal bridges hardly comprise any problem. Low mass, on the other hand, leads to challenges in terms of soundproofing and thermal comfort. In such cases, depending on the required standards, construction demands increased expenditure.

For a long time, pest infestation and rot were considered the achilles' heel of timber construction. In the recent past, this led to an increasing use of chemical wood preservatives that are considered questionable from ecological and health perspectives. Structural measures make chemical wood preservation completely unnecessary, without reducing the longevity of buildings. For many construction tasks, fire safety was also considered a significant obstacle to building with wood. However, it has now become generally accepted that, while wood tends to burn well, its fire behaviour can be described as "friendly" – it burns slowly and in a predictable manner. The latest findings point out the importance of including the carbon layer resulting from burn-up into calculations. It slows down the burning process and has thus led to a renewed evaluation of fire behaviour. An enormous advantage of wood is its characteristics with regards to comfort. The perception of wood as a warm material, its scent and its vivid appearance are reasons why many people say they feel comfortable in timber buildings [9].

Typical construction methods
The available options for the assembly of individual elements that are limited in size define the character of building with wood. Well into the 19th century, monomaterial-based and craftsmanship-related connection types such as mortise-tenon or lap joints prevailed. Craftsmanship-based joinery methods require significant amounts of labour. From the second half of the 20th century onwards, they were increasingly replaced by engineering-based connections featuring metal components such as angles or perforated plates. The currently available means of digital production, supported by CAD and CNC, allow the

processing of wood in an economical manner, leading to monomaterial connections. These can be considered advancements of craftsmanship-based joinery. On the one hand, digitally supported, precise production is contributing to a renaissance of historic methods of joinery, such as with treenails, wood pegs or wood dowels. On the other hand, the assembly of wood boards or square timber by use of wood connectors into planar construction materials without using glue or adhesives is also possible. Numerous products are based on dovetail joints, pegs and dowels as connectors for high-performance monomaterial building components. Such developments provide timber construction with a new impetus in terms of building technology, aesthetic quality and circularity. Further, timber construction comprises a variety of methods that also find use for hybrid forms. This includes timber frame construction for the creation of plate elements consisting of sheathing applied to frames, or plywood and cross-laminated timber products for solid plate elements, or traditional solid timber building methods including log construction, which has become seldom. For ceiling construction, many different systems are available as well: traditional beam ceilings and prefabricated stacked timber or dowel laminated elements spanning one direction only, non-directional solid cross-laminated timber ceilings, or timber concrete composite ceilings. Common construction types are usually comprised of layers. Every layer performs a specific function, e.g. support, insulate, air tightness, fire protection, back ventilation, soundproofing, weather protection. This idea of layering qualifies timber construction as efficient, yet also complex.

Circular potential
Building with wood displays a very good carbon footprint. For one, this is due to the fact that wood can store carbon during its growth [10]. Secondly, only a limited amount of energy is required for its production, compared to building with steel or concrete. Beyond the question of carbon emissions, wood is likely the building material with the greatest potential in terms of circularity. This is mostly due to the construction logic inherent to the material. The typically used connectors support the assembly and disassembly of timber components in a relatively simple and damage-free manner. Old half-timbered buildings were and still are easy to disassemble, once infill has been removed. In principle, the same is true for current timber construction. Aside from the described monomaterial and form-fit connections, screws are still one of the most frequently used connectors in timber construction. Thus, disassembly is much simpler than in the case of concrete or masonry buildings, even if it requires a certain degree of effort. This also supports relatively easy transport. The simple repair and modification of building components or even entire buildings coincides with the opportunity for reusing building components such as beams, ceilings and floors. Since timber can be easily processed, the required adaptation of building components can take place easily as well. In the case of wood, reuse by newly deploying a construction component generally requires testing whether it still fulfils the parameters of standardised fresh wood. History demonstrates that this is possible. The potential of reclaimed waste wood has also become a research topic [11]. Should the reuse of wood construction components be impossible, the the next step would be to use them as a source of lower-quality material. This includes reclaiming smaller pieces of wood or those with small cross sections from waste wood components, or further processing smaller pieces of wood into insulation material or wood-based material panels (not sorted by type). This often coincides with reducing wood elements in size and, effectively, cascading use [12]. This cascading use terminates with composting and thereby reintroducing the material into its natural cycle, unless the wood is contaminated by chemical additives, such as adhesives or coatings, or contains impurities such as nails. This natural process is always preferable to incineration, in order to ensure that nutrients remain in the natural cycle. Proper production and assembly sorted by type are key to achieving this.

Notes
[1] Collins 2004, p. 19–56
[2] Gerbig/Greschat 2015
[3] Volz 2003, p. 31ff.
[4] Stuttgarter Nachrichten 2018
[5] Holzbau 2021
[6] German Federal Environment Agency 2020
[7] Beck-O'Brien et al. 2020, p. 15 and p. 46
[8] see note 3
[9] Selberherr 2020, p. 15
[10] ibid.
[11] Reversibel Bauen 2019
[12] ibid.

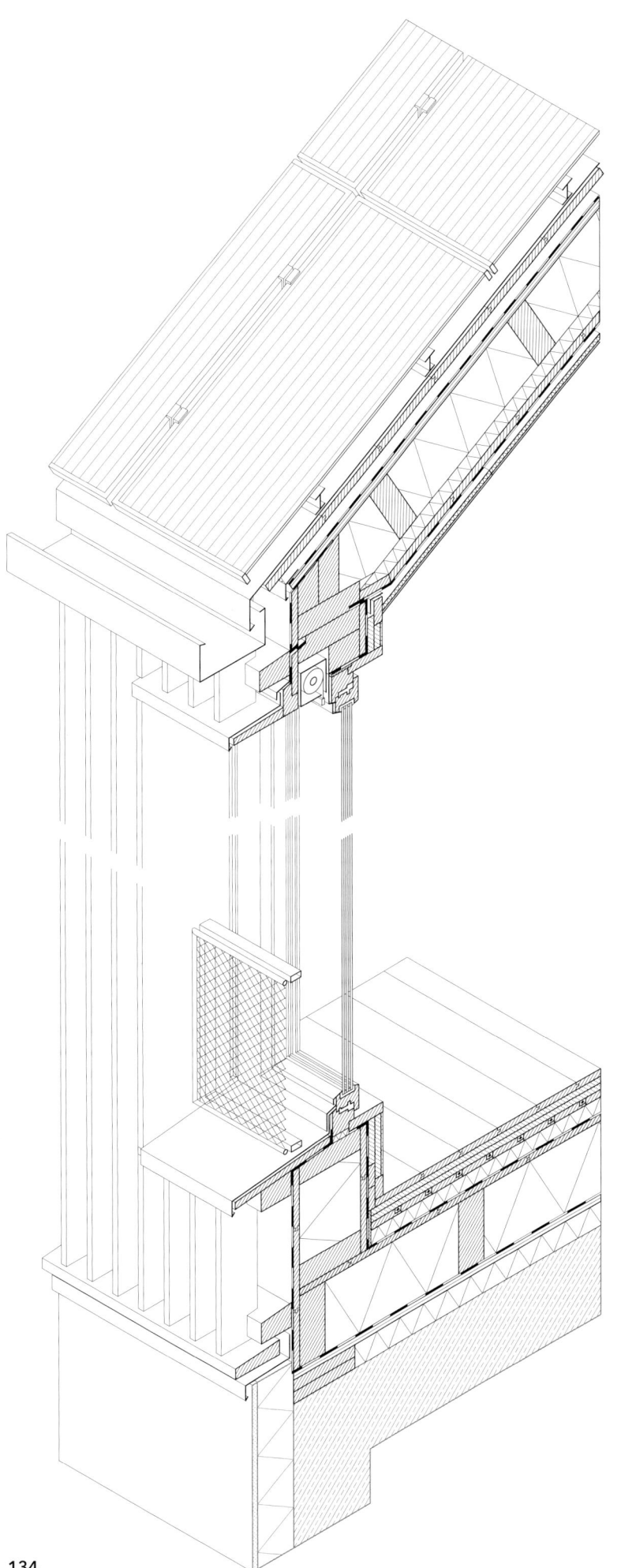

Timber: Modular Construction

Protect
standing seam roofing
timber slats

Insulate
seagrass insulation

Support
solid construction-grade timber posts

Seal
PE film
sarking membrane

Cover
reclaimed wood floorboards
loam render on loam construction panel
sheep wool felt on cellulose waste substrate panel

Isometric illustration scale 1:20
Vertical section scale 1:20

1 Roof construction:
   Photovoltaic (PV) system clamped to framing
   metal framing, clamped to standing seam
   standing seam roofing, clamped to anchor clip
   min. 24 mm solid timber sheathing, screw connections
   min. 40 mm counterbattens / back ventilation, screw connections
   sarking membrane, airtight, breathable, min. 30 cm overlap, clamped
   solid timber sheathing, screw connections
   solid construction grade timber rafters as module border
   solid construction grade timber purlins, set between rafters, max. 60 cm off centre, dovetail joint to rafters, bolt connections
   loose infill insulation, seagrass
   seagrass insulation, loose infill between battens
   diagonal sheathing, bracing effect, screw connections
   PE film, diffusion-resistant, min. 30 cm overlap, attached to wall plate
   solid timber sheathing, screw connections
   textile cover, sheep wool felt, attached to cellulose substrate panel
2 Sheet steel insect screen, corrosion-proof, screw connections to sheathing and counterbattens

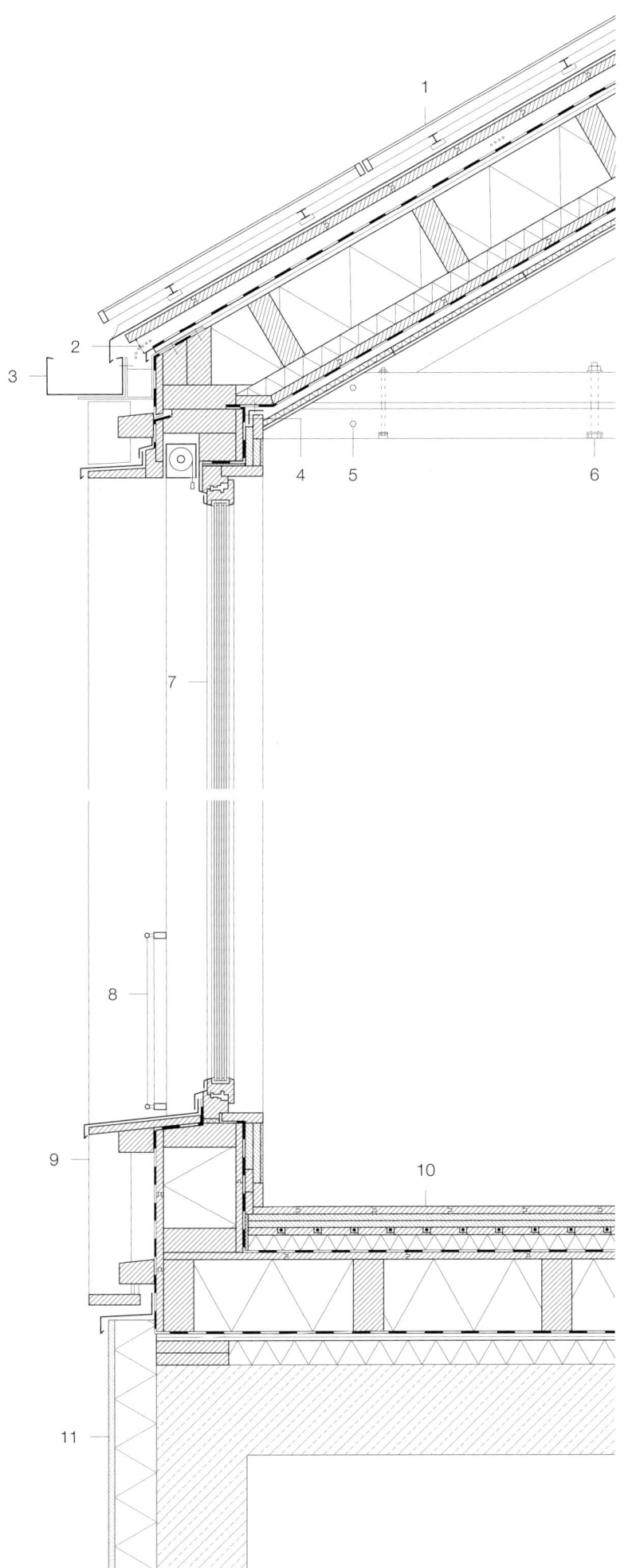

3 Box gutter, corrosion-proof, screw connections
4 Steel angle, corrosion-proof, as connector between modules, screw connections
5 Horizontal connection of modules along tie beam (horizontal border), bolt and nut connections, visible from the interior
6 Vertical connection of modules along tie beam (horizontal border), bolt and nut connections, visible from the interior
7 Windows: Triple glazing in wood frame with aluminium cover, reused
plastic dry sealant, sorted by type
8 Stainless steel mesh fall protection, clamped, set into stainless steel frame
9 Exterior wall construction:
Reclaimed wood slats, sunburnt, screw connections
solid timber lattice, living organic protective layer, bolt connections to timber frame
sarking membrane, windproof, breathable, min. 30 cm overlap
solid timber sheathing, screw connections
solid construction grade timber frame, bolt connections
loose infill insulation, seagrass
diagonal sheathing, stiffening, screw connections
PE film, diffusion-resistant, min. 30 cm overlap
solid timber battens/installation layer, screw connections
loam construction panel with double loam render layer, reinforced, screw connections
10 Floor construction:
Reclaimed wood floorboards, concealed tongue-and-groove screw connections
solid timber sheathing, screw connections
inlaid 2-ply loam construction panel
solid timber sheathing, lateral groove and heat diffuser plates to install underfloor heating pipes
cork impact soundproofing / installation layer
PE film, diffusion-resistant, min. 30 cm overlap
diagonal sheathing, stiffening, screw connections
solid construction-grade timber beam, bolt connections
loose infill insulation, seagrass
PE film, diffusion-resistant, min. 30 cm overlap
solid timber sheathing, screw connections
solid timber plates, screw connections to existing ceiling
loose infill insulation, seagrass
existing ceiling
11 Existing building

Focus on Wood

Isometric illustration, roof
Isometric illustration, wall/plinth
not to scale

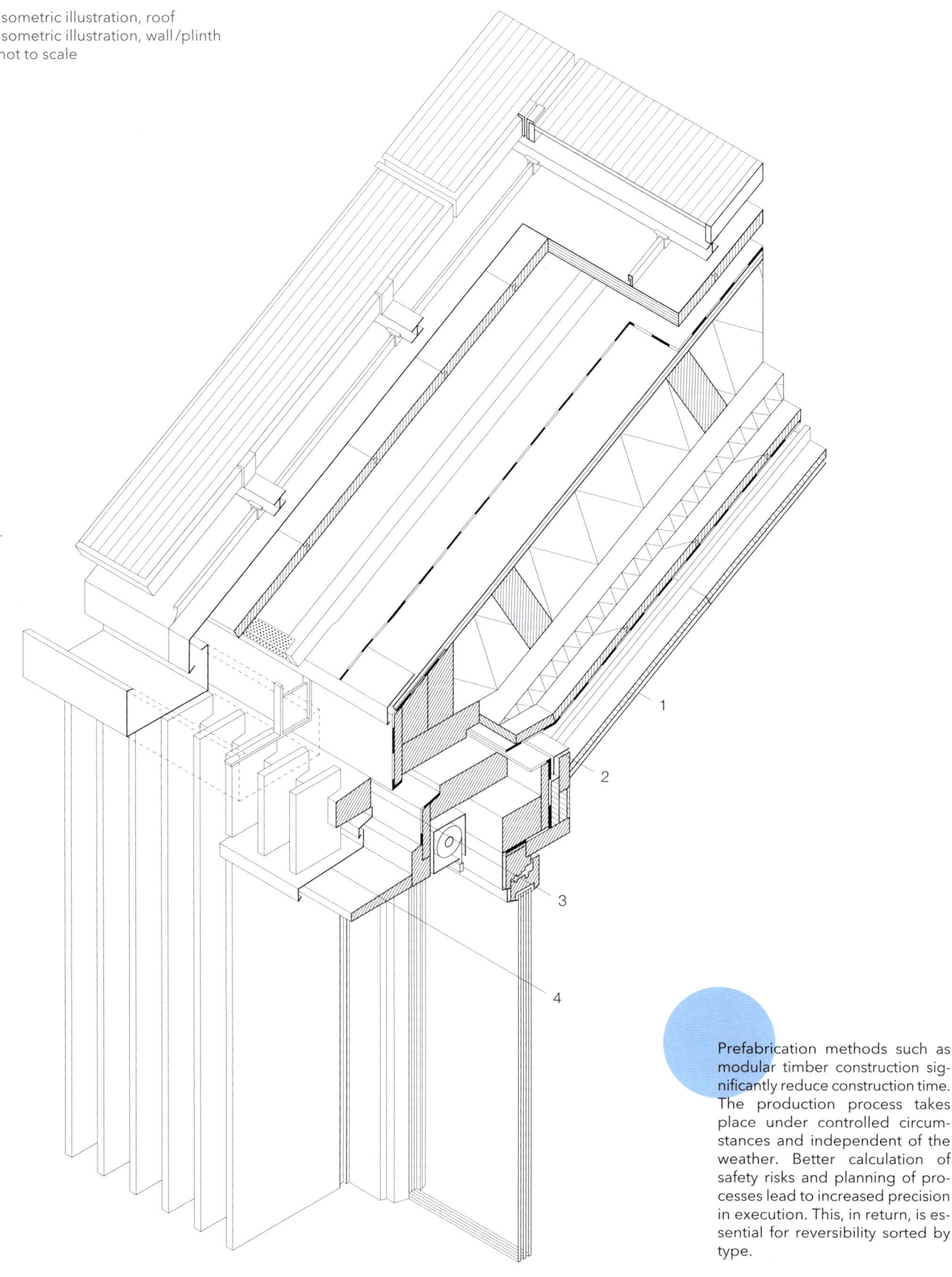

Prefabrication methods such as modular timber construction significantly reduce construction time. The production process takes place under controlled circumstances and independent of the weather. Better calculation of safety risks and planning of processes lead to increased precision in execution. This, in return, is essential for reversibility sorted by type.

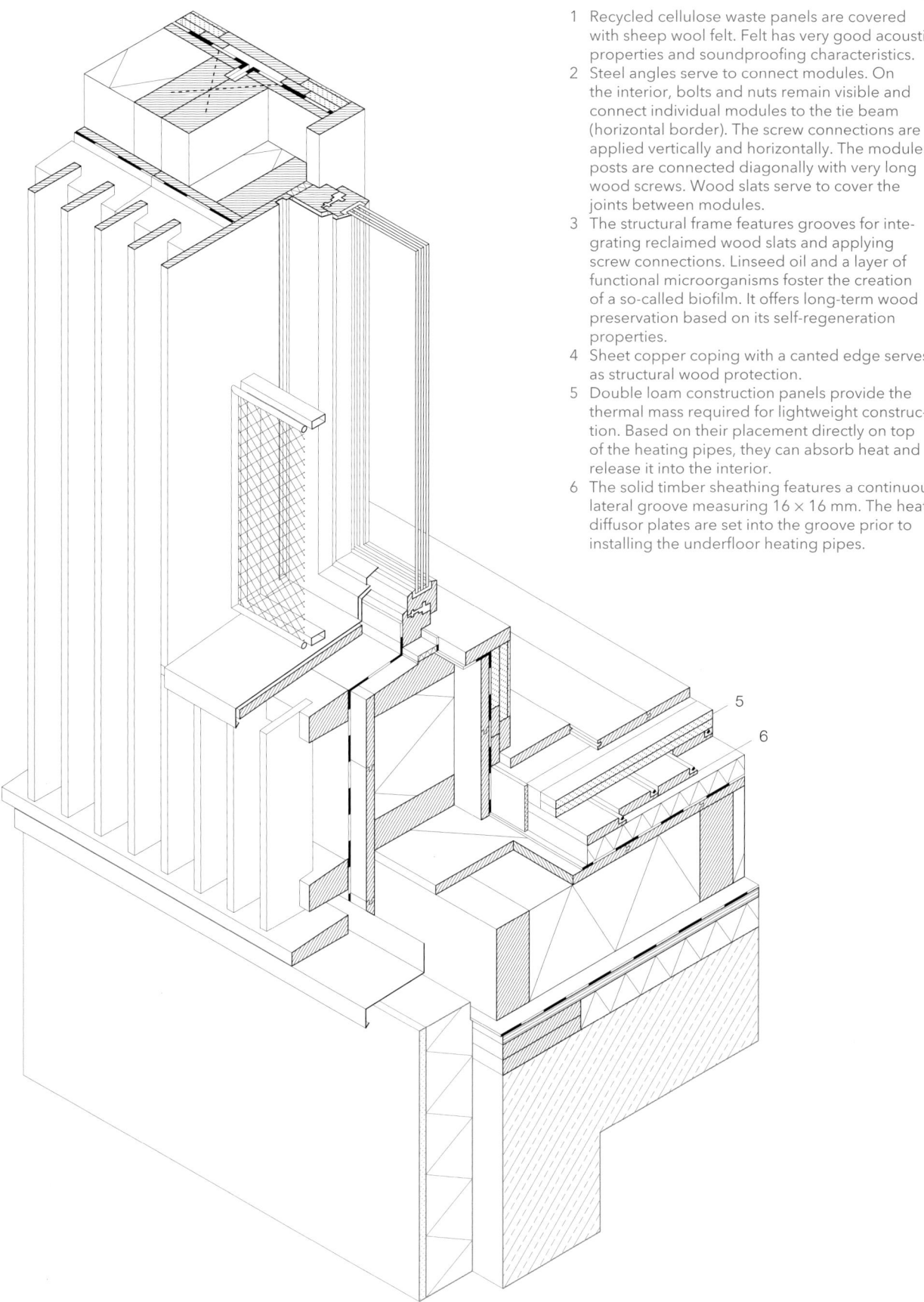

1. Recycled cellulose waste panels are covered with sheep wool felt. Felt has very good acoustic properties and soundproofing characteristics.
2. Steel angles serve to connect modules. On the interior, bolts and nuts remain visible and connect individual modules to the tie beam (horizontal border). The screw connections are applied vertically and horizontally. The module posts are connected diagonally with very long wood screws. Wood slats serve to cover the joints between modules.
3. The structural frame features grooves for integrating reclaimed wood slats and applying screw connections. Linseed oil and a layer of functional microorganisms foster the creation of a so-called biofilm. It offers long-term wood preservation based on its self-regeneration properties.
4. Sheet copper coping with a canted edge serves as structural wood protection.
5. Double loam construction panels provide the thermal mass required for lightweight construction. Based on their placement directly on top of the heating pipes, they can absorb heat and release it into the interior.
6. The solid timber sheathing features a continuous lateral groove measuring 16 × 16 mm. The heat diffusor plates are set into the groove prior to installing the underfloor heating pipes.

Focus on Wood

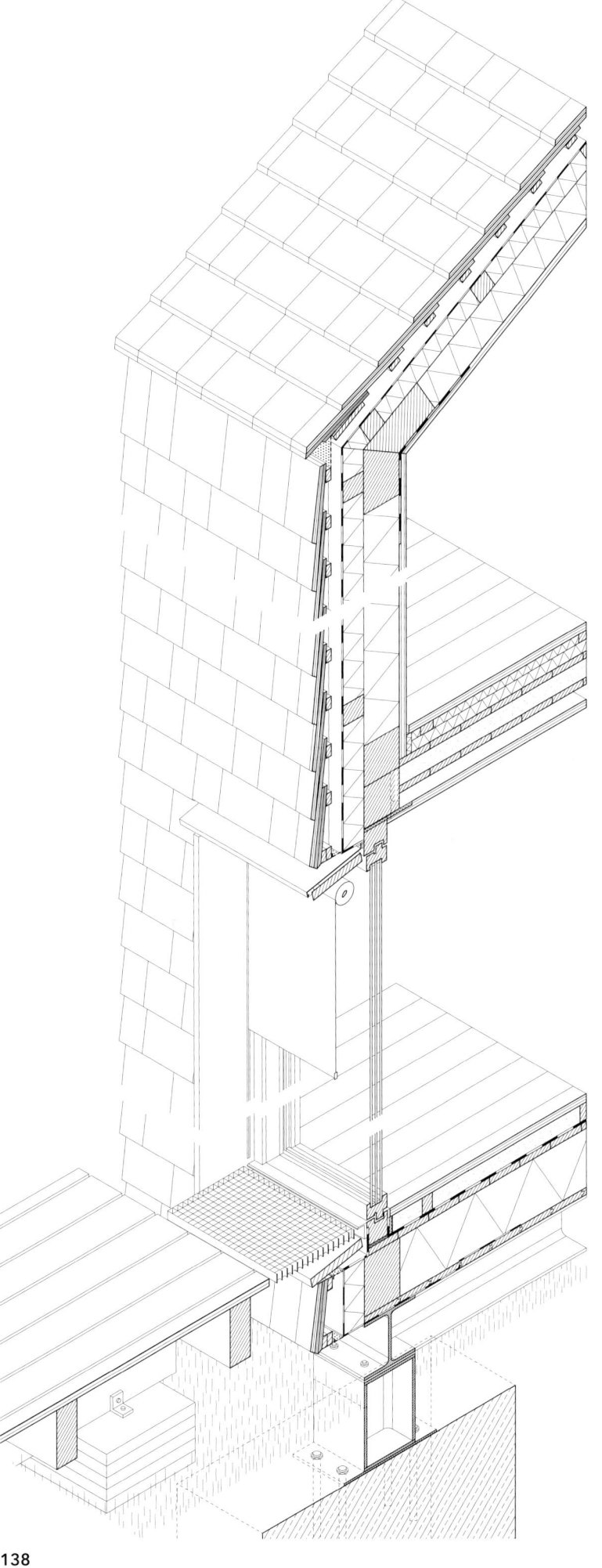

Timber: Frame Construction I

Protect
wood shingles

Insulate
softwood fibre panels
seagrass insulation

Support
solid construction timber frame
timber beams
prefabricated concrete point foundations, reused
structural grid, wide flange steel beams

Seal
PE film
sarking membrane

Cover
reclaimed wood floorboards
solid timber sheathing

Isometric illustration scale 1:20
Vertical section scale 1:20

1 Roof construction:
   Wood shingles, three layers, dowel connections
   solid timber battens, screw connections
   min. 40 mm counterbattens, screw connections/ back ventilation
   sarking membrane, windproof, breathable, min. 30 cm overlap
   softwood fibre panel thermal insulation, lignin-bonded, inlaid between wood spacers
   half-timbered solid construction timber roof, bolt connections
   loose infill insulation, seagrass
   solid timber panel, diagonal sheathing with dovetail joints, screw connections
2 Exterior wall construction:
   Wood shingles, three layers, dowel connections
   solid timber battens, screw connections
   min. 40 mm counterbattens, screw connections/ back ventilation
   sarking membrane, windproof, breathable, min. 30 cm overlap
   softwood fibre panel thermal insulation, lignin-bonded, inlaid between wood spacers
   half-timbered frame, solid construction timber
   loose infill insulation, seagrass

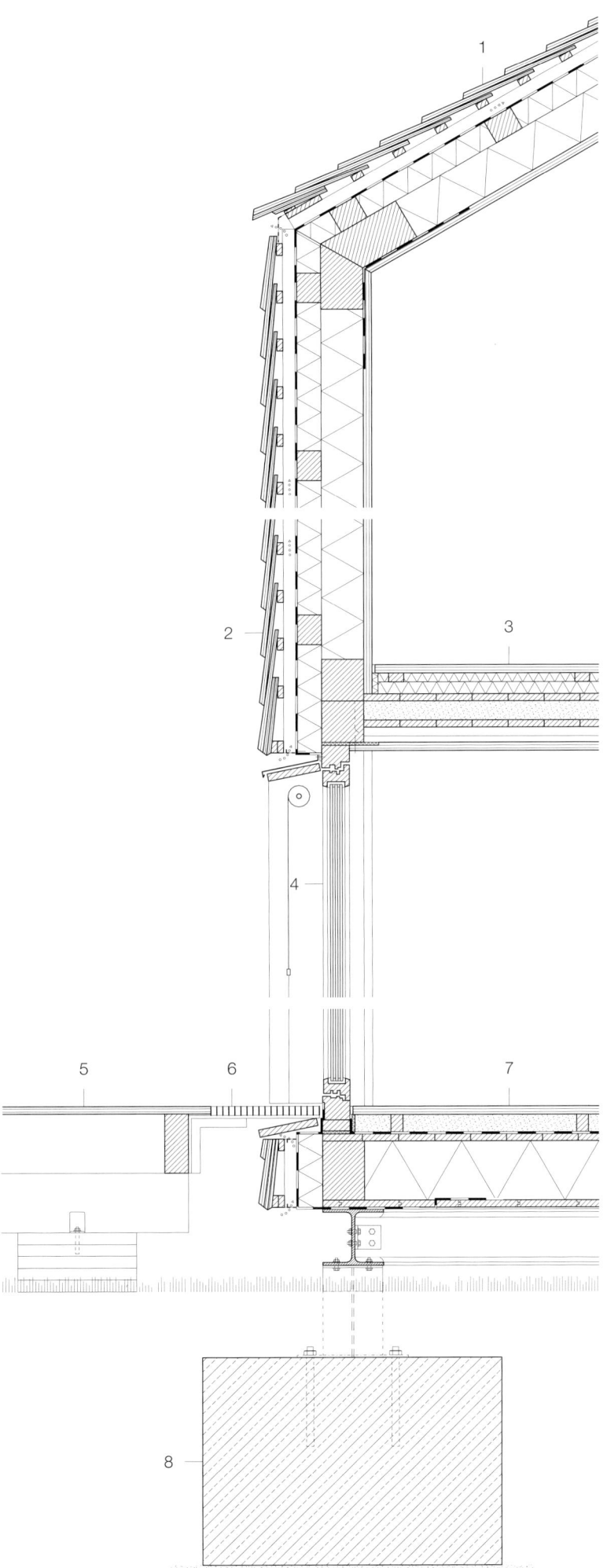

solid timber panel, diagonal sheathing with dovetail joints, screw connections
3 Ceiling slab construction:
Reclaimed wood floorboards, concealed tongue-and-groove screw connections
solid timber battens, screw connections
inlaid soft wood fibre panel, lignin-bonded
softwood fibre panel levelling insulation, lignin-bonded, loose layout
solid timber sheathing, screw connections
solid construction timber ceiling beam, dovetail joints to ring beam
inlaid mass fill, sand, to improve soundproofing
insert board, set on top of battens, screw connections
solid timber sheathing, screw connections
4 Windows: Triple glazing in wood frame, reused plastic dry sealant, sorted by type
5 Terrace decking, rift sawn, planed, untreated, screw connections
twin solid construction timber beam
concrete paver point foundations, reused, bolt connections
6 Metal grille, corrosion-proof
7 Floor construction:
Reclaimed wood floorboards, concealed tongue-and-groove screw connections
solid timber battens, screw connections
sand infill, to increase building mass
PE film, diffusion-resistant, min. 30 mm overlap
solid timber sheathing, screw connections
solid construction timber beam, dovetail joints
infill insulation, seagrass
solid timber sheathing, screw connections
wide flange steel beam, corrosion-proof, screw connections
8 Prefabricated reinforced concrete point foundation, reused, frost-free foundation
sand base layer

Focus on Wood

Isometric illustration, roof
Isometric illustration, plinth/floor
not to scale

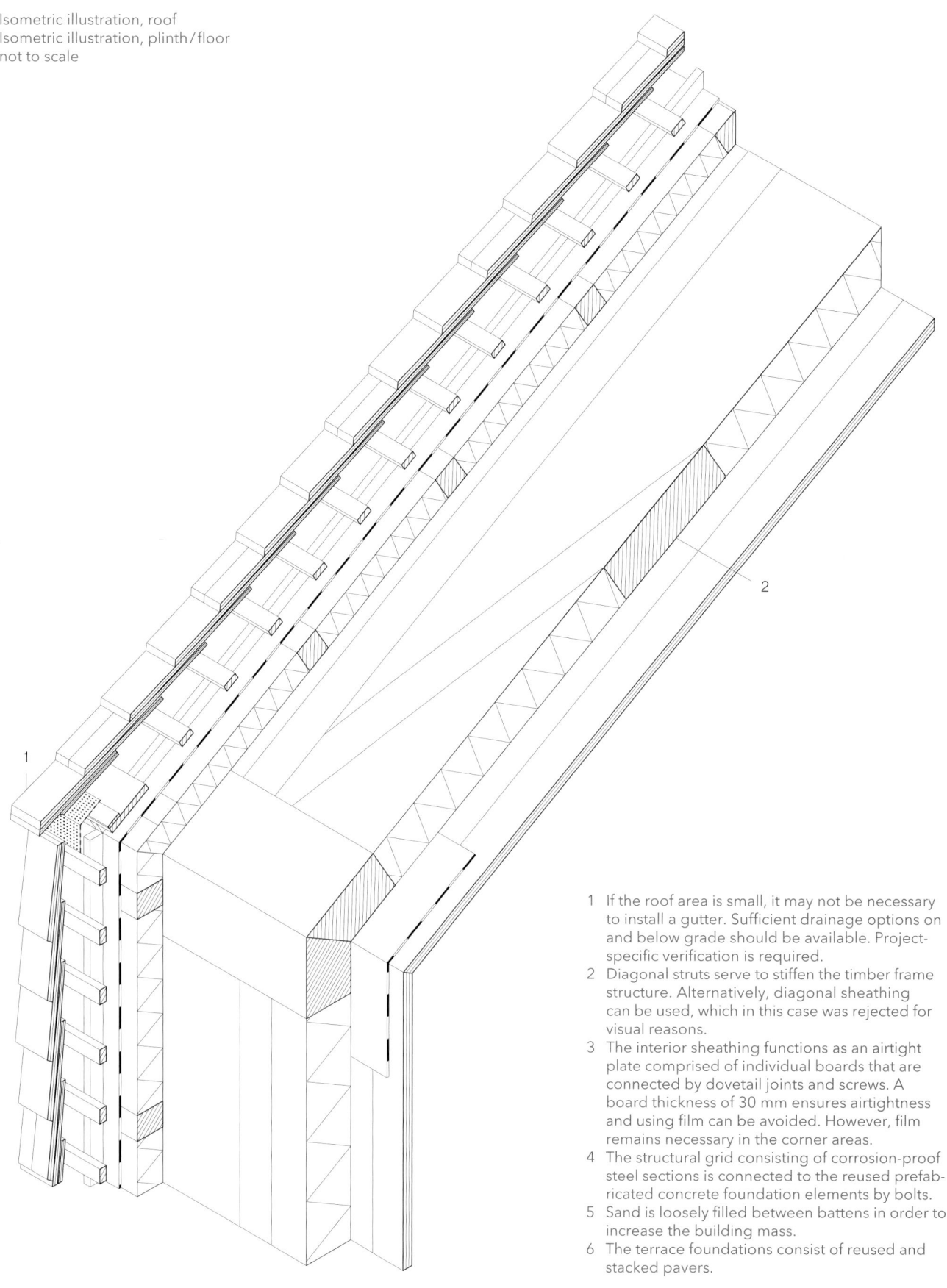

1 If the roof area is small, it may not be necessary to install a gutter. Sufficient drainage options on and below grade should be available. Project-specific verification is required.
2 Diagonal struts serve to stiffen the timber frame structure. Alternatively, diagonal sheathing can be used, which in this case was rejected for visual reasons.
3 The interior sheathing functions as an airtight plate comprised of individual boards that are connected by dovetail joints and screws. A board thickness of 30 mm ensures airtightness and using film can be avoided. However, film remains necessary in the corner areas.
4 The structural grid consisting of corrosion-proof steel sections is connected to the reused prefabricated concrete foundation elements by bolts.
5 Sand is loosely filled between battens in order to increase the building mass.
6 The terrace foundations consist of reused and stacked pavers.

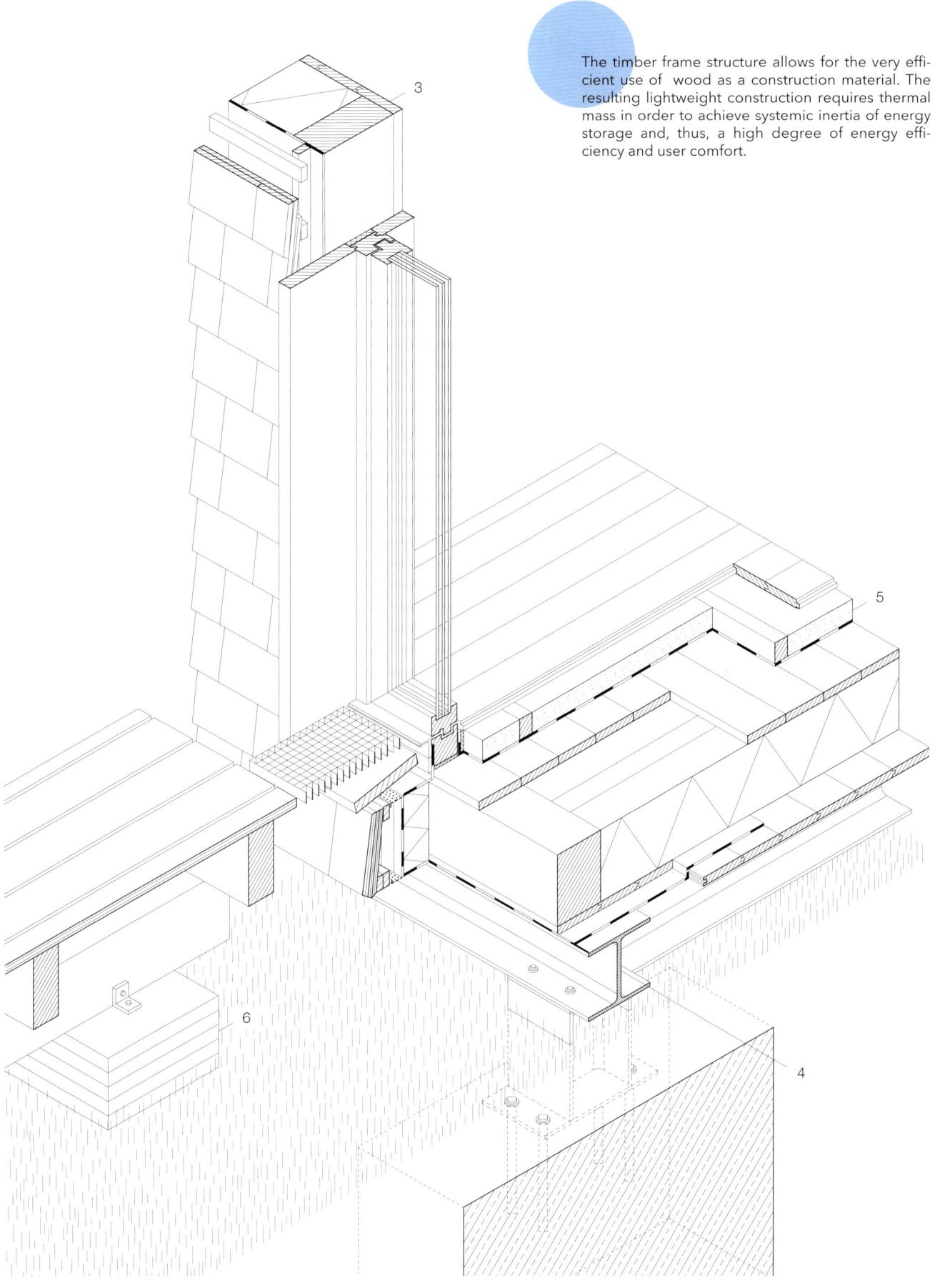

The timber frame structure allows for the very efficient use of wood as a construction material. The resulting lightweight construction requires thermal mass in order to achieve systemic inertia of energy storage and, thus, a high degree of energy efficiency and user comfort.

Focus on Wood

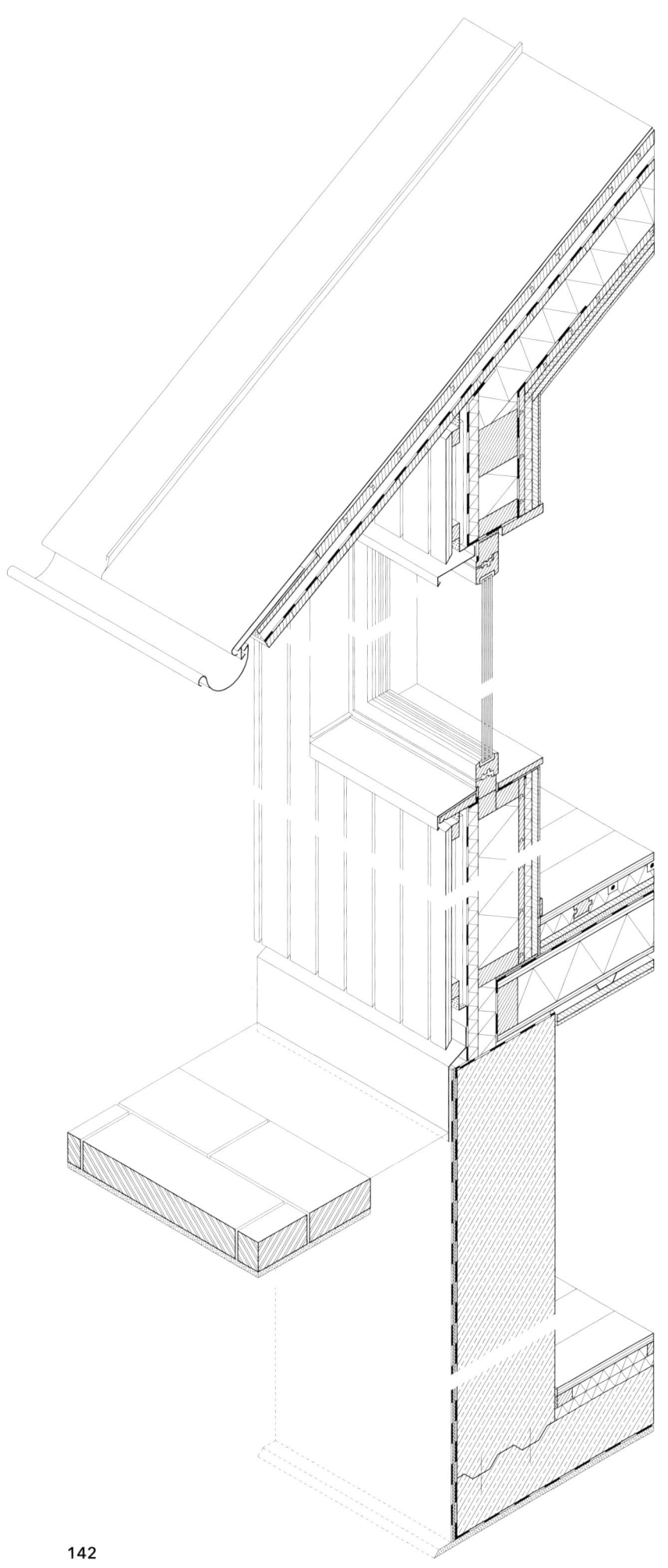

Timber: Frame Construction II

Protect
standing seam roofing
wood sheathing, back ventilation

Insulate
reed cane insulation
hemp insulation

Support
solid construction timber posts
timber beams, rafters
prefabricated reinforced concrete elements, reused

Seal
EPDM sealant
PE film
sarking membrane
diagonal sheathing, airtight

Cover
reclaimed wood floorboards
loam render on loam construction panel

Isometric illustration   scale 1:20
Vertical section   scale 1:20

1 Roof construction:
  Standing seam roofing, clamped to anchor clip
  polypropylene fabric, breathable, attached to sheathing
  min. 24 mm solid timber sheathing, screw connections
  min. 40 mm counterbattens, screw connections / back ventilation
  sarking membrane, windproof, breathable, min. 30 cm overlap
  solid timber sheathing, screw connections
  solid construction timber rafters, birdsmouth joint to wall plate and ridge beam, screw connections
  inlaid reed cane insulation, loose layout
  min. 30 mm solid timber diagonal sheathing, airtight, screw connections
  solid timber sheathing, screw connections
  inlaid hemp insulation / installation layer
  loam construction panel with double loam render layer, reinforcement, screw connections

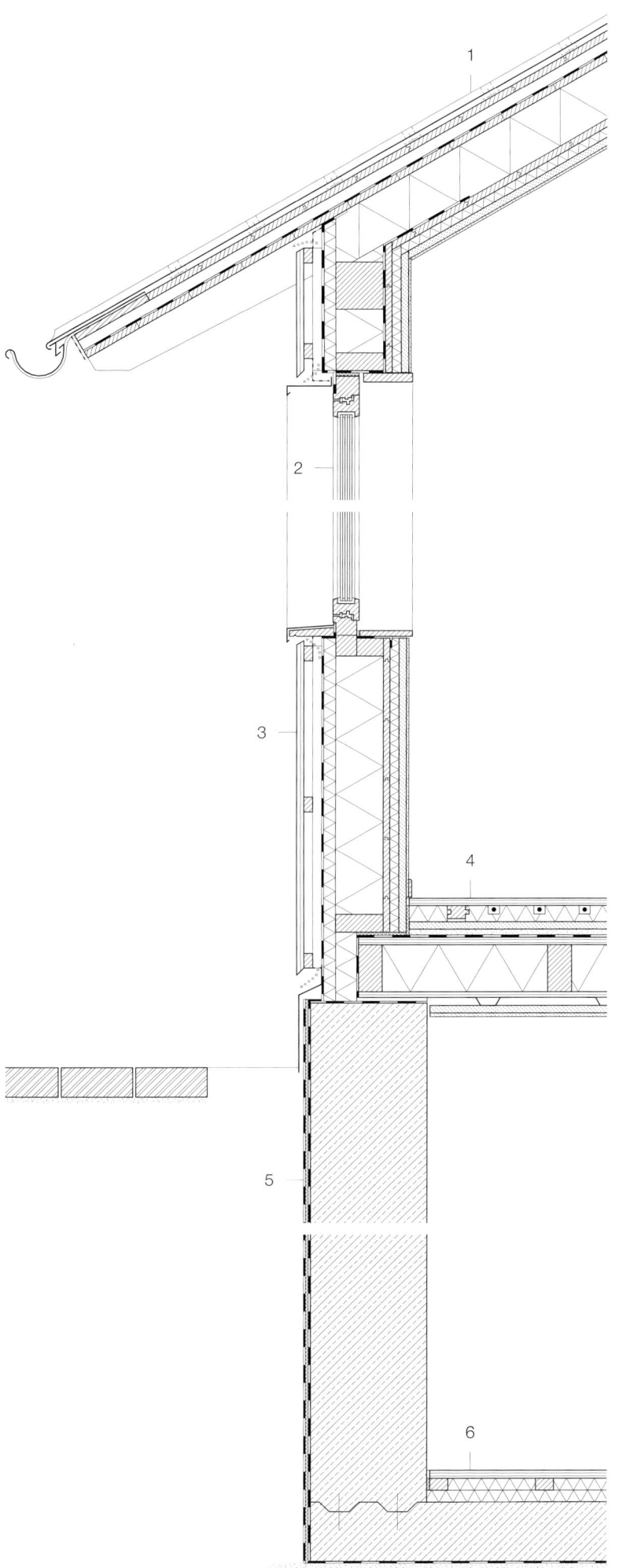

2 Windows: triple glazing in wood frame, reused plastic dry sealant, sorted by type
3 Exterior wall construction:
  Solid timber sheathing, visible screw connections
  solid timber battens, screw connections
  min. 30 mm counterbattens, screw connections / back ventilation
  sarking membrane, windproof, breathable, min. 30 cm overlap
  softwood fibre thermal insulation
  solid construction timber frame, bolt connections
  inlaid hemp thermal insulation
  min. 30 mm solid timber, diagonal sheathing, airtight, screw connections
  solid timber battens, screw connections
  inlaid hemp thermal insulation / installation layer
  loam construction panel with double loam render layer, reinforcement, screw connections
4 Floor construction:
  Reclaimed wood floorboards, concealed tongue-and-groove screw connections
  solid timber battens, screw connections
  inlaid softwood fibre panel, lignin-bonded, as installation panel with grooves and heat diffusor panels for installing underfloor heating pipes
  heavy loam construction panel, loose layout
  mass fill, sand, to improve soundproofing
  PE film, diffusion-proof, min. 30 cm overlap
  solid timber sheathing, screw connections
  solid construction timber ceiling beam, bolt connections
  inlaid hemp thermal insulation
  solid timber sheathing, screw connections
  hung ceiling system:
  sheet steel mounting channel, galvanised, screw connections
  loam construction panel with double render layer, reinforcement, screw connections
5 EPDM sealant layer against ground moisture, membrane-to-membrane weld, clamped
  dimpled sheet, min. 30 cm overlap
  EPDM sealant layer against ground moisture, membrane-to-membrane weld, loose layout
  prefabricated reinforced concrete wall element, reused, bolt connections to floor slab, frost-free foundation
6 Floor construction:
  Reclaimed wood floorboards, concealed tongue-and-groove screw connections
  solid timber battens, screw connections
  inlaid softwood fibre panel, lignin-bonded
  softwood fibre panel levelling insulation, lignin-bonded
  prefabricated reinforced concrete floor slab, reused
  EPDM sealant layer against ground moisture, membrane-to-membrane weld, loose layout
  sand base layer

Isometric illustration, roof
Isometric illustration, plinth/floor
not to scale

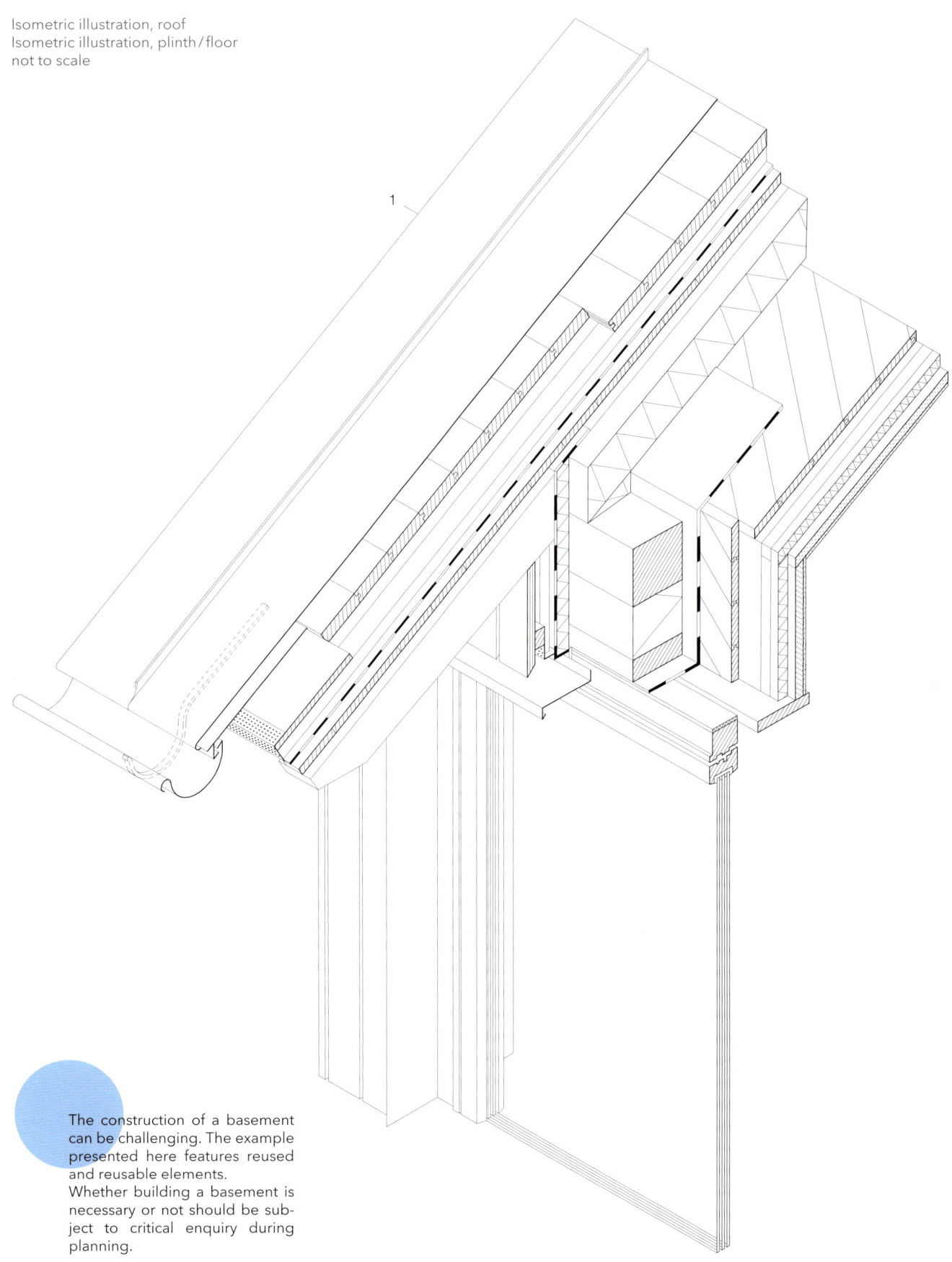

The construction of a basement can be challenging. The example presented here features reused and reusable elements.
Whether building a basement is necessary or not should be subject to critical enquiry during planning.

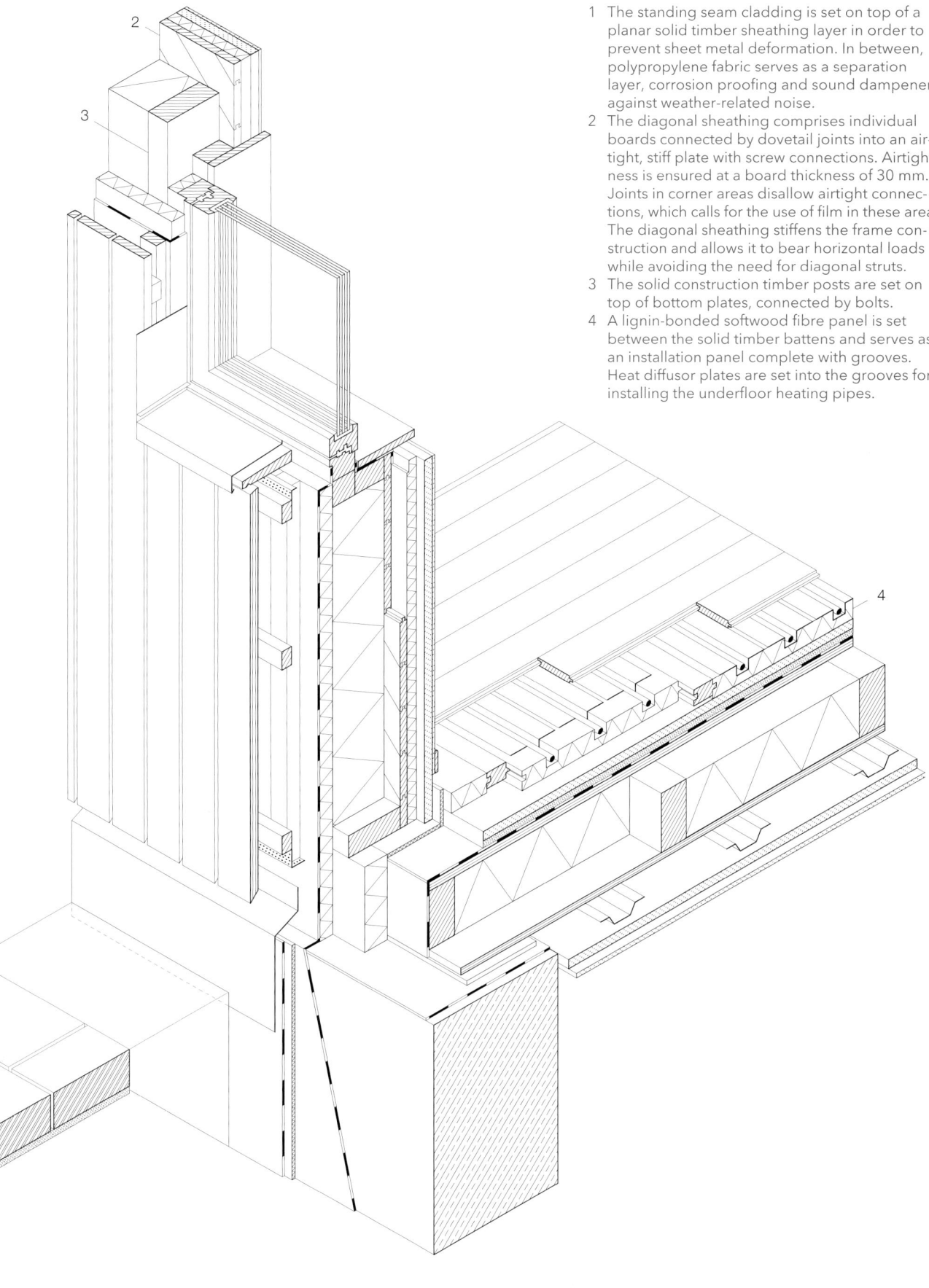

1. The standing seam cladding is set on top of a planar solid timber sheathing layer in order to prevent sheet metal deformation. In between, polypropylene fabric serves as a separation layer, corrosion proofing and sound dampener against weather-related noise.
2. The diagonal sheathing comprises individual boards connected by dovetail joints into an airtight, stiff plate with screw connections. Airtightness is ensured at a board thickness of 30 mm. Joints in corner areas disallow airtight connections, which calls for the use of film in these areas. The diagonal sheathing stiffens the frame construction and allows it to bear horizontal loads while avoiding the need for diagonal struts.
3. The solid construction timber posts are set on top of bottom plates, connected by bolts.
4. A lignin-bonded softwood fibre panel is set between the solid timber battens and serves as an installation panel complete with grooves. Heat diffusor plates are set into the grooves for installing the underfloor heating pipes.

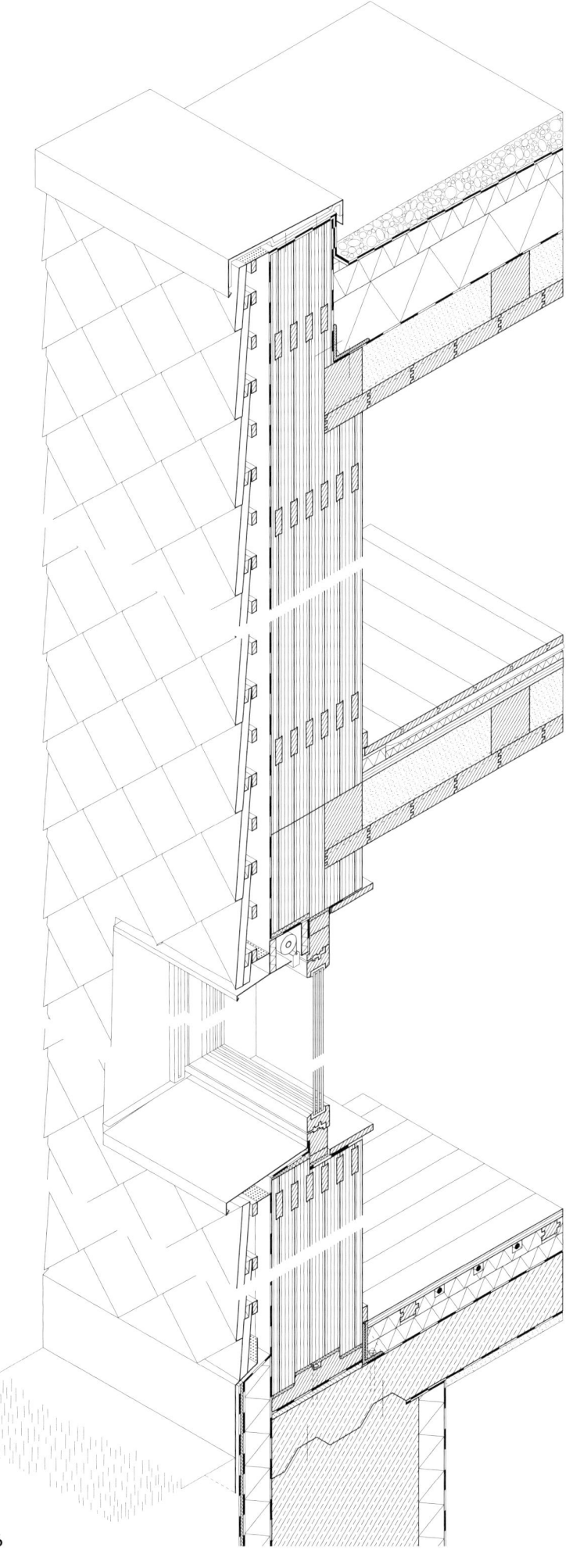

Timber: Solid Construction I

Protect
parapet coping, gravel fill
reprocessed plastic shingles,
back-ventilated

Insulate
solid timber wall
foam glass panels
foam glass gravel

Support
solid timber wall
timber beams
prefabricated reinforced concrete foundation

Seal
EPDM sealant layer
sarking membrane

Cover
reclaimed wood floorboards
solid timber panel, visible on the interior

Isometric illustration   scale 1:20
Vertical section   scale 1:20
Section, ceiling slab   scale 1:20

1 Roof edge: Parapet coping, canted, clamped to anchor clip
   anchor clip, screw connections to solid timber wall
2 Roof construction:
   Gravel fill
   EPDM sealant, membrane-to-membrane weld, 2-ply, attached to parapet, clamped by anchor clip
   foam glass panel insulation, 2 % to falls, loose layout
   foam glass panel insulation, loose bond
   PE film, diffusion-resistant, attached to parapet, clamped by anchor clip
   solid construction timber roof beam, bolt connections
   mass infill, sand, to improve soundproofing
   solid timber plate, finger joints, screw connections
3 Exterior wall construction
   reprocessed plastic shingles, screw connections
   solid timber battens, screw connections

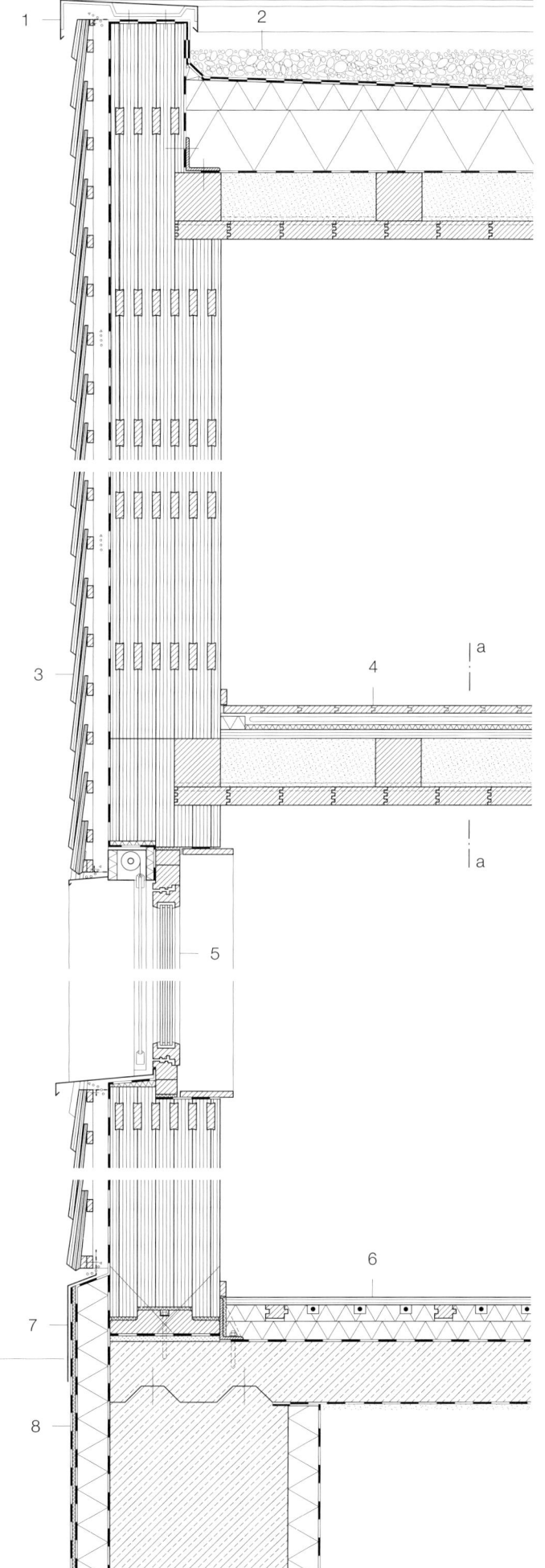

min. 40 mm counterbattens, screw connections/
back ventilation
sarking membrane, windproof, breathable,
min. 30 cm overlap
solid stacked timber wall, internal transverse
spline joints
4   Ceiling slab construction:
Reclaimed wood floorboards, concealed
tongue-and-groove screw connections
solid timber battens, screw connections
inlaid softwood fibre panel, lignin-bonded, as
installation panel with grooves and heat diffusor
plates to install underfloor heating pipes
softwood fibre panel levelling insulation, lignin-
bonded
solid timber sheathing, screw connections to
ceiling beam
solid construction timber beam, screw connections
mass infill, sand, to improve soundproofing
solid timber panel, finger joints, screw connections
5   Windows:
Triple glazing in wood frame, reused

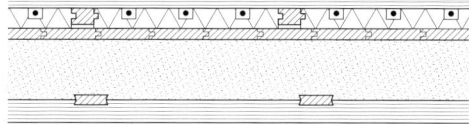

aa

plastic dry sealant, sorted by type
6   Floor construction:
Reclaimed wood floorboards, concealed
tongue-and-groove screw connections
solid timber battens, screw connections
inlaid softwood fibre panel, lignin-bonded, as
installation panel with grooves and heat diffusor
plates to install underfloor heating pipes
softwood fibre levelling insulation, lignin-
bonded
PE film, diffusion-resistant, min. 30 cm overlap
prefabricated reinforced concrete floor slab
element, reused
EPDM sealant against ground moisture,
membrane-to-membrane weld, min. 30 cm
overlap, loose layout, clamped between
foundation and floor slab
sand base layer
7   Sheet steel plinth coping, galvanised, screw
connections
8   EPDM sealant against ground moisture,
membrane-to-membrane weld, clamped by
mounting track
dimpled sheet, min 30 cm overlap
EPDM sealant against ground moisture,
membrane-to-membrane weld, attached to
foam glass insulation panel
prefabricated reinforced concrete foundation
element, bolt connections to floor slab, frost
free foundation
sand base layer

Focus on Wood   147

Isometric illustration
wall / ceiling
Isometric illustration
plinth / floor
not to scale

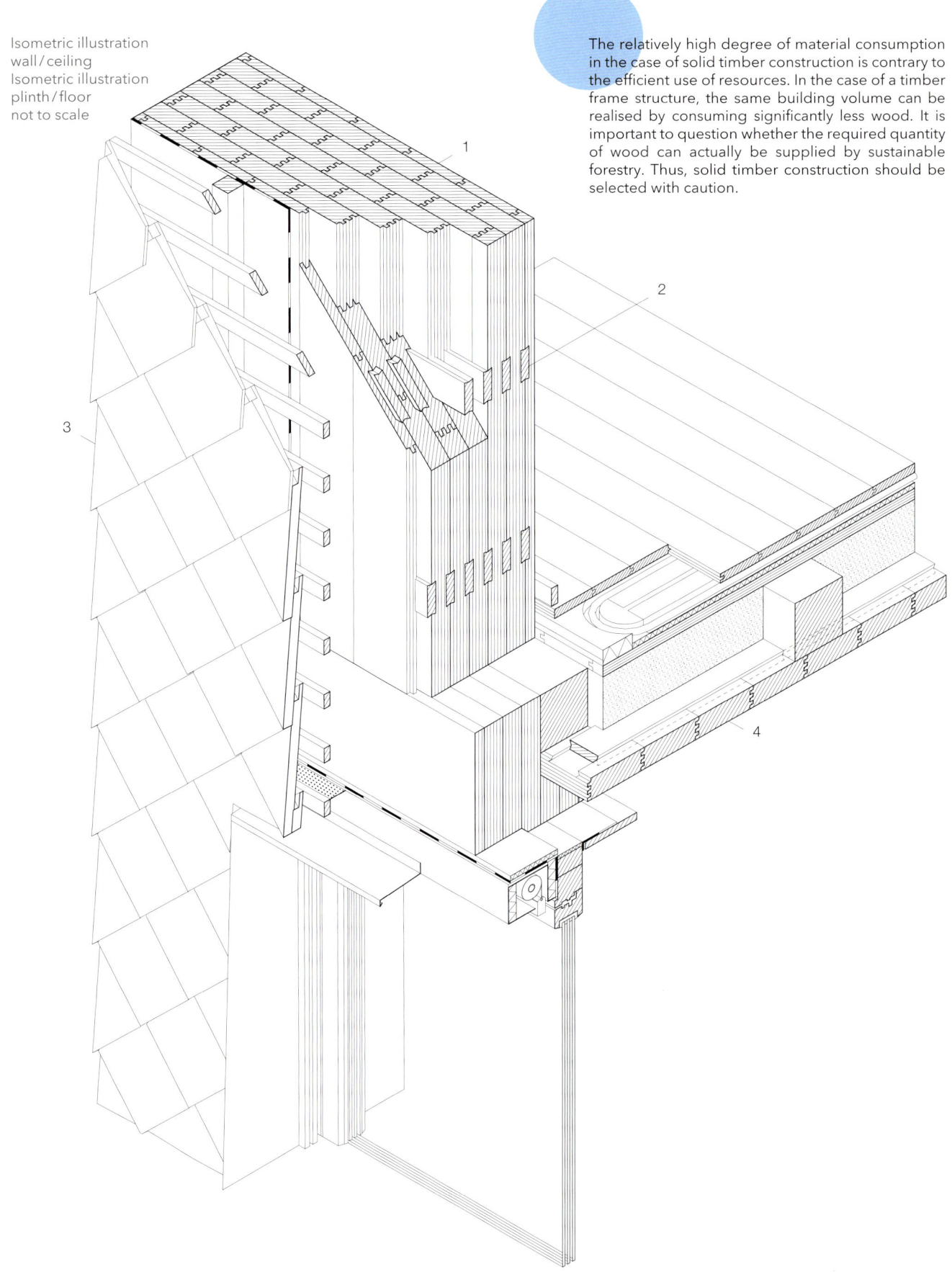

The relatively high degree of material consumption in the case of solid timber construction is contrary to the efficient use of resources. In the case of a timber frame structure, the same building volume can be realised by consuming significantly less wood. It is important to question whether the required quantity of wood can actually be supplied by sustainable forestry. Thus, solid timber construction should be selected with caution.

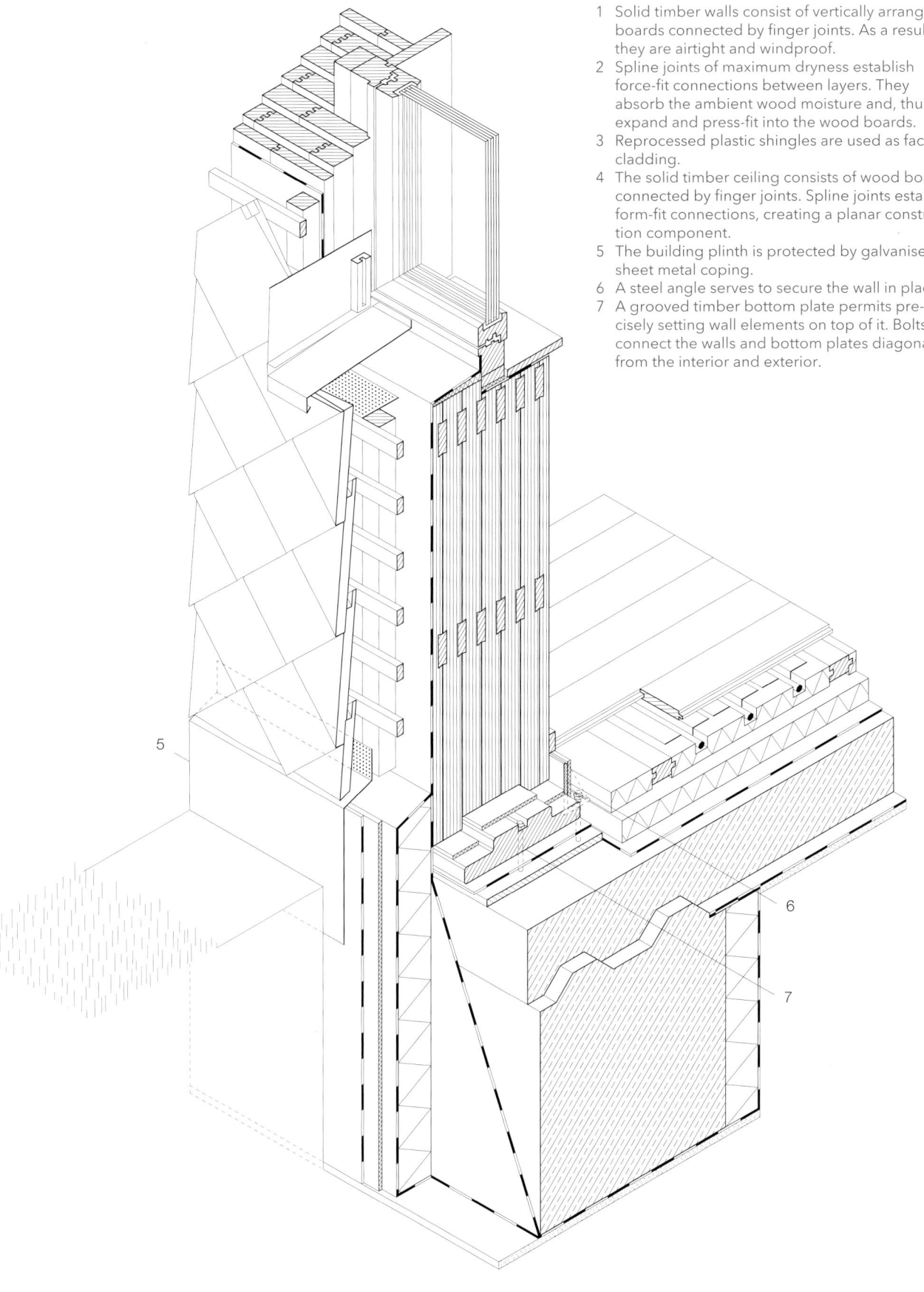

1 Solid timber walls consist of vertically arranged boards connected by finger joints. As a result, they are airtight and windproof.
2 Spline joints of maximum dryness establish force-fit connections between layers. They absorb the ambient wood moisture and, thus, expand and press-fit into the wood boards.
3 Reprocessed plastic shingles are used as facade cladding.
4 The solid timber ceiling consists of wood boards connected by finger joints. Spline joints establish form-fit connections, creating a planar construction component.
5 The building plinth is protected by galvanised sheet metal coping.
6 A steel angle serves to secure the wall in place.
7 A grooved timber bottom plate permits precisely setting wall elements on top of it. Bolts connect the walls and bottom plates diagonally from the interior and exterior.

Focus on Wood

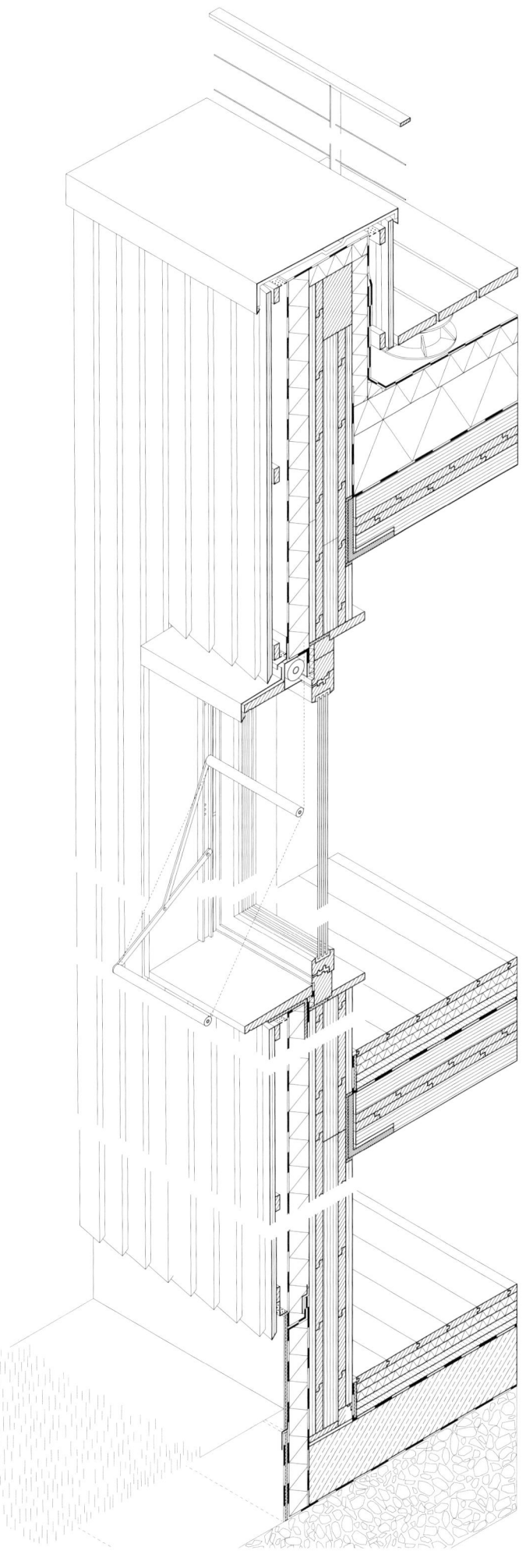

Timber: Solid Construction II

Protect
parapet coping
clapboard, back-ventilated
terrace decking

Insulate
foam glass panels
hemp lime panels
softwood fibre panels
foam glass gravel

Support
solid timber construction
prefabricated reinforced concrete element,
foundation / floor slab, reused

Seal
EPDM sealant
PE film sarking membrane

Cover
reclaimed wood floorboards
solid timber wall, visible on the interior

Isometric illustration   scale 1:20
Vertical section   scale 1:20

1 Roof edge:
   Parapet coping, canted, clamped to anchor clip
   anchor clip, screw connections to solid
   timber wall
2 Emergency overflow
3 Roof construction:
   Terrace decking, rift sawn, planed, untreated,
   screw connections
   solid timber battens, screw connections
   plastic levelling pedestals, sorted by type,
   height-adjustable
   EPDM roof sealant, 2-ply, membrane-to-
   membrane weld, attached to parapet, clamped
   beneath wind paper and battens
   foam glass insulation panel, 2 % to falls, loose
   layout
   foam glass insulation panel, loose bond
   PE film, diffusion-resistant, min. 30 cm overlap,
   attached to parapet
   solid timber ceiling, multiple layers, hardwood
   dowel connectors, bolt connections to solid
   timber wall bracket

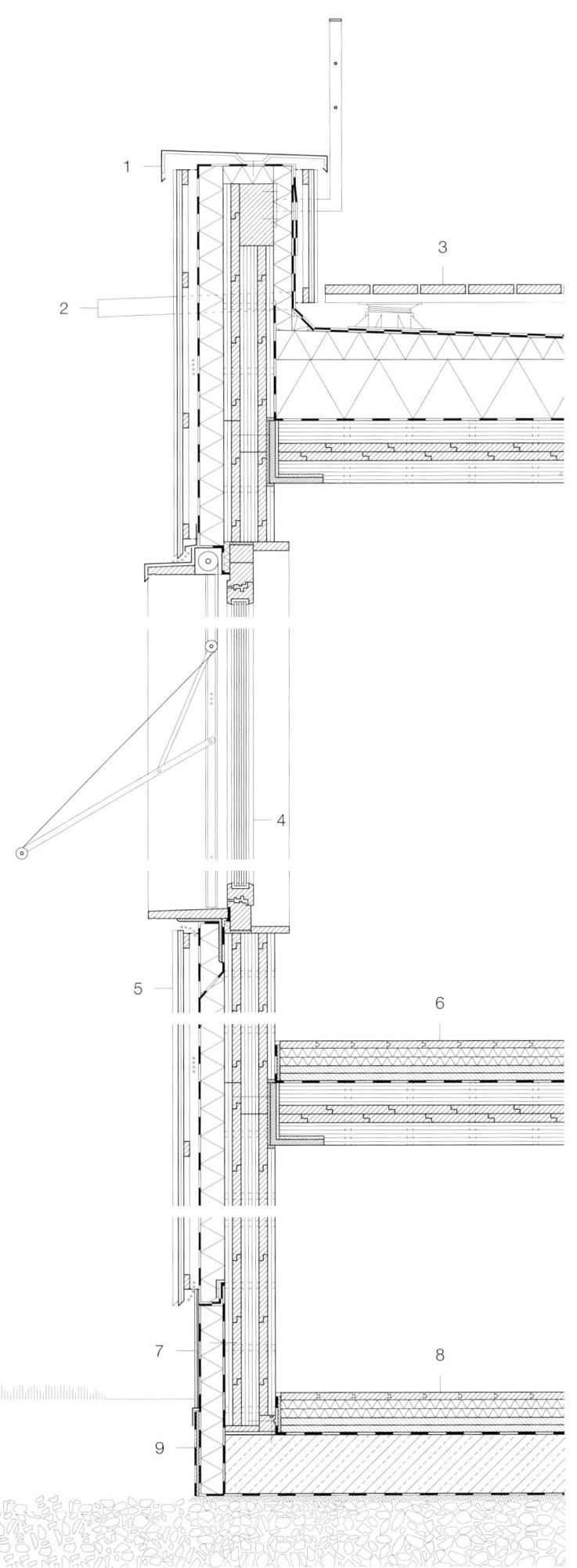

4 Window: triple glazing in wood frame, reused plastic dry sealant, sorted by type
5 Exterior wall construction:
solid timber clapboard siding, vertical, screw connections
min. 20 mm solid timber battens, screw connections
min. 30 mm counterbattens, screw connection / back ventilation
sarking membrane, windproof, breathable, min. 30 cm overlap, clamped
hemp lime insulation panel, insulation dowel connections
solid timber wall, multiple layers, hardwood dowel connections
6 Ceiling slab construction:
Reclaimed wood floorboards, concealed tongue-and-groove screw connections
solid timber battens, screw connections
inlaid double softwood fibre panel levelling insulation, lignin bonded
heavy double loam construction panel, loose layout
kraft paper trickle protection, min. 30 cm overlap, loose layout
multilayer solid timber ceiling, hardwood dowel connections, bolt connections to solid timber wall bracket
7 Splash guard, lime render on render substrate panel, plinth height min. 30 cm above ground level
8 Floor construction:
Reclaimed wood floorboards, concealed tongue-and-groove screw connections
solid timber battens, screw connections
inlaid double softwood fibre panel levelling insulation, lignin-bonded
heavy double loam construction panel, loose layout
PE film, diffusion-resistant, min. 30 cm overlap
prefabricated reinforced concrete element floor slab foundation, reused
EPDM sealant layer against ground moisture, membrane-to-membrane weld, min. 30 cm overlap, clamped
sand base layer
structural perimeter insulation
foam glass gravel fill, min. 80 cm deep
9 EPDM sealant layer against ground moisture, membrane-to-membrane weld, clamped by mounting track
dimpled sheet, min. 30 cm overlap
EPDM sealant layer against ground moisture, membrane-to-membrane weld, attached to foam glass insulation panel
prefabricated reinforced concrete floor slab foundation, reused

Focus on Wood

Isometric illustration, roof
Isometric illustration, plinth / floor
not to scale

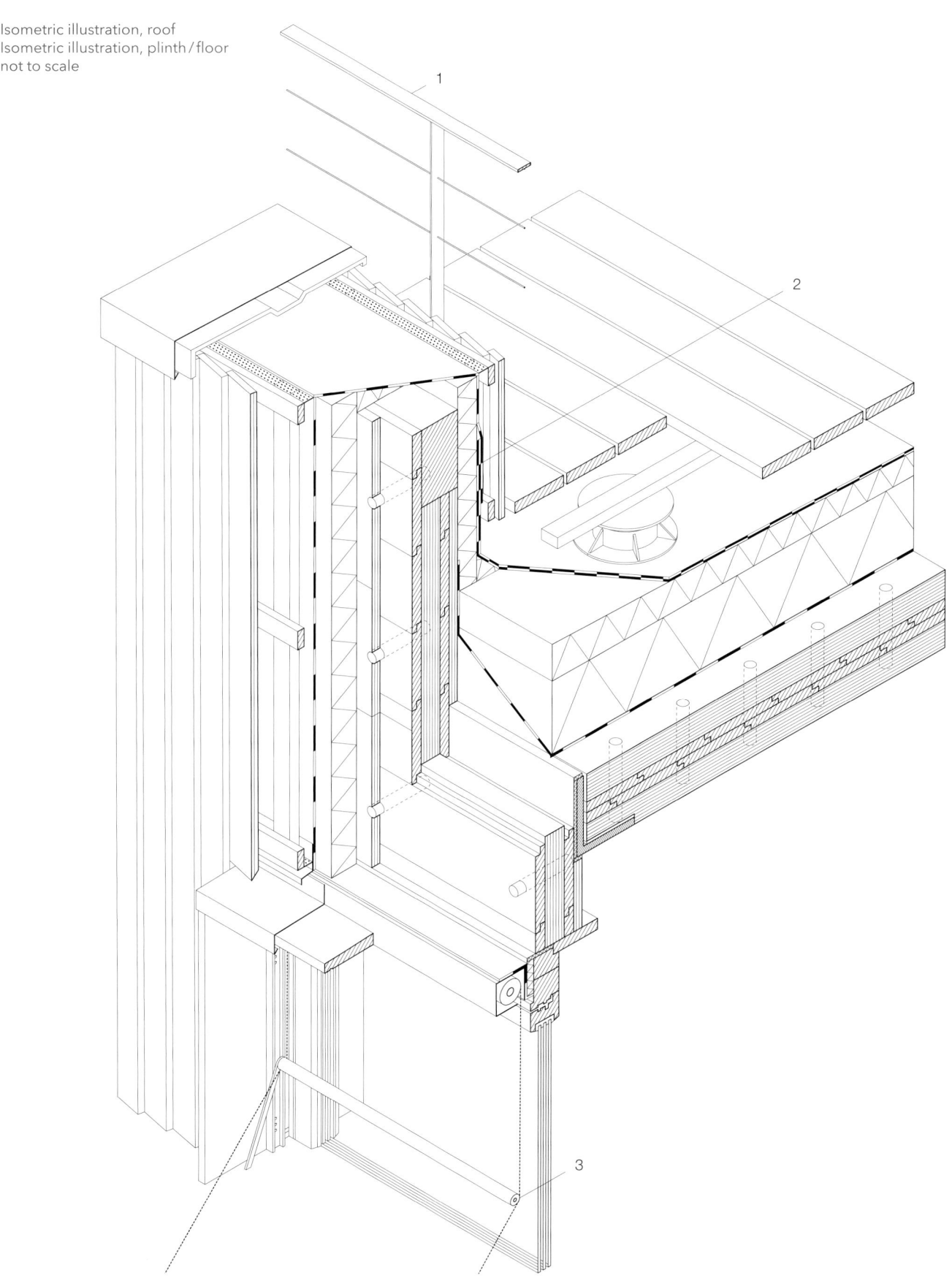

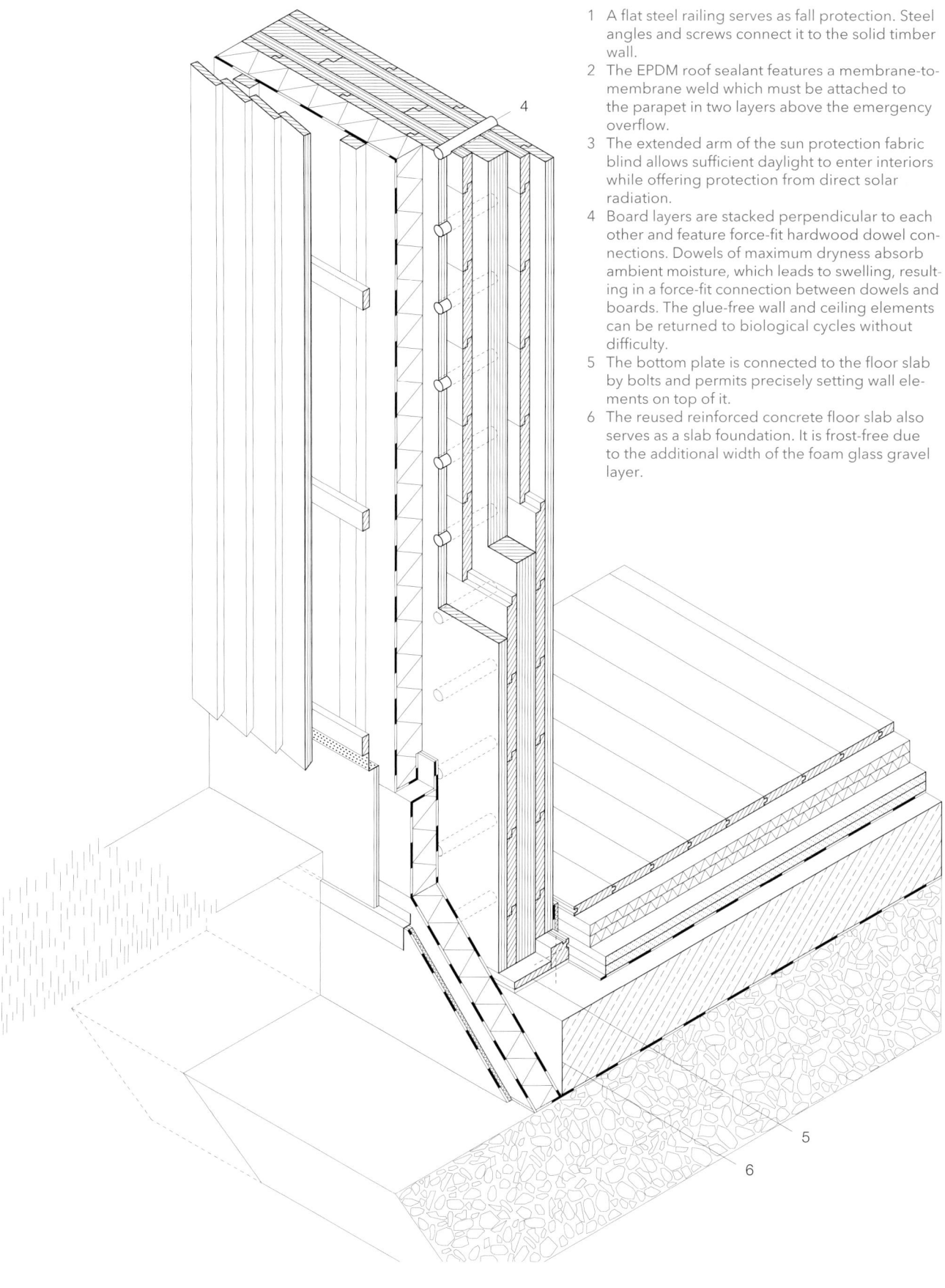

1. A flat steel railing serves as fall protection. Steel angles and screws connect it to the solid timber wall.
2. The EPDM roof sealant features a membrane-to-membrane weld which must be attached to the parapet in two layers above the emergency overflow.
3. The extended arm of the sun protection fabric blind allows sufficient daylight to enter interiors while offering protection from direct solar radiation.
4. Board layers are stacked perpendicular to each other and feature force-fit hardwood dowel connections. Dowels of maximum dryness absorb ambient moisture, which leads to swelling, resulting in a force-fit connection between dowels and boards. The glue-free wall and ceiling elements can be returned to biological cycles without difficulty.
5. The bottom plate is connected to the floor slab by bolts and permits precisely setting wall elements on top of it.
6. The reused reinforced concrete floor slab also serves as a slab foundation. It is frost-free due to the additional width of the foam glass gravel layer.

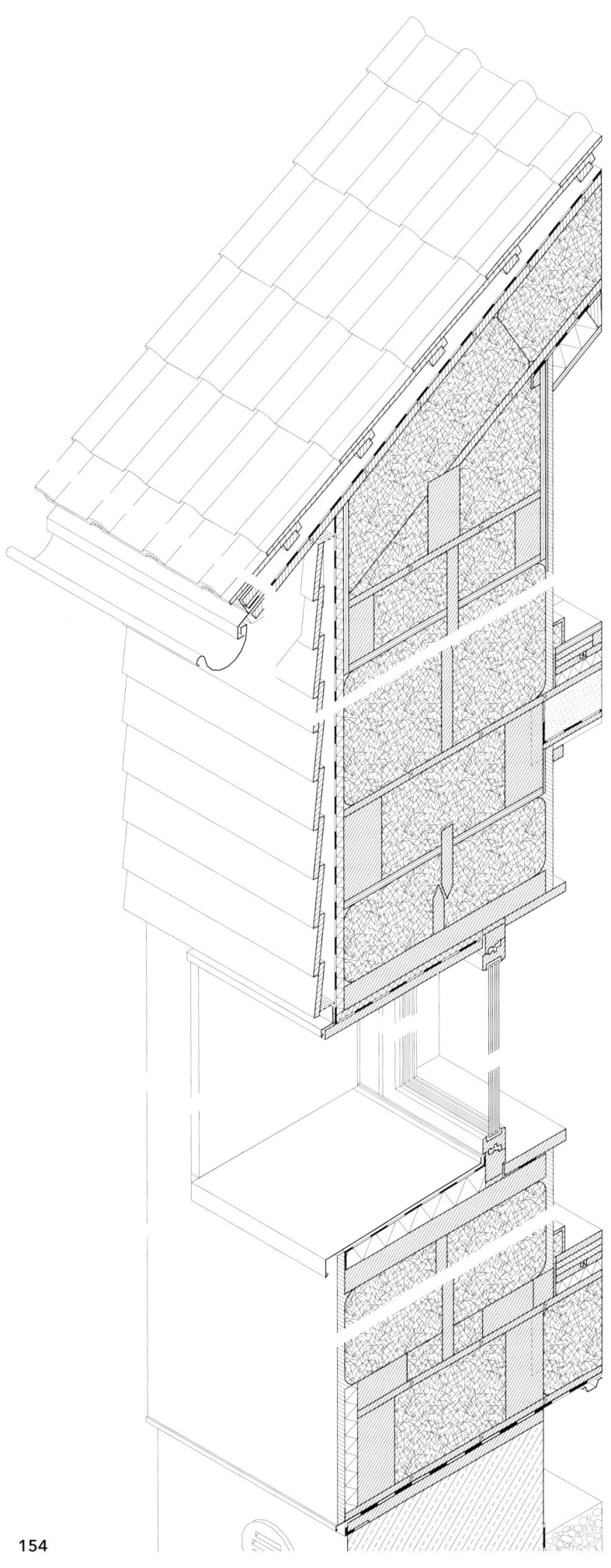

Straw: Solid Construction

Protect
pantiles
solid timber clapboard siding on battens
lime render finish layer

Insulate
straw, tied into bales
straw panels

Support
wood beams, rafters
straw bale with hazel rod reinforcement
timber frame ring beam
prefabricated reinforced concrete foundation element, reused

Seal
EPDM sealant
PE film
sarking membrane

Cover
reclaimed wood floorboards
loam render
solid timber sheathing

Isometric illustration   scale 1:20
Vertical section   scale 1:20

1 Roof construction:
  Pantile, supported by battens, screw connections
  solid timber battens, screw connections
  min. 40 mm counterbattens, screw connections/back ventilation
  sarking membrane, windproof, breathable, min. 30 cm overlap
  solid timber sheathing, screw connections
  solid construction timber rafters, birdsmouth joints to wall plate/ridge beam, screw connections
  inlaid straw bale, bound, loose layout
  solid timber panel, diagonal siding with dovetail joints, screw connections
  PE film, diffusion-resistant, min. 30 cm overlap, attached to bottom plate
  solid timber battens, screw connections
  inlaid straw panel insulation, clamped/

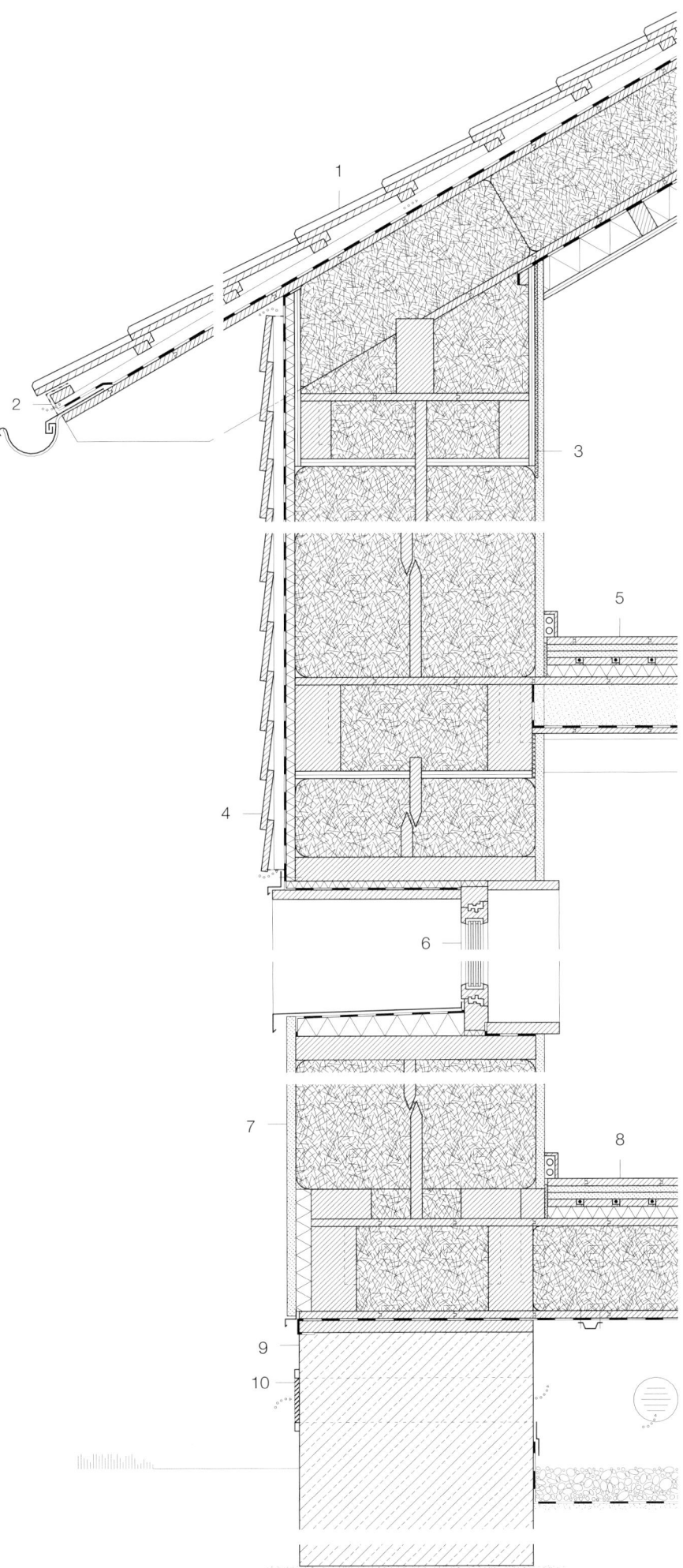

1. Roof construction:
   installation layer
   solid timber sheathing, visibly exposed, screw connections
2. Sheet metal insect screen, corrosion proof, screw connections to sheathing and counterbattens
3. Timber frame ring beam
4. Exterior wall construction:
   Solid timber clapboard, screw connections
   min. 40 mm solid timber battens, screw connections / back ventilation
   sarking membrane, windproof, breathable, min. 30 cm overlap
   softwood fibre panel, lignin-bonded, screw connections
   straw bale, load-bearing, bound, hazel rod reinforcement
   loam render primer, metal mesh reinforcement
   loam fine render
5. Ceiling slab construction:
   Reclaimed wood floorboards, concealed tongue-and-groove screw connections
   solid timber battens, screw connections
   inlaid double loam construction panel
   solid timber sheathing with lateral groove and heat diffusor plates to install underfloor heating pipes
   cork impact soundproofing / installation level
   solid timber sheathing, screw connections
   solid construction timber beam, dovetail joint to ring beam
   mass infill, sand, to improve soundproofing
   PE film, diffusion-proof, min. 30 cm overlap
   solid timber sheathing, screw connections
6. Window: Triple glazing in wood frame, reused plastic dry sealant, sorted by type
7. Exterior wall construction:
   Lime render finish
   lime render primer, metal mesh reinforcement
   straw bale, load-bearing, bound, hazel rod reinforcement
   loam render primer, grass fibre mat reinforcement
   loam fine render
8. Floor construction:
   Reclaimed wood floorboards, concealed tongue-and-groove screw connections
   solid timber battens, screw connections
   inlaid double loam construction panel
   solid timber sheathing with lateral groove and heat diffusor plates to install underfloor heating pipes
   cork impact soundproofing / installation layer
   solid timber sheathing, screw connections
   solid construction timber beam, dovetail joint to ring beam
   inlaid straw bale, bound, loose layout
   solid timber sheathing, screw connections
   EPDM sealant against ground moisture, membrane-to-membrane weld, clamped
   metal clamps to fix sealant membrane in place, bottom screw connections
9. Prefabricated reinforced concrete foundation element, reused, top edge min. 50 cm above ground level, frost-free foundation
10. Ventilation grille for crawlspace ventilation

Isometric illustration, roof
Isometric illustration, plinth/floor
not to scale

1 Hazel rods serve as reinforcement for straw bales and establish force-fit connections with the timber ring beam.
2 A prefabricated wood frame is required to set windows into walls. It is dimensioned in order to take the settling of straw into account. It is smaller than the shell construction opening in the straw bale wall.
3 The ducts for electrical installations are situated behind the baseboard. The wall-mount installation permits maintenance from the interiors and adaptation to changing user requirements.
4 The building plinth features a protective layer of lime render.
5 The ceiling beams are connected by dovetail joints and bolts to the ring beam.

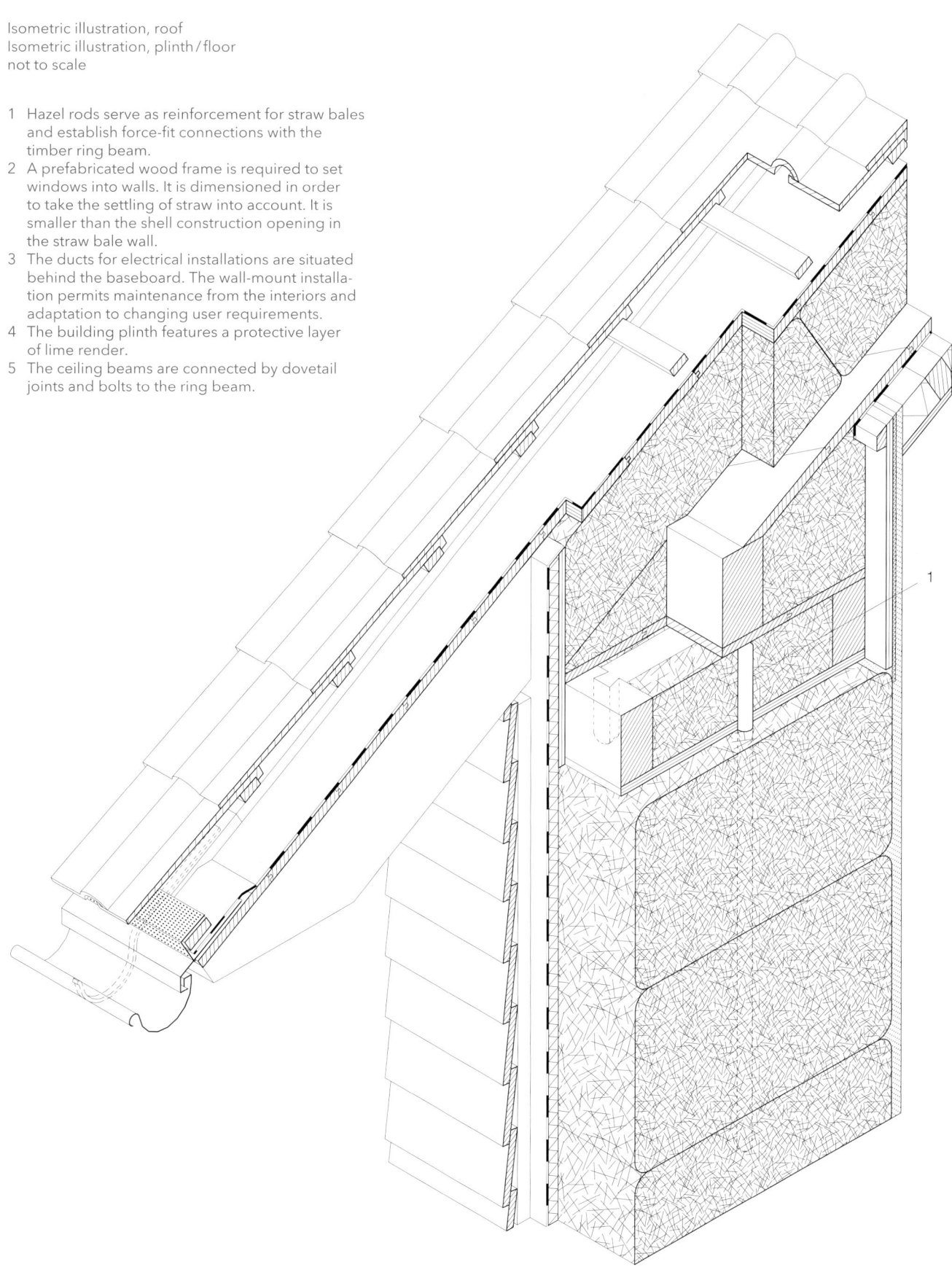

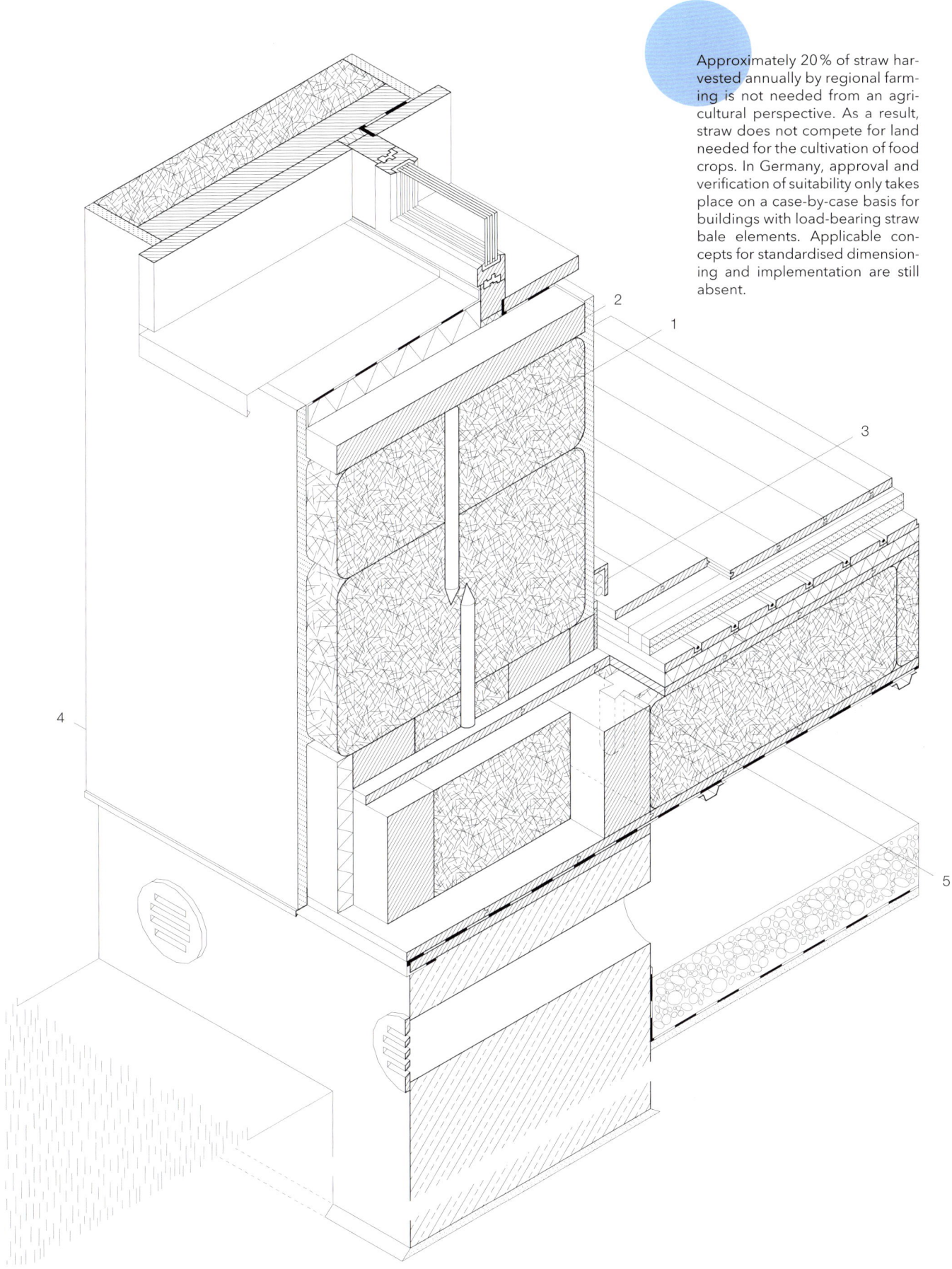

Approximately 20% of straw harvested annually by regional farming is not needed from an agricultural perspective. As a result, straw does not compete for land needed for the cultivation of food crops. In Germany, approval and verification of suitability only takes place on a case-by-case basis for buildings with load-bearing straw bale elements. Applicable concepts for standardised dimensioning and implementation are still absent.

Focus on Wood

Focus on Masonry

The term masonry describes the assembly of natural or artificial stone blocks into a structural building component [1]. The assembly follows a certain bond with related offset vertical joints between the blocks or bricks. In brickwork, vertical mortar layers are described as perpends and horizontal mortar joints are called beds.

Masonry as a technique originates in the simple layering of unhewn or undressed stone found in the landscape. The stability of resulting masonry walls was only achieved by intelligently deciding on how to set which stones. The handling of stones by craftspeople led to improving the distribution of loads from one stone to the next. This way, the joint surfaces coming into contact with each other were continuously optimised until, eventually, mortar as a formable and hardening joint filler was developed [2]. Masonry wall components featuring mortar joints result in a high degree of stability with relatively low effort required in terms of craftsmanship. They also enable the use of artificial blocks produced in economically feasible ways that are, however, unsuitable for dry masonry construction. In its industrialsed form, this type of masonry is still in use today [3].

Masonry construction types can be found all across the globe. For the most part and from an historical point of view, they comprised the dominant construction type for prestigious and ecclesiastical architecture. The pyramids of Giza or Greek temple structures are well-known early examples of the application of masonry techniques. Since antiquity, brick arches or vaults enabled the creation of extensive spaces with only a limited number of columns, or even completely column-free. The development of fired brick led to the production of uniform and calculable construction blocks that permitted easy handling and transportation (Fig. 1).

Raw material deposits

The components used in masonry include stone, blocks or brick and – except in the case of dry masonry – mortar. A distinction is made between natural and artificial stone or blocks. As a result, different material variations are available on the market.

Currently, natural stone plays only a limited role, due to the costly sourcing, production and handling of the raw material for modern masonry purposes. Most of all, it is used for facades or specialised building elements based on specific design intentions. Industrially produced artificial stone or blocks created by firing (brick, clinker) or pressing (sand lime blocks), on the other hand, permit combinations based on a modular system of block dimensions. In the recent past, industrially developed artificial stone materials have become available that display a diverse range of characteristics. This includes lightweight, porous blocks with good thermal insulation properties or heavy blocks that can bear heavy loads [4].

Masonry structures can be created from fired clay brick as well as blocks consisting of dried loam, sand lime, concrete or aerated concrete. Fired clay brick is available in a large variety of colours, especially in reddish and yellowish hues, depending on the geological composition.
Air-dried loam brick (see Focus on Loam, p. 206ff.) is most likely the world's oldest craftsmanship-based construction block type, originating in Egypt around 14 000 BC.

The method of firing brick to improve its characteristics and to preserve it is likely known since 5000 BC [5].
Sand lime blocks are artificially created by subjecting blocks comprise lime and quartz-containing sand to vapour pressure [6].
Aerated concrete blocks comprise steam-hardened, fine porous concrete and include quartz-containing sand in combination with cement, lime and gypsum. The characteristic porosity provides the block with a

1 Construction industry masonry block formats

low bulk density, which is achieved by adding a foaming agent during the production process. Aerated concrete blocks comprise up to 80% air.

Concrete blocks contain non-reinforced concrete (see Focus on Concrete, p. 174ff.) that can be used for construction tasks similar to other masonry block applications. Due to the prevailing manual production methods, common until a few decades ago for the creation of masonry structures, masonry blocks are optimised for handling by an individual in terms of dimensions and weight. Current masonry systems, due to economic reasons, focus on large-format blocks that require setting with technological means. In addition to the typically rectangular blocks, artificial units in different formats are available as specialised construction blocks.

Characteristics

Masonry is capable of bearing vertical compressive forces perpendicular to bed joints. Traditional masonry walls, however, due to their system characteristics, are hardly capable of bearing tensile loads. Masonry walls can only bear shear and bending stress caused by lateral forces, for instance due to wind loads, based on friction resulting from the intrinsic weight of the wall in combination with the adhesive tension strength and shear resistance of joints [7]. In general, these characteristics differ strongly according to the block material, the mortar performance and its adhesion capacity, as well as the processing method.

A range of specialised brick types exists. Brick fired at very high temperatures is described as clinker. During the firing process, so-called sintering sets in that leads to the closure of brick pores. This is why clinker is suitable most of all for facing shells and exposed facade surfaces. Some types of clinker display strong load-bearing characteristics, while others such as lightweight, vertically perforated brick with air pockets feature good thermal insulation properties. The firing process can influence the structure, hardness and visual appearance of the brick.

In the case of masonry with mortar joints, mortar serves as a binding agent and establishes force-fit and form-fit connections to the brick or block. Different mortar materials display different characteristics. Commonly used industrial mortar typically contains cement as well as synthetic aggregate in order to increase performance. Trass lime, lime and loam mortar can be simply removed by hammering it from masonry brick or block. Today, these mortar types are rarely used, due to their limited hardness compared to mortar with cement content. However, they can contribute to a greater degree of circularity in masonry construction.

Typical construction methods

Masonry walls consist of multiple strata of layered blocks or brick according to different types of bond. Layering can lead to single-brick walls (stretcher course) or walls with multiple brick shells placed side-by-side and bonded by alternating header and stretcher bricks (e.g. English bond or English cross bond). Masonry walls can be employed as singular wall plates or in more complex configurations featuring a multitude of wall and corner connections. The realised types comprise either single or multiple layers of load-bearing or non-load-bearing masonry walls. In the case of multilayer construction types, masonry is suitable as a load-bearing layer as well as a weatherproof material for facade exteriors. In the case of single-layer construction types, highly insulated brick or blocks are used that simultaneously perform insulating and load-bearing functions.

To this day, the standard dimensions for artificial blocks are based on octametric measuring systems: All block dimensions are oriented on a grid module of one-eighth of a metre (12.5 cm) [8].

Structural systems that bear compressive forces match the material characteristics of brick. The load-bearing behaviour of a masonry structure is influenced by the specific way the material is used. In the case of interlaced arrangements of masonry walls in the plan view, they can display high performance characteristics, even if their thickness is limited. Instead of interlaced wall arrangements, wall plates or columns can be created as construction components. In such cases, the specific characteristics of the individual materials used (type of block or mortar) play increased roles for construction and load-bearing capacity [9].

Aside from using brick for the construction of walls, it is also suitable for creating arch, dome or vault structures. Masonry wall openings can be created with arches or with horizontal lintels. In Europe, for the most part and due to cost concerns, industrially produced lintels are used. Alternative materials, such as e.g. concrete, often replace brick in such cases. Brick is rarely used for ceiling structures in modern masonry construction and is mostly used in the form of inserted or semi-finished elements of industrially prefabricated ceiling systems.

Circular potential

In 2018 in Germany about 219 million tonnes of mineral construction waste was produced. 27.3 % of this figure comprised concrete, ceramic and most of all, masonry construction waste [10]. The amount of demolished masonry walls per year is estimated at 25 million tonnes [11].

In principle, due to the longevity and robustness of the material, most blocks or bricks are reusable. However, reuse is difficult, due to the large amount of manual labour required for separation sorted by type, or due to difficult to separate mortar joints.

From an historic perspective, reusing natural stone and brick from dismantled masonry buildings for creating new buildings is a common practice (see "History of the Building Culture of Reuse", p. 30ff). Dry masonry wall constructions can be dismantled by hand. The only limitations for reusing blocks are their form and dimensions. Historic lime and loam mortar permit the separation of blocks and mortar in a relatively simple way, thereby enabling the reuse of blocks from masonry walls with mortar joints. While the development of cement mortar led to improving the performance of wall structures, it also significantly complicated the separation of blocks or bricks and mortar. Today, demolished masonry, as a form of mineral construction waste, mostly enters cascading use in different forms at decreasing levels of quality. Depending on the characteristics of the demolition material, typical fields of application include fill material for road construction, such as base or frost protection layers, fill material for civil engineering purposes, or concrete aggregate [12].

Similar to other mineral deconstruction materials, masonry rubble offers only limited recyclability at stable levels of product quality. Recent industrial developments reflect the aim of increasing the share of demolition material within new products. For brick, the potential of using recyclate sorted by type ranges from 20 to 60 %, depending on the source material. Specific research projects even achieved recyclate shares of up to 70 % [13].

Aside from using alternative mortar types, such as lime or trass lime mortar, masonry construction sorted by type consisting of loam block, loam mortar or dry masonry assembled in the form of interlocking systems contributes to the high-quality reuse of load-bearing masonry structures. In the case of masonry facade exteriors, a range of products is now available that employ mechanical fastening systems. Such systems permit reclamation of individual blocks or brick sorted by type and, thus, loss-free reuse. They even function without requiring any mortar [14].

Notes
[1] Dierks/Wormuth 2012, p. 65
[2] Moro 2021, p. 254–258
[3] ibid., p. 258
[4] Hillebrandt et al. 2018, p. 69
[5] Pfeifer/Ramcke 2001, p. 10
[6] see note 1
[7] see note 5, p. 92–95
[8] Bundesverband Kalksandsteinindustrie e.V. 2019
[9] Belz 1999, p. 86
[10] Kreislaufwirtschaft Bau/Bundesverband Baustoffe – Steine und Erden e.V. 2021
[11] Müller 2018, p. 234–235
[12] Martens/Goldmann 2016, p. 358
[13] see note 4, p. 69
[14] see note. 4, p. 50

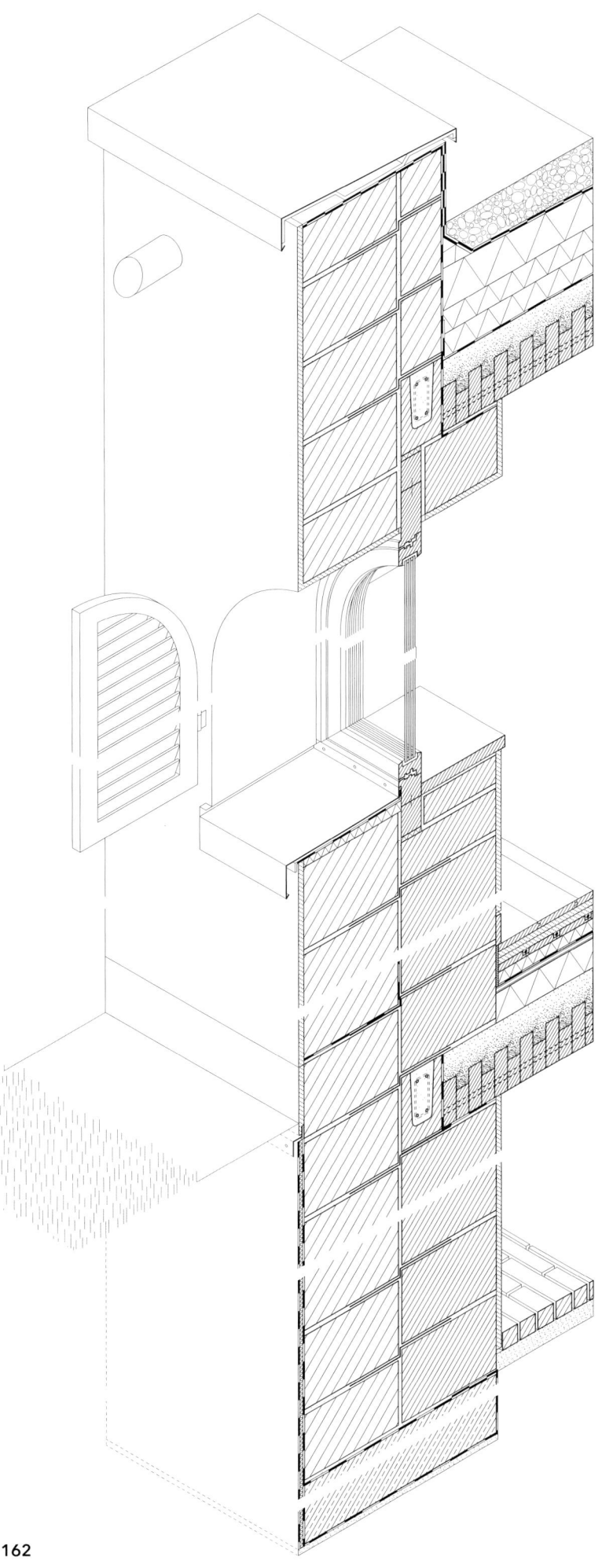

Masonry: Solid Construction

Protect
parapet coping, gravel fill
lime render

Insulate
vertically perforated brick, insulating
foam glass panels
cork panels

Support
vertically perforated brick, load-bearing
ring beam, U-shell
timber loam composite ceiling
prefabricated reinforced concrete element
foundation, reused

Seal
EPDM sealant membrane
PE film

Cover
reclaimed wood floorboards
lime render

Isometric illustration   scale 1:20
Vertical section   scale 1:20

1   Roof edge:
    Parapet coping, canted, clamped to anchor clip
    anchor clip, screw connections to masonry wall
2   Roof construction:
    Gravel fill
    EPDM sealant, membrane-to-membrane weld, 2-ply, attached to parapet, clamped by anchor clip
    foam glass insulation panel, 2 % to falls, loose layout
    double foam glass insulation panels
    PE film, diffusion-resistant, min. 30 cm overlap, attached to parapet, clamped by anchor clip
    timber loam composite ceiling
    rammed earth on cross-laminated timber ceiling
    stacked timber, dowel connections
3   Wood folding shutters, point by point screw connections
4   Windows: triple glazing in wood frame, reused
    plastic dry sealant, sorted by type
5   Exterior wall construction:
    Lime render

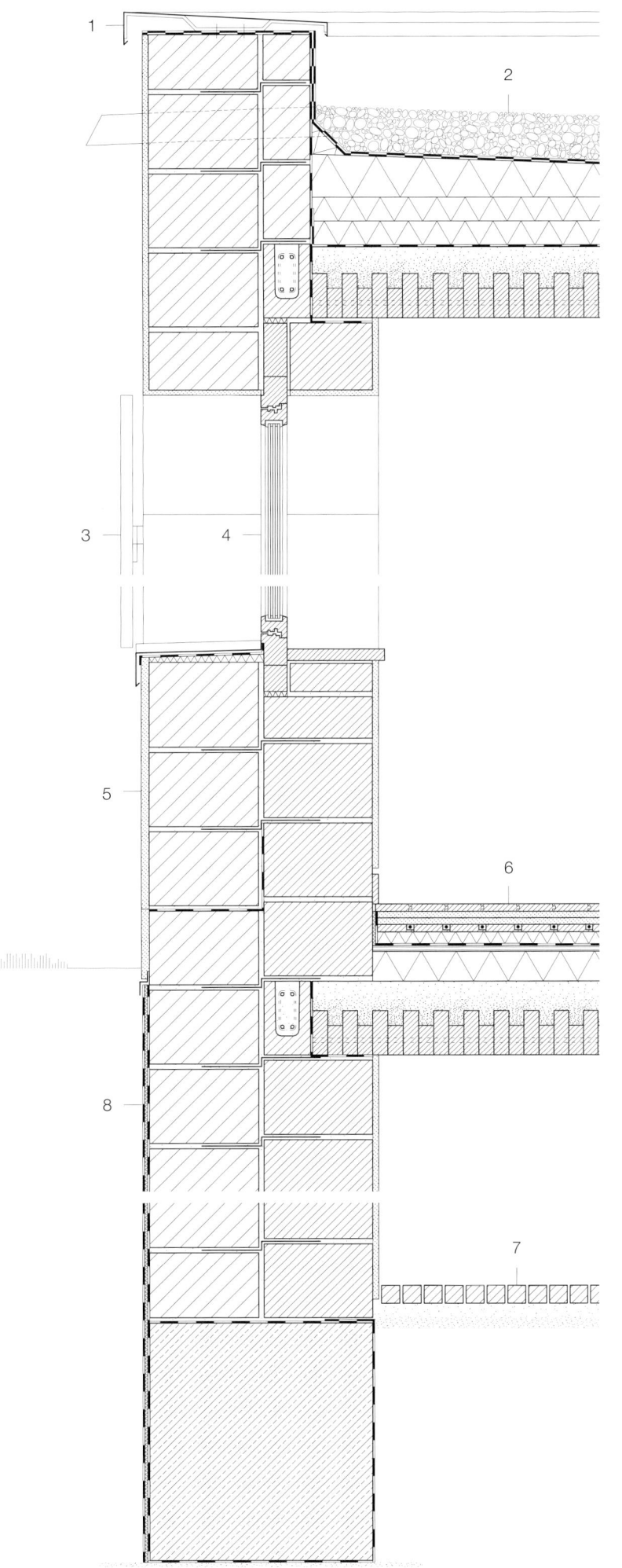

1 Sheet metal flashing
2 Green roof construction
3 Window sill
4 Wood-aluminium window
5 Wall construction:
 vertically perforated brick, insulating, mortar joints
 lime mortar joints
 flat steel anchor, embedded in mortar bed
 vertically perforated brick, load-bearing, mortar joints
 lime render
6 Floor construction:
 Reclaimed wood floorboards, concealed tongue-and-groove screw connections
 solid timber battens, screw connections
 inlaid double loam construction panel
 solid timber sheathing with lateral grooves and heat diffusor panels to install underfloor heating pipes
 cork impact soundproofing/installation layer
 PE film, diffusion-resistant, min. 30 cm overlap, clamped
 solid timber panel, diagonal sheathing with dovetail joints, screw connections
 solid timber battens, screw connections
 inlaid cork insulation, loose fill
 timber loam composite ceiling:
 Rammed earth on cross-laminated timber ceiling
 stacked timber, dowel connections
7 Basement construction:
 Paving stones, loose layout
 min. 40 mm crushed stone fill
 gravel base/capillary break layer
8 EPDM sealant layer against ground moisture, membrane-to-membrane weld, clamped
 dimpled sheet, min. 30 cm overlap
 EPDM sealant layer against ground moisture, membrane-to-membrane weld, loose layout
 vertically perforated brick, insulating, mortar joints
 lime mortar joints
 flat steel anchor, embedded in mortar bed
 vertically perforated brick, load-bearing, mortar joints
 lime render

Focus on Masonry 163

Isometric illustration, roof
Isometric illustration, plinth/floor
not to scale

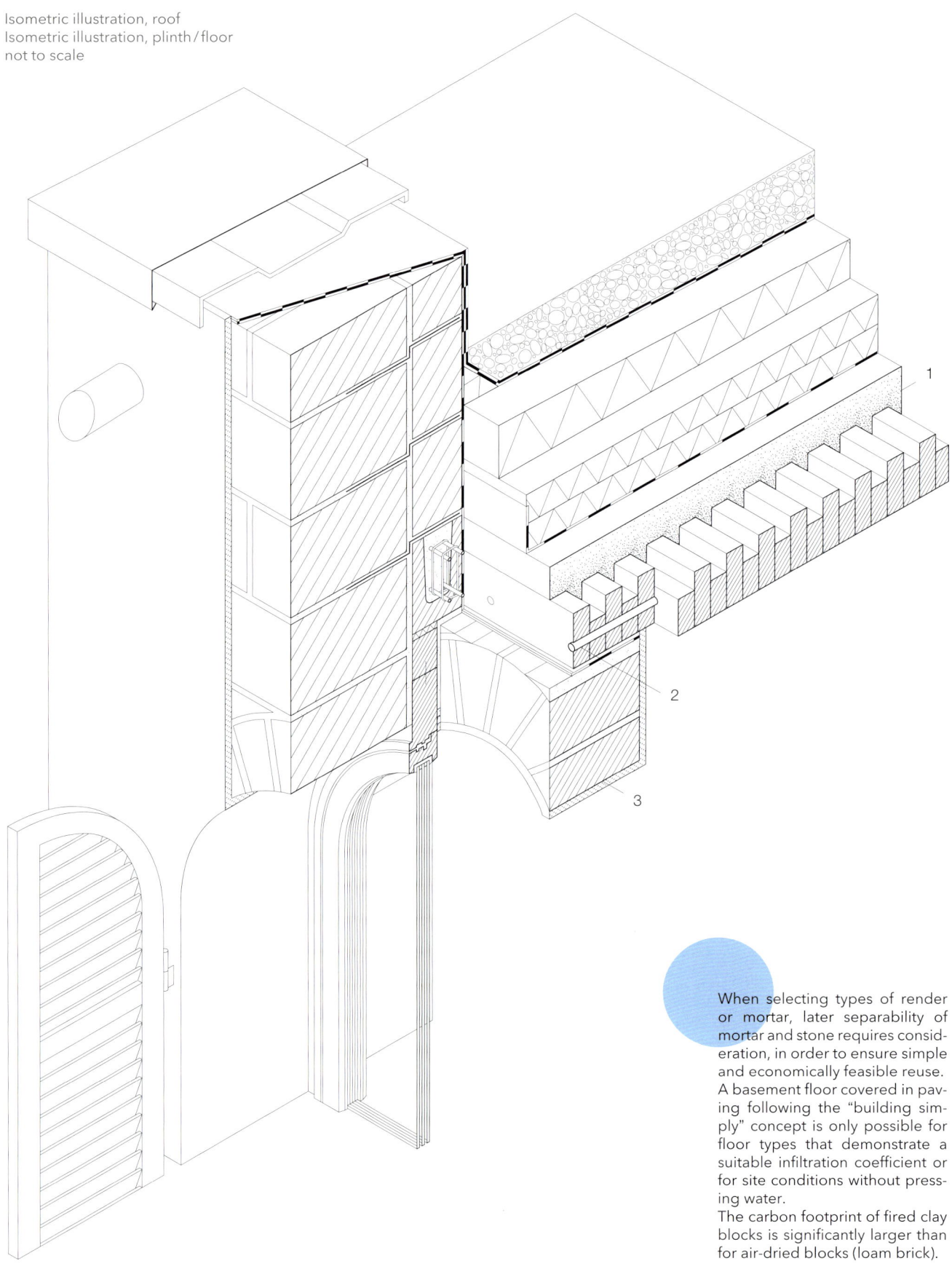

When selecting types of render or mortar, later separability of mortar and stone requires consideration, in order to ensure simple and economically feasible reuse. A basement floor covered in paving following the "building simply" concept is only possible for floor types that demonstrate a suitable infiltration coefficient or for site conditions without pressing water.
The carbon footprint of fired clay blocks is significantly larger than for air-dried blocks (loam brick).

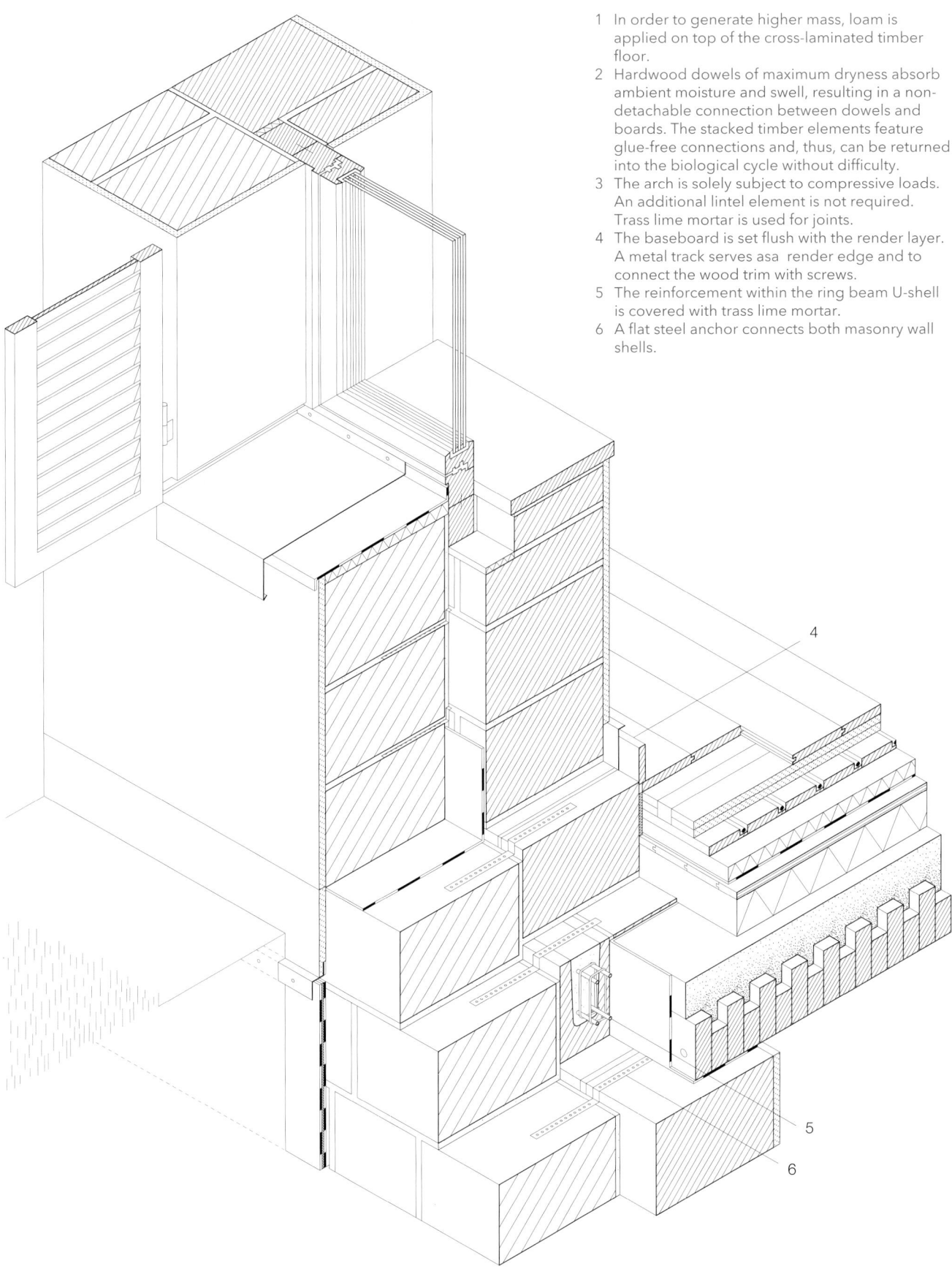

1. In order to generate higher mass, loam is applied on top of the cross-laminated timber floor.
2. Hardwood dowels of maximum dryness absorb ambient moisture and swell, resulting in a non-detachable connection between dowels and boards. The stacked timber elements feature glue-free connections and, thus, can be returned into the biological cycle without difficulty.
3. The arch is solely subject to compressive loads. An additional lintel element is not required. Trass lime mortar is used for joints.
4. The baseboard is set flush with the render layer. A metal track serves as a render edge and to connect the wood trim with screws.
5. The reinforcement within the ring beam U-shell is covered with trass lime mortar.
6. A flat steel anchor connects both masonry wall shells.

Focus on Masonry

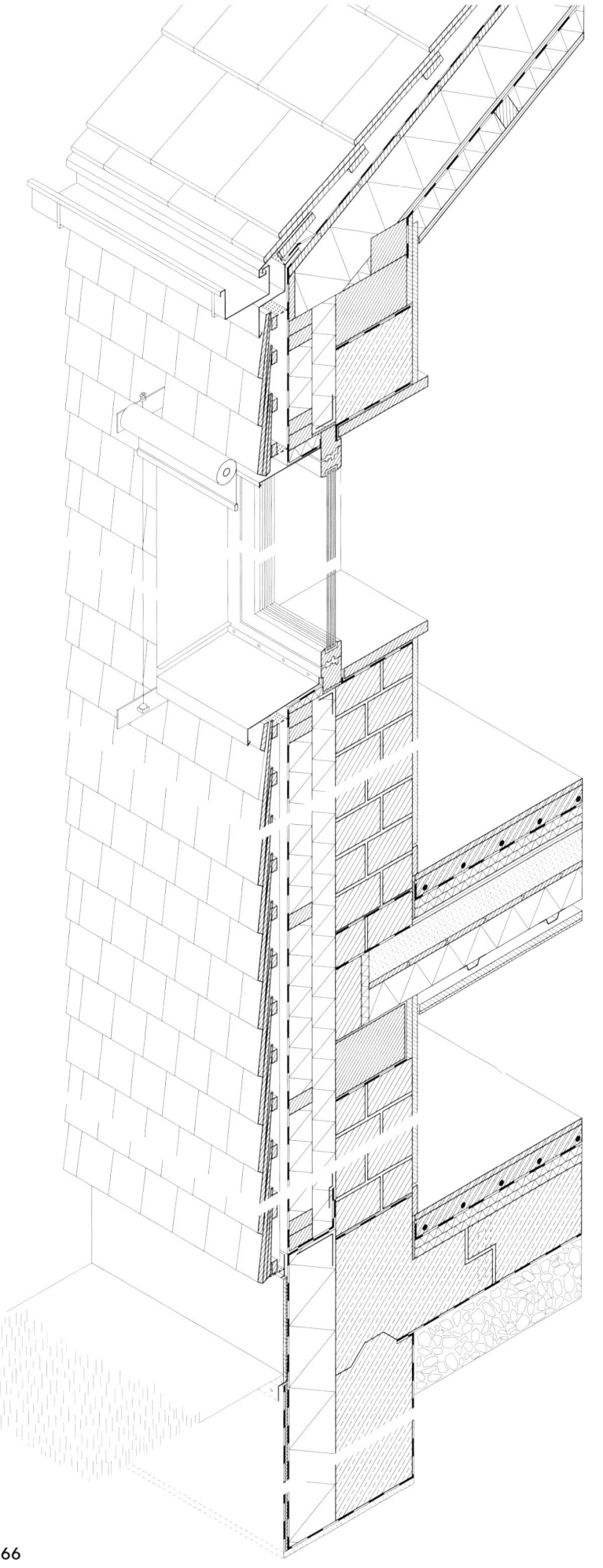

Masonry: Loam Blocks

Protect
slate roofing
wood shingles, back-ventilated
plinth render, lime

Insulate
reed cane insulation, hemp insulation panels
foam glass fill

Support
loam blocks
solid construction timber ring beam
timber beam, rafters
prefabricated reinforced concrete element
foundation, reused

Seal
EPDM sealant
PE film, sarking membrane

Cover
lime render
loam render
loam terrazzo

Isometric illustration   scale 1:20
Vertical section   scale 1:20
Section floor slab   scale 1:20

1 Roof construction:
  Slate roofing, screw connections
  solid timber battens, screw connections
  min. 40 mm counterbattens, screw connections/ back ventilation
  sarking membrane, windproof, breathable, min. 30 cm overlap
  solid timber sheathing, screw connections
  solid construction timber rafters, birdsmouth joints to wall plate/ridge beam, bolt connections
  inlaid reed cane insulation
  solid timber panel diagonal siding with dovetail joints, screw connections
  PE film, diffusion-resistant, min. 30 cm overlap, attached to bottom plate
  solid timber battens, screw connections
  inlaid hemp insulation/installation layer
  solid timber sheathing, screw connections
2 Insect screen, sheet steel, corrosion-proof, screw connections to sheathing and counterbattens
3 Solid construction timber ring beam
4 Textile sun protection, exterior

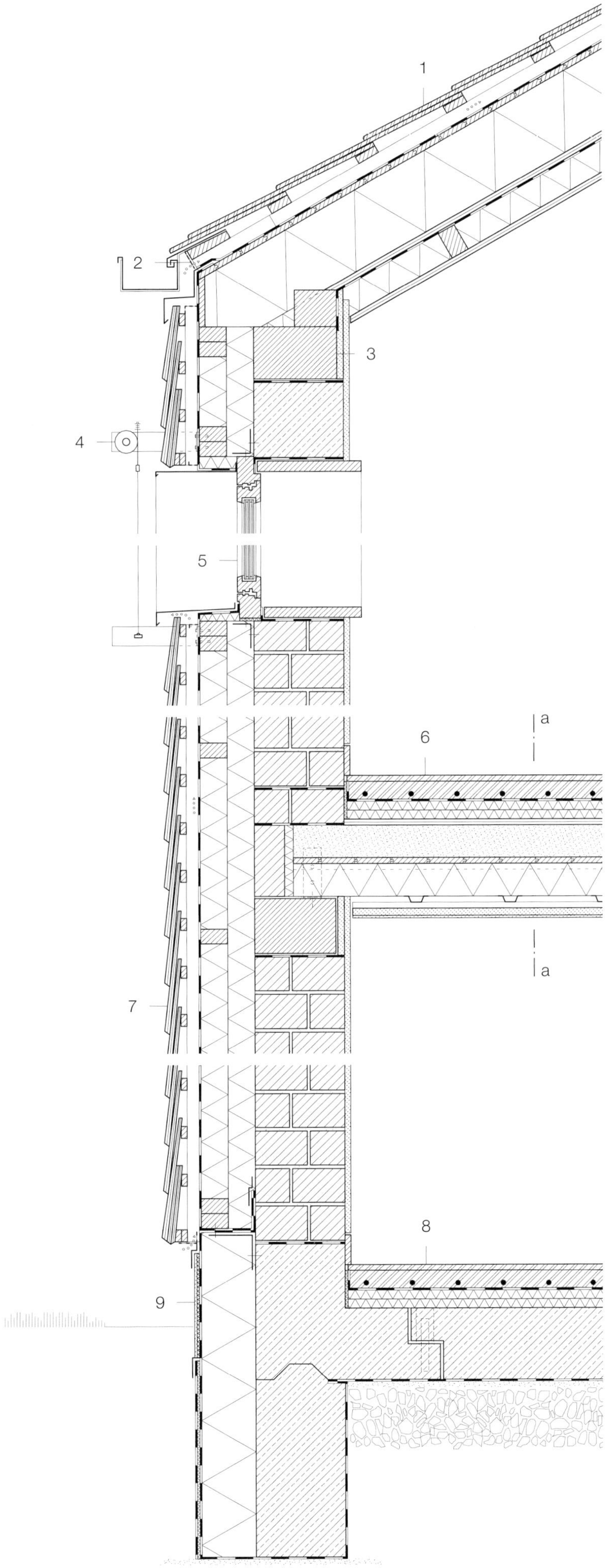

5 Window: Triple glazing in wood frame, reused dry plastic sealant, sorted by type
6 Floor construction:
Loam terrazzo top layer, sanded
loam terrazzo base layer, embedded underfloor heating pipes
PE film, diffusion-resistant, min. 30 cm overlap
cork rigid impact soundproofing
softwood fibre panel levelling insulation, lignin-bonded
solid timber sheathing, screw connections
solid construction timber beam
mass fill, sand, to improve soundproofing
insert board, set on top of battens, screw connections
hemp insulation, infilled between beams
hung ceiling system:
sheet steel mounting channel, galvanised, screw connections
loam construction panel with double render layer, reinforcement, screw connections
7 Exterior wall construction:
Wood shingles, three layers, dowel connections
solid timber battens, screw connections
min. 40 mm counterbattens, screw connections/back-ventilated
sarking membrane, windproof, breathable, min. 30 cm overlap

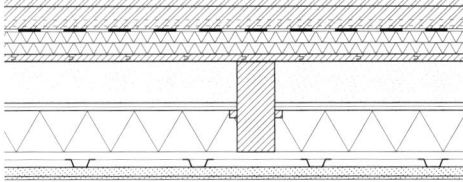

aa

twin mullion transom structure, screw connections
inlaid hemp insulation panel
min. 24 cm loam block masonry wall, trass lime mortar joints
double loam render layer with grass fibre reinforcement
8 Floor construction:
Loam terrazzo top layer, sanded
loam terrazzo base layer, embedded underfloor heating pipes
PE film, diffusion-resistant, min. 30 cm overlap
cork rigid soundproofing panel
softwood fibre levelling insulation panel, lignin-bonded
prefabricated reinforced concrete floor slab element, reused
EPDM sealant layer against ground moisture, membrane-to-membrane weld, min. 30 cm overlap, clamped between foundation and floor slab
sand base layer
foam glass gravel fill perimeter insulation
9 Lime render/splash water protection on render substrate panel, plinth height min. 50 cm above ground level (for loam construction materials)

Focus on Masonry

Isometric illustration, roof
Isometric illustration, ceiling / wall
not to scale

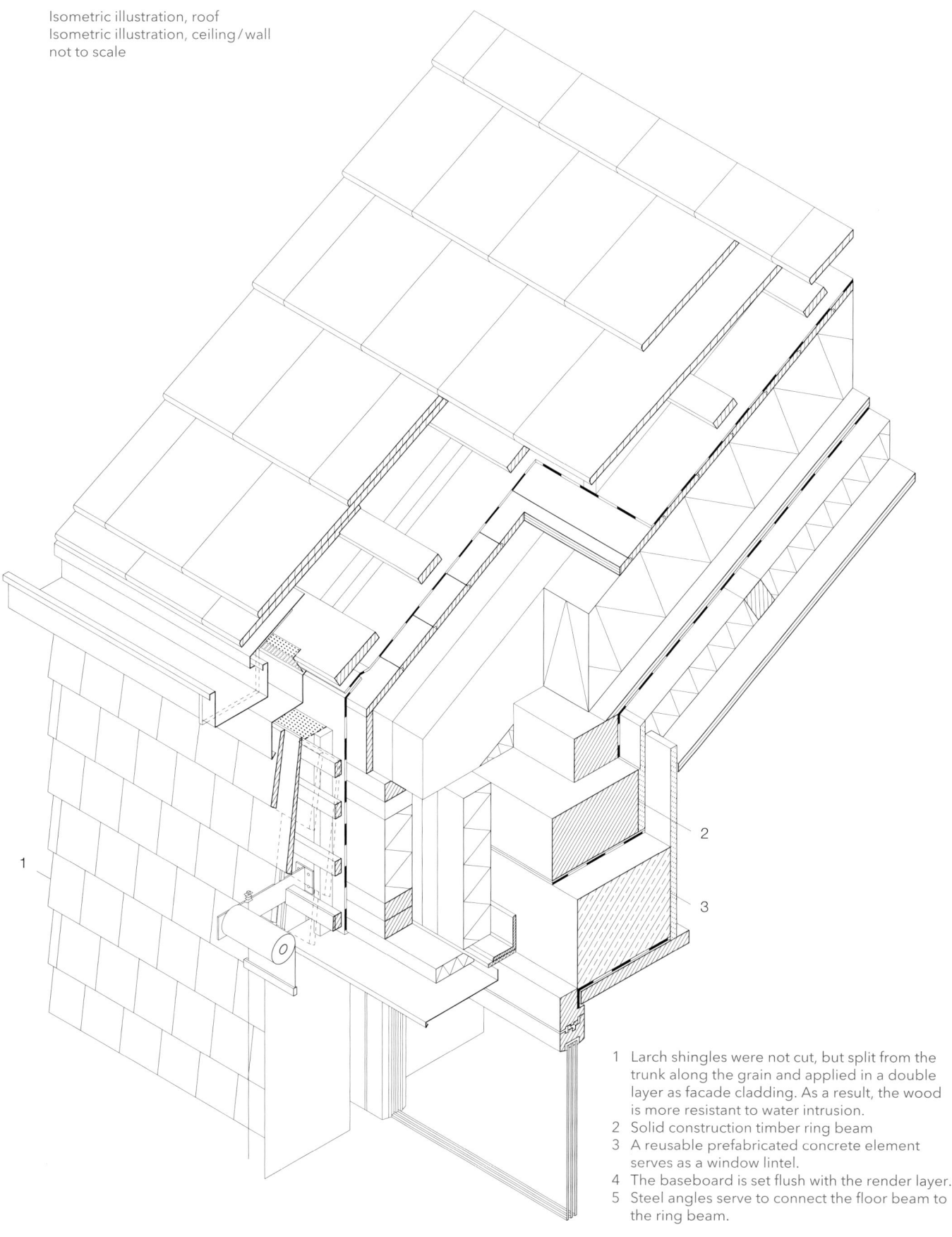

1 Larch shingles were not cut, but split from the trunk along the grain and applied in a double layer as facade cladding. As a result, the wood is more resistant to water intrusion.
2 Solid construction timber ring beam
3 A reusable prefabricated concrete element serves as a window lintel.
4 The baseboard is set flush with the render layer.
5 Steel angles serve to connect the floor beam to the ring beam.

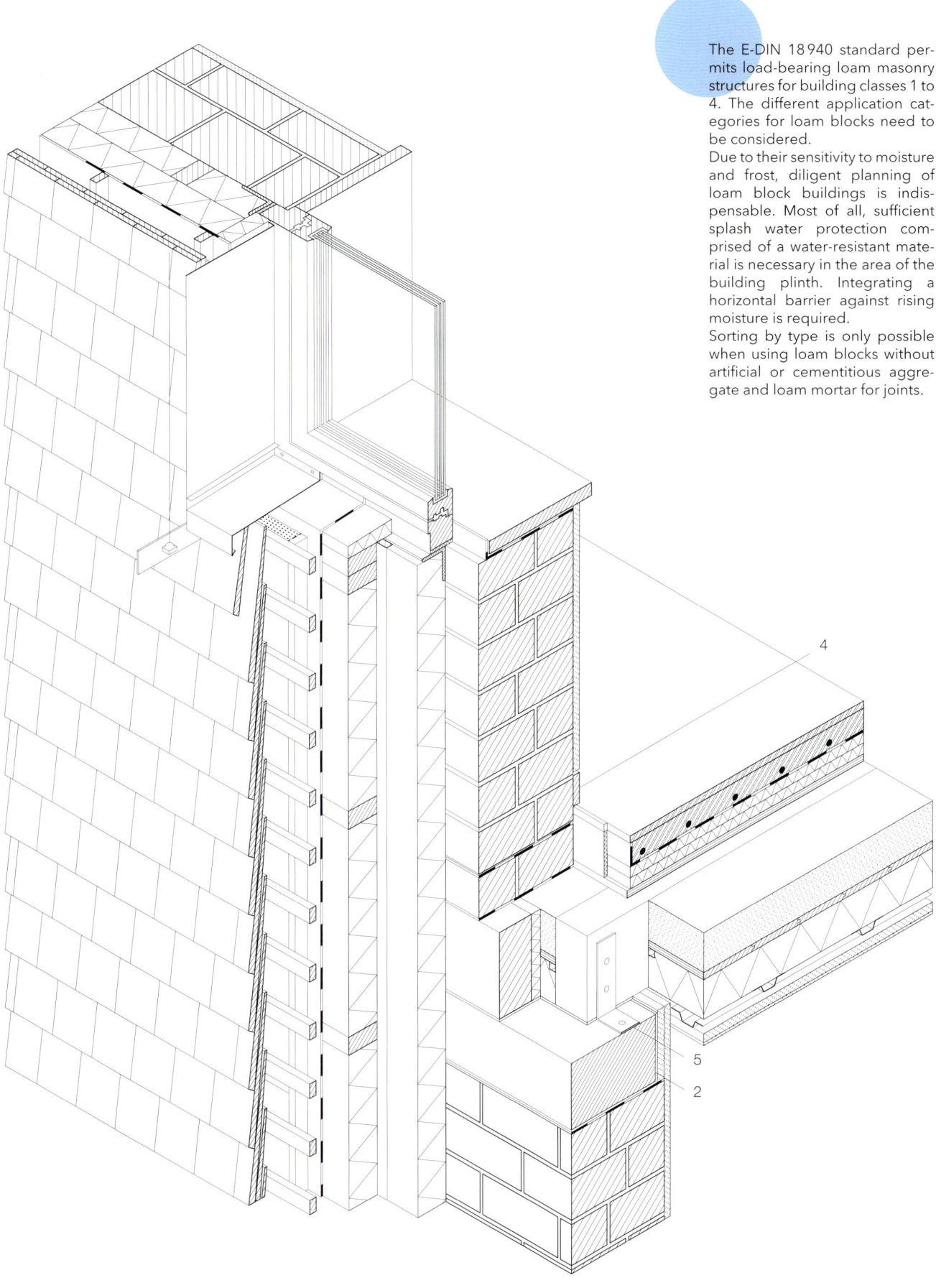

The E-DIN 18940 standard permits load-bearing loam masonry structures for building classes 1 to 4. The different application categories for loam blocks need to be considered.

Due to their sensitivity to moisture and frost, diligent planning of loam block buildings is indispensable. Most of all, sufficient splash water protection comprised of a water-resistant material is necessary in the area of the building plinth. Integrating a horizontal barrier against rising moisture is required.

Sorting by type is only possible when using loam blocks without artificial or cementitious aggregate and loam mortar for joints.

Focus on Masonry

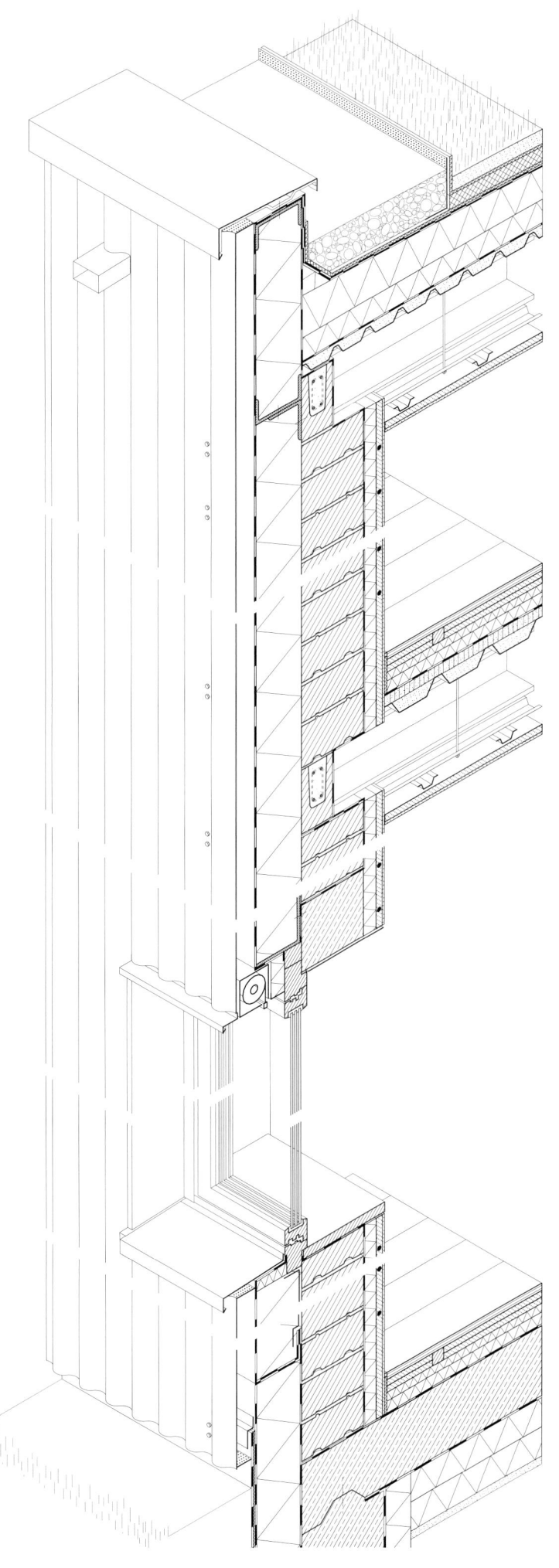

Masonry: Sand Lime Blocks

Protect
parapet coping, extensive greening
corrugated sheet steel, galvanised

Insulate
reed cane insulation
foam glass panels

Support
sand lime modular wall blocks
ring beam U-shell
wide flange steel beam, corrosion-proof
prefabricated reinforced concrete foundation element, reused

Seal
EPDM sealant membrane
PE film, sarking membrane

Cover
reclaimed wood floorboards
loam render on loam construction panel

Vertical section   scale 1:20
Isometric illustration   scale 1:20

1 Roof edge
  parapet coping, canted, clamped to anchor clip
  anchor clip, screw connections to topmost stud wall steel channel
  lightweight metal stud parapet construction
2 Extensive greening
  gravel perimeter strip
  vegetation substrate fill layer
  polypropylene filter fleece, loose layout
  HDPF drainage element, sorted by type or stainless steel, loose layout, reused
  plastic storage protection mat, sorted by type, loose layout
  EPDM sealant membrane, membrane-to-membrane weld, 2-ply, attached to parapet, clamped
  foam glass insulation panels, min. 2 % to falls, loose layout
  foam glass panel insulation, loose bond
  PE film, diffusion-resistant, min. 30 cm overlap, attached to parapet, clamped
  corrugated sheet metal, galvanised, point by point screw connections to wide flange steel beam, corrosion-proof
  mass infill, sand, to improve soundproofing, in

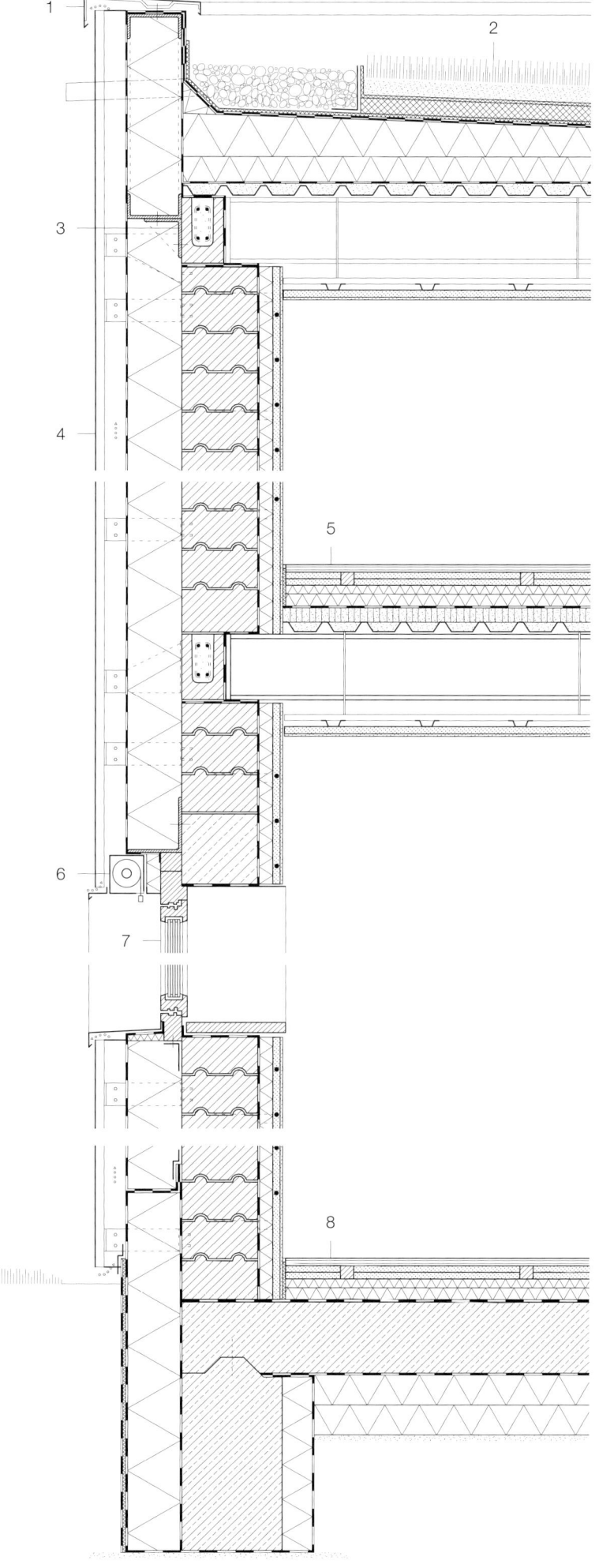

1 cardboard honeycomb set between corrugated sheet metal ribs
wide flange steel beam, corrosion-proof, supported by masonry wall
hung ceiling system:
Sheet steel framing, corrosion-proof, screw connections
sheet steel mounting channel, corrosion-proof, screw connections
loam render on loam construction panel, screw connections
3 Structural steel angle, corrosion-proof, as anchor to lightweight metal stud wall, thermal separation, bolt connections
4 Corrugated sheet steel, galvanised, screw connections
steel bracket framing, galvanised, bolt connections to masonry wall
40 mm back ventilation
sarking membrane, windproof, breathable, min. 30 cm overlap
reed cane mat insulation, anchored to masonry wall, insulation dowel connections
sand lime modular construction block, interlocked
solid timber battens, screw connections
inlaid reed cane mineral wool insulation / installation layer
loam render on loam construction panel, integrated heating pipes, screw connections
5 Reclaimed wood floorboards, concealed tongue-and-groove screw connections
solid timber battens, screw connections
inlaid double loam construction panel
cork impact soundproofing
cork levelling layer
PE film, diffusion-resistant, min. 30 cm overlap
corrugated sheet metal, galvanised, point-to-point screw connections to wide flange steel beam, corrosion-proof
mass infill, sand, to improve soundproofing, in cardboard honeycomb set between corrugated sheet metal ribs
wide flange steel beam, corrosion-proof, supported by masonry wall
hung ceiling system:
Sheet steel framing, corrosion-proof, screw connections
sheet steel mounting channel, corrosion-proof, screw connections
double loam render layer on loam construction panel, reinforcement, screw connections
6 Textile sun protection
7 Window: Triple glazing in wood frame, reused plastic dry sealant, sorted by type
8 Reclaimed wood floorboards, concealed tongue-and-groove screw connections
solid timber battens, screw connections
inlaid double loam construction panel
double softwood fibre panel impact soundproofing, lignin-bonded
PE film, diffusion-resistant, min. 30 cm overlap, clamped
prefabricated reinforced concrete floor slab element, reused
sealant membrane against ground moisture
EPDM layer, membrane-to-membrane weld, loose layout
double foam glass panel perimeter insulation, loose layout
sand base layer

Focus on Masonry

Isometric illustration, roof
Isometric illustration, plinth/floor
not to scale

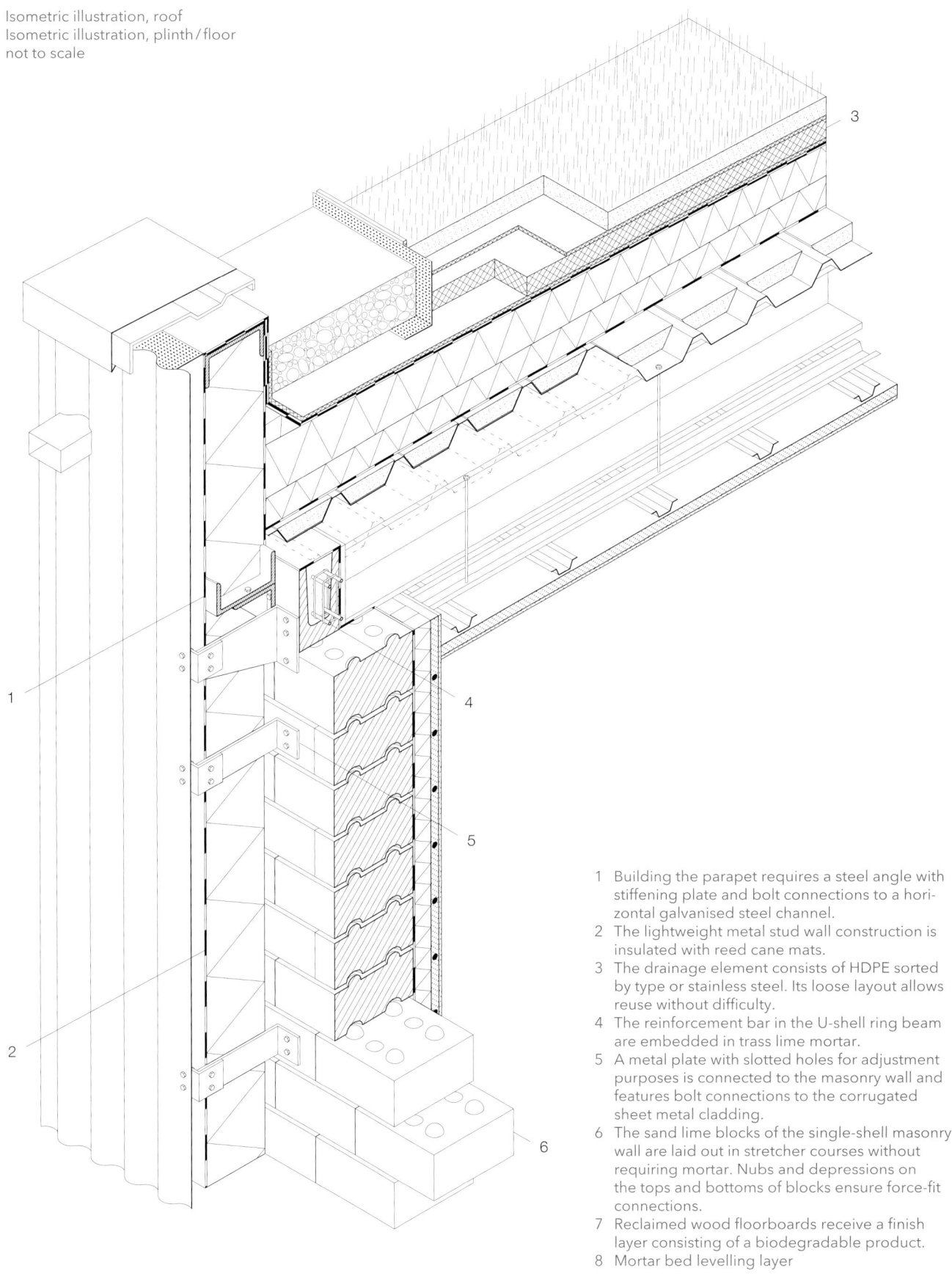

1. Building the parapet requires a steel angle with stiffening plate and bolt connections to a horizontal galvanised steel channel.
2. The lightweight metal stud wall construction is insulated with reed cane mats.
3. The drainage element consists of HDPE sorted by type or stainless steel. Its loose layout allows reuse without difficulty.
4. The reinforcement bar in the U-shell ring beam are embedded in trass lime mortar.
5. A metal plate with slotted holes for adjustment purposes is connected to the masonry wall and features bolt connections to the corrugated sheet metal cladding.
6. The sand lime blocks of the single-shell masonry wall are laid out in stretcher courses without requiring mortar. Nubs and depressions on the tops and bottoms of blocks ensure force-fit connections.
7. Reclaimed wood floorboards receive a finish layer consisting of a biodegradable product.
8. Mortar bed levelling layer

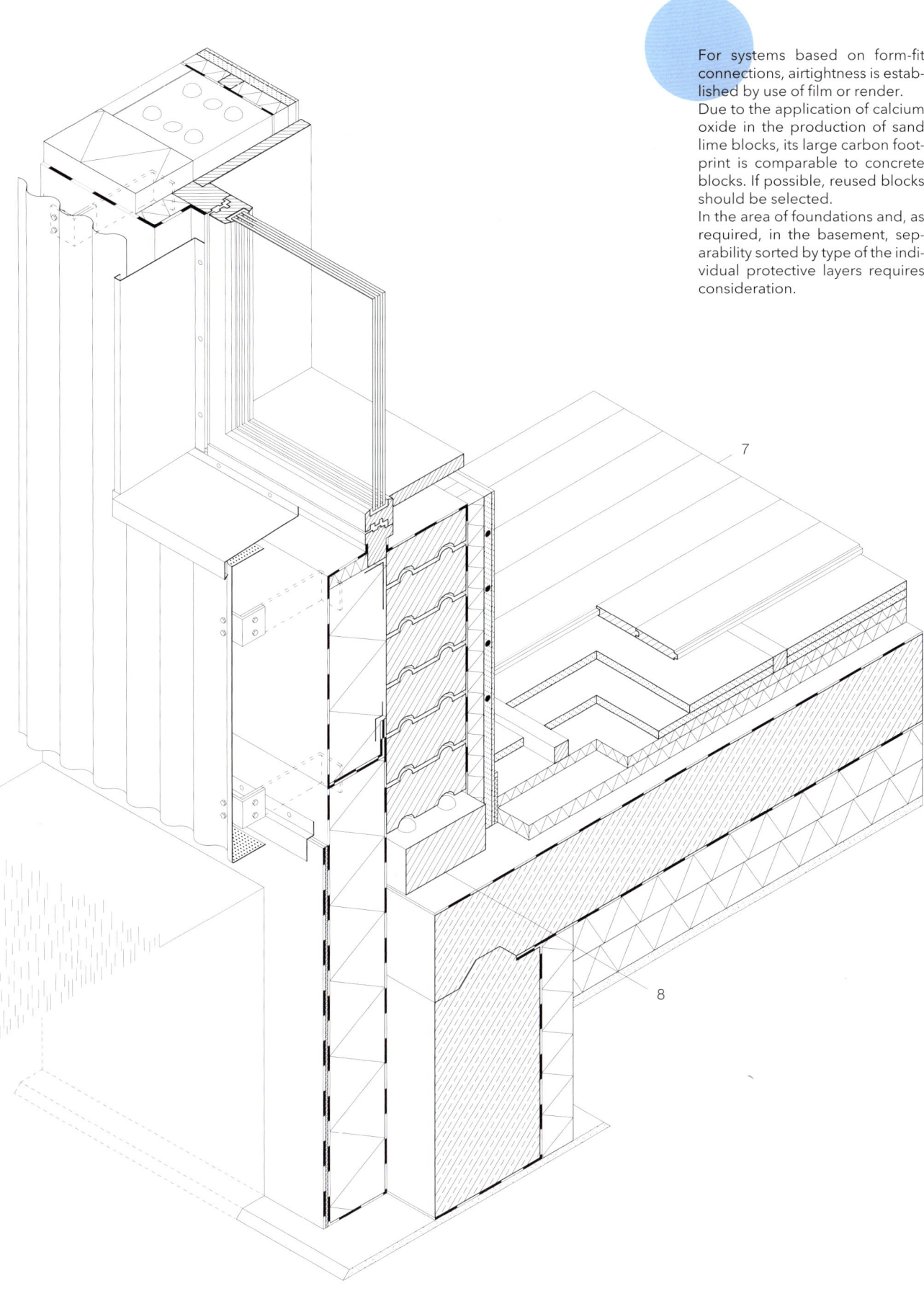

For systems based on form-fit connections, airtightness is established by use of film or render.
Due to the application of calcium oxide in the production of sand lime blocks, its large carbon footprint is comparable to concrete blocks. If possible, reused blocks should be selected.
In the area of foundations and, as required, in the basement, separability sorted by type of the individual protective layers requires consideration.

Focus on Masonry

Focus on Concrete

More than 5000 years ago and possibly by chance, composite materials similar to concrete were developed that consisted of mineral binding agents and small pebble aggregate. The discovery of mineral binding agents that achieve solidity not by drying, but through chemical bonds with water led to a revolution in the construction field. With a view to construction sorted by type, the concept of a binding agent – a material with the purpose of connecting many different particles within a heterogeneous particulate composite material [1] – almost seems paradoxical.

Characteristics

The likely most important mineral binding agent for the construction industry was developed in the mid-19th century with the invention of Portland cement. Mixed with stone particles of different sizes and by adding water, cement reacts and hardens into a material resembling stone. Completely enveloping the stone particles with hardened cement paste results in the creation of concrete as a composite material – as of today one of the most frequently used construction materials in the world. In simple terms, to produce cement, clay and limestone are crushed, ground and fired

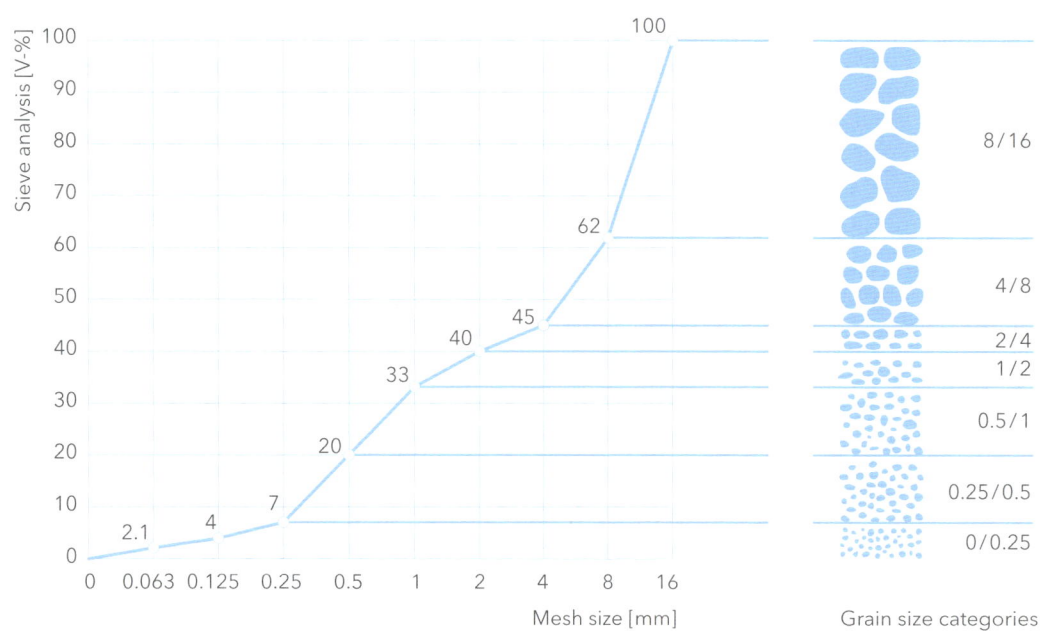

1 Grain size distribution curve for concrete aggregate

into cement clinker at temperatures of about 1450 °C. By discharging CO_2, this leads to the creation of the main component of cement: Calcium oxide (CaO). Aside from the tremendous amount of energy required for firing clinker, most of all the chemical reaction is the reason why cement production causes about 8 % of human induced global greenhouse gases. The high degrees of strength and solidity, the plastic formability, the fireproofing characteristics and the high resistance to moisture qualify concrete as a high-performance construction material. Given the current mindset in the building sector, replacing it seems hardly possible.

Raw material deposits
The raw materials required for the production of concrete include cement, sand, large grains of rock and water. These are (currently) available almost worldwide. The relatively simple production of this mix of materials in connection to the previously described characteristics of concrete are the reason why the use of the material is steadily increasing globally. Concrete is employed in civil engineering, for the construction of roads and tunnels, for architectural construction and all conceivable building types. It is problematic because of aspects related to the availability of sand and gravel, as well as CO_2 emissions caused during the production of cement. In the long run, these aspects will lead to a significant increase in prices and, thus, to a more sparing use of the material. Not least because of the looming shortage of raw materials, the recycling process for concrete has become central to building materials research. Aside from the already described ecological problems surrounding cement as a binding agent, aggregates such as sand and gravel are to be viewed critically as well. Both belong to the most heavily mined resources by mass worldwide. The extraction process occupies large, otherwise agriculturally used areas, leading to massive interventions in the existing ecological balance. With a view to the gigantic amounts of sand in the world's deserts, the idea that sand is scarce seems rather odd. This is, however, due to the fact that the size and form of sand grains is relevant (Fig. 1). Desert sand and its round shape and homogeneous grain size is the result of wind erosion. For concrete, less round and more angular sand grains are required that permit compact distribution due to diverse grain sizes and polygonal forms. This allows a reduction in the amount of required cement (see "The linear system", p. 15f.).

Typical construction methods
In the construction field, concrete mostly finds use as in-situ concrete, meaning it is poured into formwork on site. Since concrete has poor tensile strength, typically reinforcement, mostly consisting of steel, serves the purpose of absorbing tensile stress in a concrete building component. Steel-reinforced concrete is, thus, capable

of bearing compressive and tensile loads. The alkaline environment of cured cement protects the embedded steel from corrosion. If construction elements are not cast on site, but instead, in a workshop or a factory, they are described as prefabricated concrete. Such elements typically include reinforcement in order to bear tensile loads. The admixture of lightweight mineral grain, including perlite, expanded clay or foam glass gravel results in the creation of concrete types with low thermal conductivity and low weight. Such concrete mixtures are categorised as lightweight concrete. Compared to typical mixtures, they display significantly lower compressive strength. Walls consisting of lightweight concrete – also described as insulating concrete – are often built with a greater thickness than walls consisting of conventional concrete. With a view to reuse or recycling, lightweight concrete should be considered problematic. Due to a high degree of porosity and the resulting limited load-bearing capacity of the concrete component, it is suitable only to a limited degree as aggregate of recycling concrete or as crushed concrete for base layers in road construction. The insulating perlite, expanded clay or foam glass gravel aggregate disallows separation sorted by type from the other concrete constituents. Thus, lightweight concrete demolition waste needs to be sent to landfill [2].

Monolithic construction comprised of lightweight concrete has the advantage that additional layers for insulation or render tend to be unnecessary. Load-bearing structure, insulation and the exterior weatherproof layer can be realised by using a single material. This contributes to details that are simple and less error-prone, resulting in robust and relatively durable buildings.

Circular potential

Conventional concrete presents its own problems in terms of being harmful to the climate or its circularity. This is offset by the diverse range of applications this extremely robust construction material permits – which represents a dilemma that researchers have been discussing for decades. When deliberating on how to produce concrete in more sustainable ways, three factors are important: Binding agents, reinforcement and aggregate. The previously described emissions-heavy cement production process urgently needs alternative binding agents. Well-researched alternatives include so-called geopolymers. Their production leads to a reduction of CO_2 emissions of up to 70 % compared to typical Portland cement [3]. Unlike cement, geopolymers, such as calcined clay – already known to the Romans – or Celitement clinker are not hydraulic binders. Instead, they belong to the inorganic binder group. These binding agents received their name due to material characteristics that are similar to rocks and minerals ("geo-") and due to a hardening process that is comparable to the polymerisation of plastics. Geopolymers demonstrate greater degrees of stability and solidity and better acid resistance than typical concrete. Currently, however, geopolymers are neither available in quantities required for large-scale technological applications, nor can they be produced in an economical manner.

The reinforcement most building elements depend on in order to absorb tensile stress currently still consists of steel. The separation of steel and concrete following the end of the life cycle of a reinforced concrete element is simple when using concrete crushers. However, such construction elements often lose their load-bearing capacity due to corroding steel reinforcement. In most cases, a sufficiently thick concrete cover ranging from 3 to 5 cm can protect steel from corrosion, while it is not required for load-bearing purposes. In order to save concrete, it would therefore be desirable to produce thinner components with an alternative reinforcement that relies on less concrete cover. Research approaches oriented on natural materials such as bamboo fibre show promising results. Textile concrete reinforced with glass fibre and basalt or carbon rovings have been in use in construction for years. Textile fabric with small mesh sizes, however, complicates separation sorted by type, compared to steel-reinforced concrete.

This is particularly the case for types of concrete reinforced with non-metallic short fibre that are very efficient and economically feasible in terms of production. In relation to the climate balance, textile or short fibre concrete is preferable to typical steel-reinforced concrete, since it consumes less concrete and the energy-intensive production of steel is omitted. In the search for alternative aggregate types, the use of recycling concrete plays an increasingly relevant role. For this type of concrete, the typical aggregate such as sand or gravel is replaced with crushed concrete to a certain extent. It is also gaining increasing political support [4]. "This approach [...] currently limits permissible exchange rates to a maximum of 45 % of grain types with a minimum grain diameter of 2 mm. The use of finer-grain recyclate, so-called crushed sand, is currently not under discussion. The potential of this approach is, thus, severely limited. Its economic feasibility and eco-friendly character are also strongly dependent on the transport distances for specific grain types. Only reprocessing close to the site of use seems to be economically efficient and unharmful to the environment" [5]. Various approaches to the selective separation of concrete components are currently being researched. For instance, at the Fraunhofer Institute for Building Physics (IBP) the principle of electrodynamic fragmentation allows selectively extracting and filtering concrete components and, thus, retrieving high-quality recyclate.

In keeping with the German Circular Economy Act (Kreislaufwirtschaftsgesetz, KrWG), the creation of waste should either be avoided completely, or waste should be recycled. In the context of producing concrete in a sustainable manner while supporting its reuse sorted by type, cascading use as e.g. fill material is not considered satisfying and, thus, only the third-best solution, according to the KrWG.

The reuse of (steel-reinforced) concrete elements, for instance within a standardised construction system, such as proposed by Le Corbusier as long ago as 1914, would lead to a drastic reduction in waste production and energy consumption. With the Maison Dom-Ino principle – a system for the industrial fabrication of houses based on a steel-reinforced concrete frame construction type – Le Corbusier created the basis for the later separation of load-bearing structure, interior finishes and facades. In a certain way, he was a pioneer of modern construction sorted by type [6].

In the case of concrete, most of all the energy required for the production of cement is of ecological and, thus, economic relevance. About 500 kWh (approx. 50 l of fuel oil for heating) are used for the production of one cubic metre of concrete. For its deconstruction, reprocessing and renewed use for construction, about 25 kWh of energy (approx. 2.6 l fuel oil for heating) is required per cubic metre [7]. Today, in terms of producing construction elements, reuse is about 50 % less expensive than manufacturing comparable new prefabricated elements [8]. If reusing concrete structures is also supposed to retain value, (steel-reinforced) prefabricated concrete elements display greater advantages than in-situ concrete constructions. From the viewpoint of sorting by type, however, (steel-reinforced) concrete as an heterogeneous material mix requires critical deliberation. Yet, viewed as a monomaterial system – as such, a near-indispensable mainstay of current construction practice – its potential for non-destructive separation of components is beneficial for construction sorted by type. This is the case particularly for detaching the building envelope from the load-bearing structure. It is desirable to establish and expand a well-functioning market for used concrete elements and a standardised system of dimensions that allows defining prefabricated elements. For this purpose, the prevailing mindset among planners engaged in processes of design and realisation needs to be advanced and adapted. As a result, the free formability of spatial enclosures so popular among architectural creators ever since the construction material concrete was first developed, would be confronted with real limitations. The way we deal with standardised concrete elements can, however, also lead to a new kind of creativity.

Notes
[1] Koenders/Weise/Vogt 2020
[2] Binder/Riegler-Floors 2018, p. 105
[3] Beton und Naturstein 2019
[4] Hillebrandt et al. 2018, p. 70
[5] Moffatt/Haist 2019, p. 34
[6] Zeumer/El Khouli/John 2014, p. 10
[7] Meyser 2011
[8] Asam 2007, p. 4

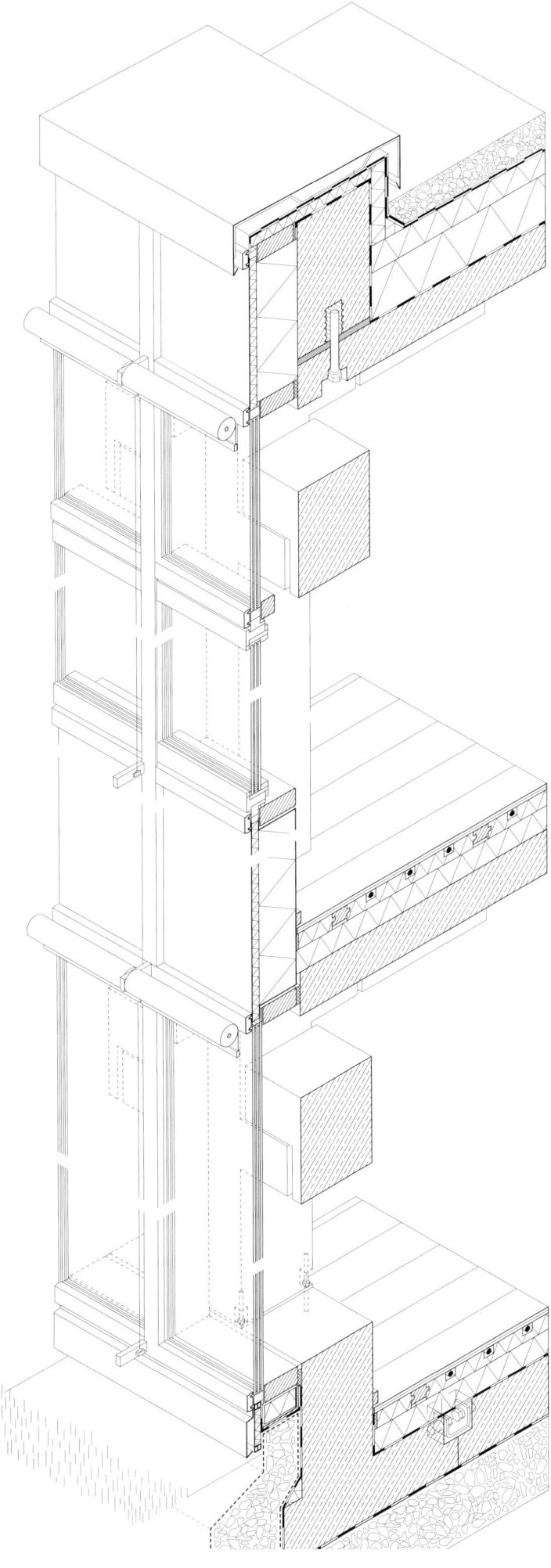

Concrete: Frame Construction

Protect
parapet coping, clay brick chips
mullion transom structure

Insulate
foam glass panels
hemp insulation panels
mullion transom structure

Support
prefabricated reinforced concrete elements, reused

Seal
EPDM sealant layer, PE film
mullion transom structure

Cover
reclaimed wood floorboards

Isometric illustration   scale 1:20
Vertical section, facade   scale 1:20
Section, floor slab   scale 1:20

1   Roof edge:
    Parapet coping, canted, clamped to anchor clip
    anchor clip, screw connections to reinforced concrete element
2   Roof construction:
    Min. 50 mm recycled clay brick chips, loose fill
    2-ply EPDM sealant, membrane-to-membrane weld, attached to parapet, clamped by anchor clip
    foam glass panel insulation, min. 2% to falls, loose layout
    foam glass panel insulation, loose layout in bond
    PE film, breathable, min. 30 cm overlap, attached to parapet, clamped
    prefabricated reinforced concrete ribbed ceiling, reused
3   Textile sun protection, exterior application, screw connections to mullion transom structure
4   Window: Triple glazing in mullion transom structure
    solid timber mullion and transom with aluminium pressure plate and cap, reused
5   Exterior wall construction:
    Sheet steel coffer system, canted, corrosion-proof
    inlaid double hemp insulation layer

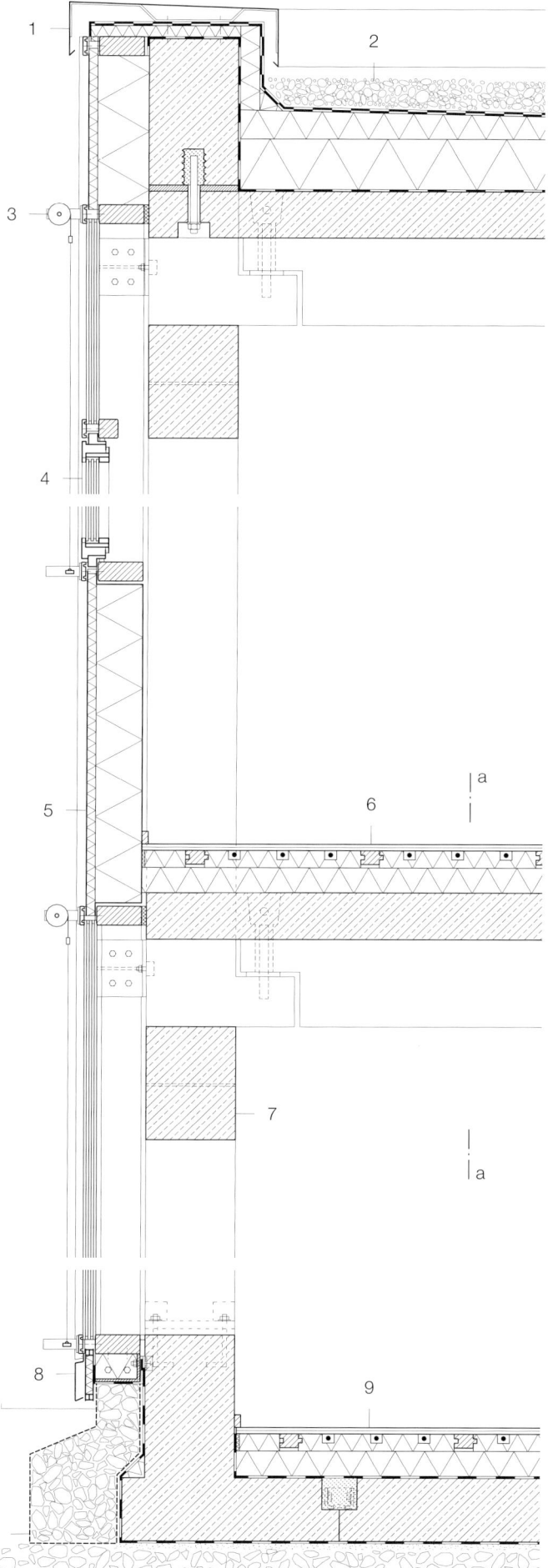

coffer section, set between and clamped by transoms
6 Ceiling construction:
   Reclaimed wood floorboards, concealed tongue-and-groove screw connections
   solid timber battens, screw connections
   inlaid softwood fibre panel, lignin-bonded, as installation panel with grooves and heat diffusor plates to install underfloor heating pipes
   softwood fibre levelling insulation, lignin-bonded
   prefabricated reinforced concrete ribbed ceiling, reused
7 Prefabricated reinforced concrete beam, with notch for elastic column bearing
8 Sheet steel plinth coping, galvanised, screw connections to mullion transom structure
9 Floor construction:
   reclaimed wood floorboards, concealed tongue-and-groove screw connections
   solid timber battens, screw connections
   inlaid softwood fibre panel, lignin-bonded, as installation panel with grooves and heat diffusor plates to install underfloor heating pipes
   softwood fibre panel levelling insulation, lignin-bonded
   PE film, diffusion-resistant, min. 30 cm overlap

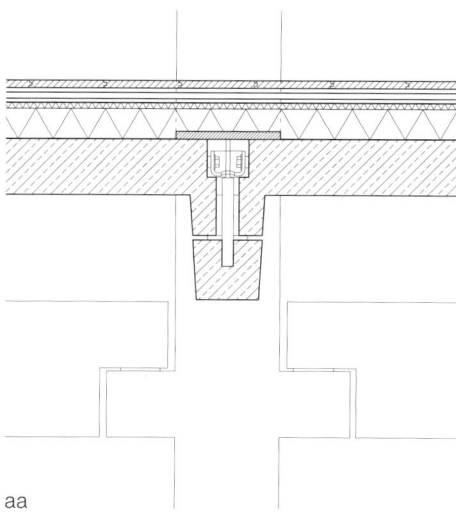

aa

   prefabricated reinforced concrete floor slab element, reused
   EPDM sealant against ground moisture, membrane-to-membrane weld, min. 30 cm overlap, loose layout
   sand base layer
   foam glass gravel perimeter insulation, loose fill
10 Foam glass gravel perimeter insulation, filled in textile fabric bags to support construction, mechanical connections to reinforced concrete
   EPDM sealant against ground moisture, membrane-to-membrane weld, loose layout
   prefabricated reinforced concrete plinth element, reused, bolt connections to floor slab

Focus on Concrete   179

Isometric illustration, roof
Isometric illustration, plinth/floor
not to scale

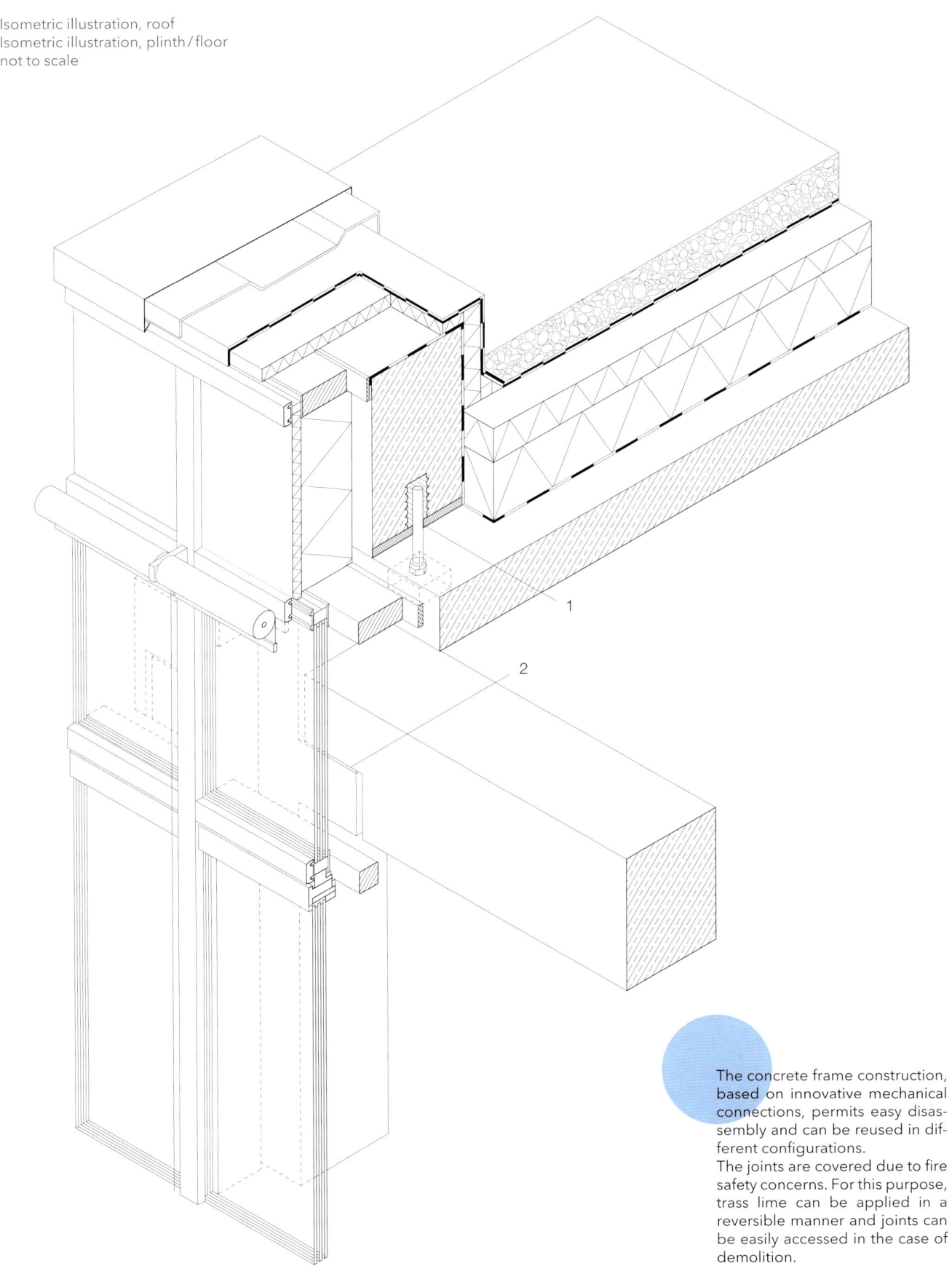

The concrete frame construction, based on innovative mechanical connections, permits easy disassembly and can be reused in different configurations.
The joints are covered due to fire safety concerns. For this purpose, trass lime can be applied in a reversible manner and joints can be easily accessed in the case of demolition.

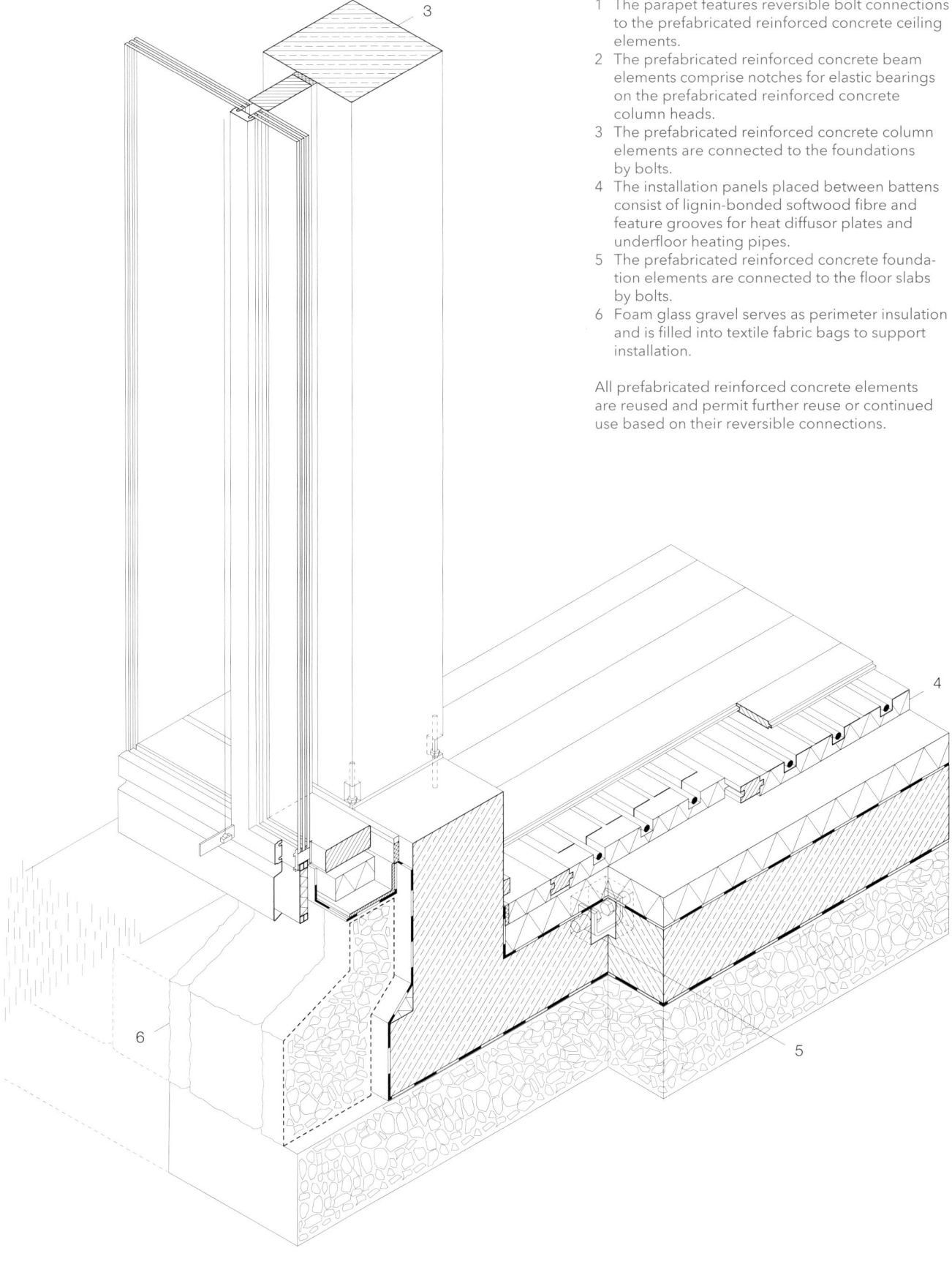

1. The parapet features reversible bolt connections to the prefabricated reinforced concrete ceiling elements.
2. The prefabricated reinforced concrete beam elements comprise notches for elastic bearings on the prefabricated reinforced concrete column heads.
3. The prefabricated reinforced concrete column elements are connected to the foundations by bolts.
4. The installation panels placed between battens consist of lignin-bonded softwood fibre and feature grooves for heat diffusor plates and underfloor heating pipes.
5. The prefabricated reinforced concrete foundation elements are connected to the floor slabs by bolts.
6. Foam glass gravel serves as perimeter insulation and is filled into textile fabric bags to support installation.

All prefabricated reinforced concrete elements are reused and permit further reuse or continued use based on their reversible connections.

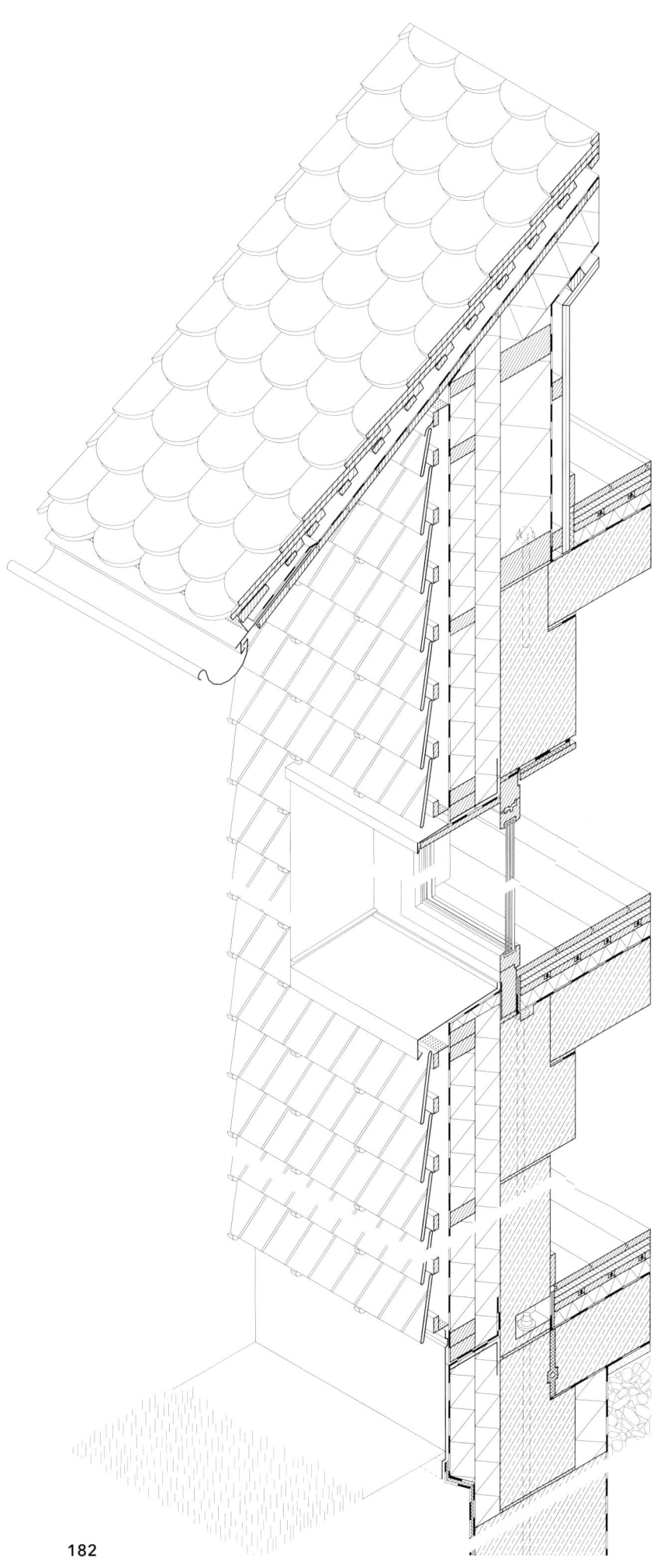

Concrete: Solid Construction

Protect
flat roof tile
compressed ventilation duct panels

Insulate
reed cane insulation
hemp insulation panels

Support
prefabricated reinforced concrete elements, reused

Seal
EPDM sealant membrane
PE film, sarking membrane

Cover
reclaimed wood floorboards
solid timber sheathing

Isometric illustration   scale 1:20
Vertical section   scale 1:20

1 Roof construction:
  Hanging flat roof tile, screw connections
  solid timber battens, screw connections
  min. 40 mm counterbattens, screw connections/
  back ventilation
  sarking membrane, windproof, breathable,
  min. 30 cm overlap
  solid timber sheathing, screw connections
  solid construction timber rafters, birdsmouth
  joints to wall plate/ridge beam, screw
  connections
  inlaid reed cane insulation, loose layout
  PE film, diffusion-resistant, min. 30 cm overlap,
  attached to wall plate, clamped
  solid timber battens, screw connections to joists
  solid timber sheathing, screw connections
2 Sheet steel insect screen, corrosion-proof, screw
  connections to sheathing and counterbattens
3 Exterior wall construction, rooftop:
  Sheathing, repurposed compressed and cut
  ventilation ducts, suspended
  solid timber battens, screw connections
  min. 40 cm counterbattens, screw connections/
  back ventilation
  sarking membrane, windproof, breathable,
  min. 30 cm overlap, welded
  solid construction timber framing

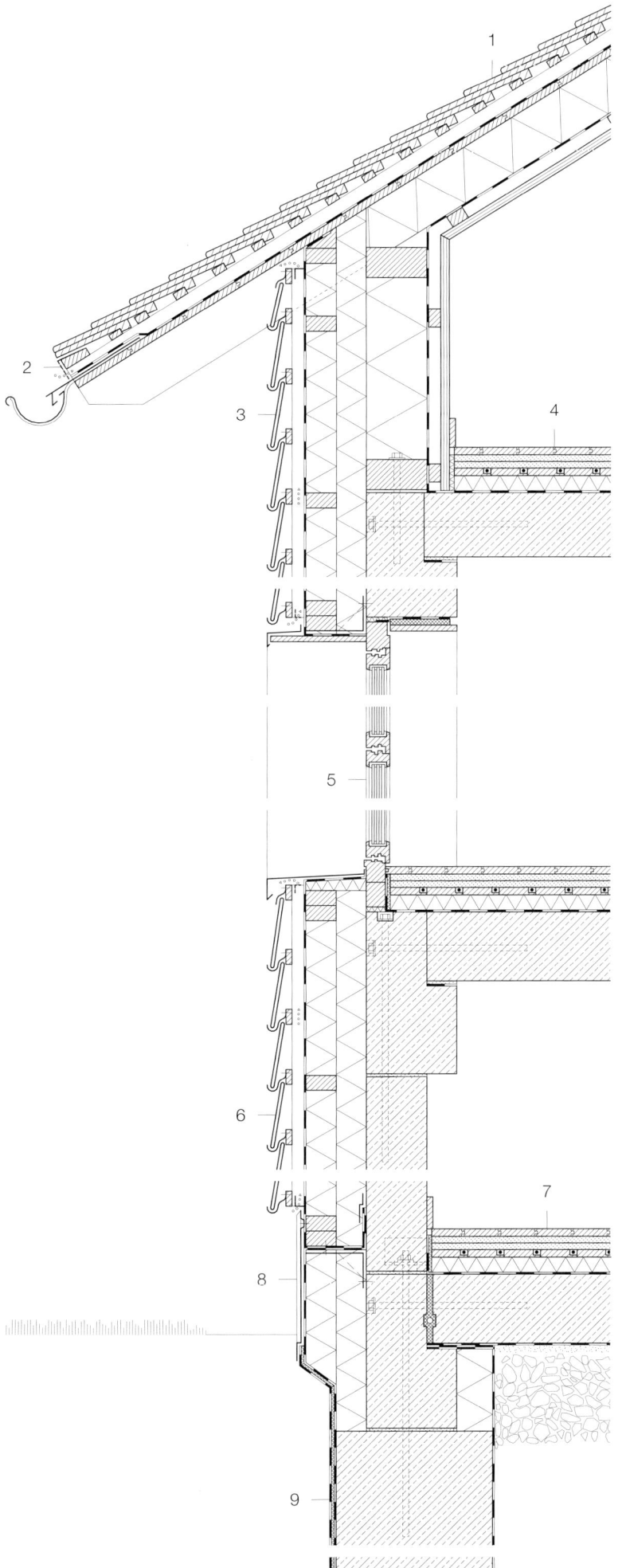

1 Roof construction:
 clay interlocking tiles, reused
 solid timber battens, screw connections
 min. 40 cm counterbattens, screw connections/
 back ventilation
 sarking membrane, windproof, breathable,
 min. 30 cm overlaps
 inlaid double hemp insulation panels
 solid construction timber framing, screw
 connections
 inlaid hemp insulation panels
 PE film, diffusion-resistant, min. 30 cm overlap,
 attached to wall plate, clamped
 solid timber battens, screw connections
 solid timber sheathing, screw connections
4 Floor construction:
 Reclaimed wood floorboards, concealed
 tongue-and-groove screw connections
 solid timber battens, screw connections
 inlaid double loam construction panel
 solid timber sheathing with lateral groove
 and heat diffusor plates to install underfloor
 heating pipes
 impact soundproofing/hemp installation layer
 PE film, diffusion-resistant, min. 30 cm overlap,
 clamped
 prefabricated reinforced concrete ceiling element, reused, bolt connections to wall bracket
5 Window: Triple glazing in wood frame, fixed
 glazing as fall protection (laminated safety glass),
 reused
 plastic dry sealant, sorted by type
6 Exterior wall construction, typical floor:
 Sheathing, repurposed compressed and cut
 ventilation duct panels, suspended
 solid timber battens, screw connections
 min. 40 cm counterbattens, screw connections/
 back ventilation
 sarking membrane, windproof, breathable,
 min. 30 cm overlaps
 solid construction timber framing
 inlaid double hemp insulation panel
 prefabricated reinforced concrete element,
 reused, bolt connections to foundation
7 Floor construction:
 Reclaimed wood floorboards, concealed
 tongue-and-groove screw connections
 solid timber battens, screw connections
 inlaid double loam construction panel
 solid timber sheathing with lateral groove and
 heat diffusor plates to install underfloor heating
 pipes
 impact soundproofing/ hemp installation layer
 PE film, diffusion-resistant, min. 30 cm overlap,
 clamped
 prefabricated reinforced concrete floor slab,
 reused
 EPDM sealant against ground moisture,
 membrane-to-membrane weld, 30 cm overlap,
 loose layout
 sand base layer
 foam glass gravel fill insulation layer
8 Sheet steel plinth coping, galvanised, canted,
 screw connections to battens
9 EPDM sealant against ground moisture,
 membrane-to-membrane weld, clamped by
 mounting track
 dimpled sheet, 30 cm overlaps
 EPDM sealant against ground moisture,
 membrane-to-membrane weld, loose layout
 prefabricated reinforced concrete foundation
 element, reused, frost-free foundation

Isometric illustration, roof
Isometric illustration, wall / ceiling
not to scale

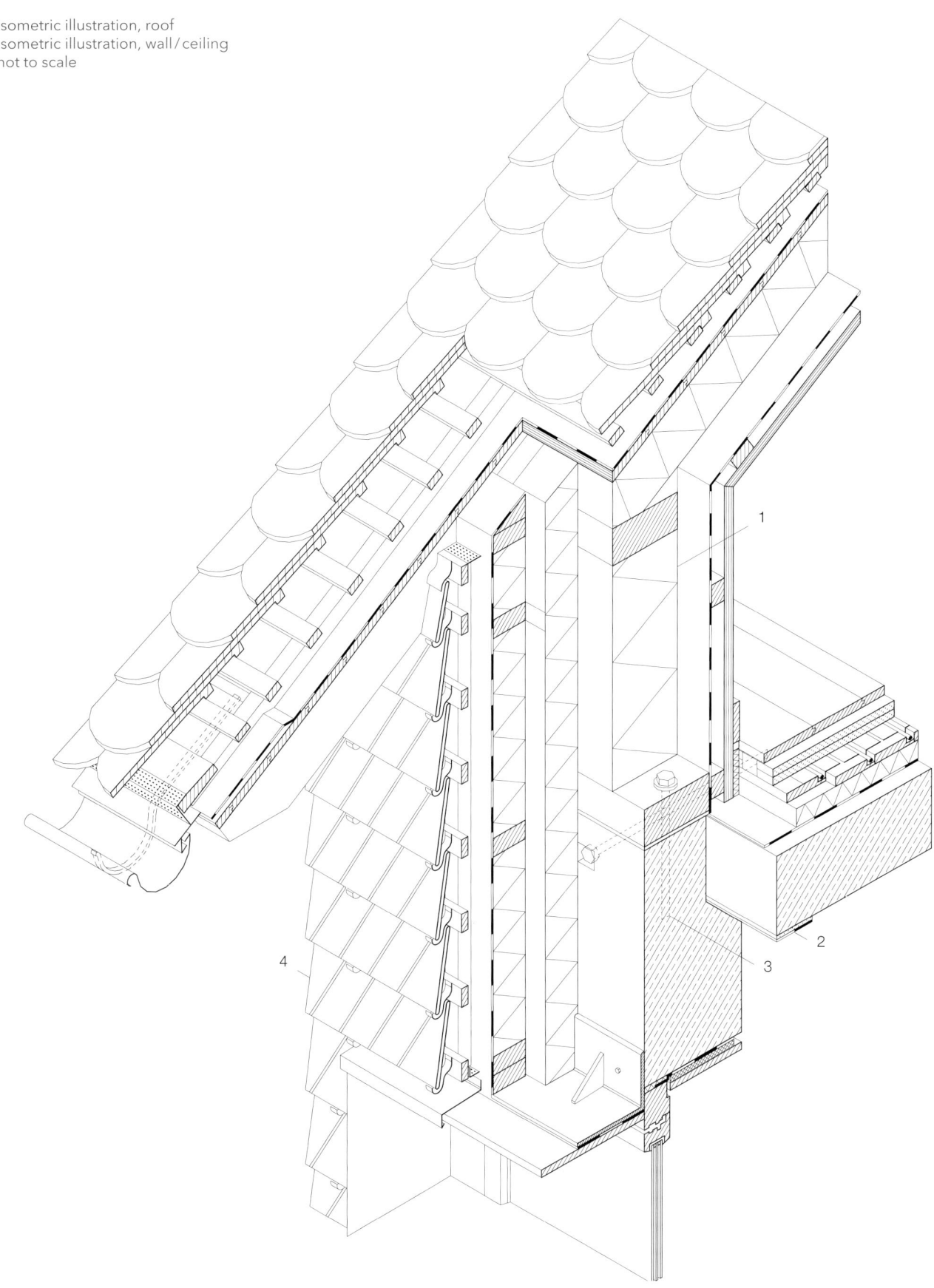

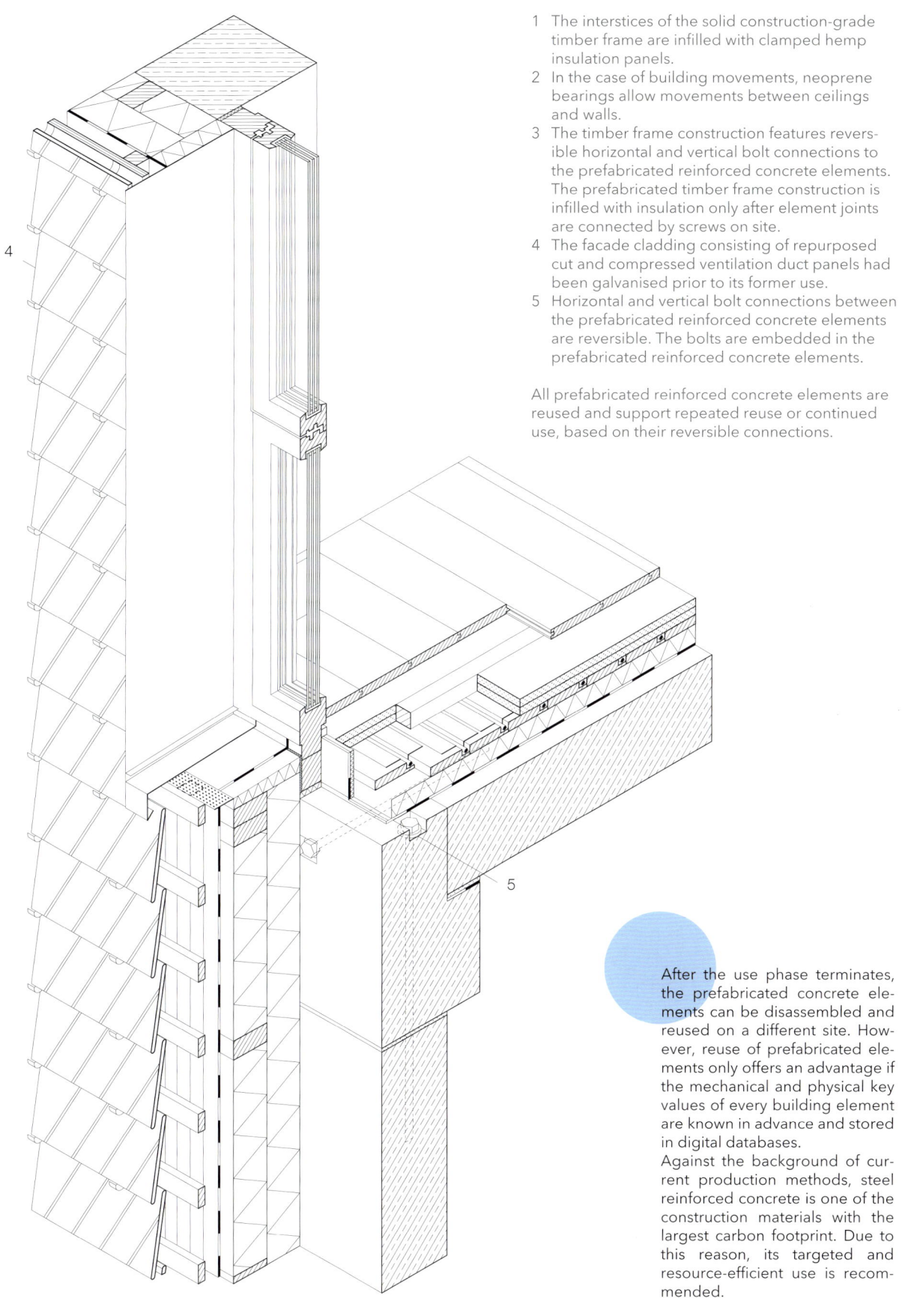

1. The interstices of the solid construction-grade timber frame are infilled with clamped hemp insulation panels.
2. In the case of building movements, neoprene bearings allow movements between ceilings and walls.
3. The timber frame construction features reversible horizontal and vertical bolt connections to the prefabricated reinforced concrete elements. The prefabricated timber frame construction is infilled with insulation only after element joints are connected by screws on site.
4. The facade cladding consisting of repurposed cut and compressed ventilation duct panels had been galvanised prior to its former use.
5. Horizontal and vertical bolt connections between the prefabricated reinforced concrete elements are reversible. The bolts are embedded in the prefabricated reinforced concrete elements.

All prefabricated reinforced concrete elements are reused and support repeated reuse or continued use, based on their reversible connections.

After the use phase terminates, the prefabricated concrete elements can be disassembled and reused on a different site. However, reuse of prefabricated elements only offers an advantage if the mechanical and physical key values of every building element are known in advance and stored in digital databases.

Against the background of current production methods, steel reinforced concrete is one of the construction materials with the largest carbon footprint. Due to this reason, its targeted and resource-efficient use is recommended.

Focus on Concrete

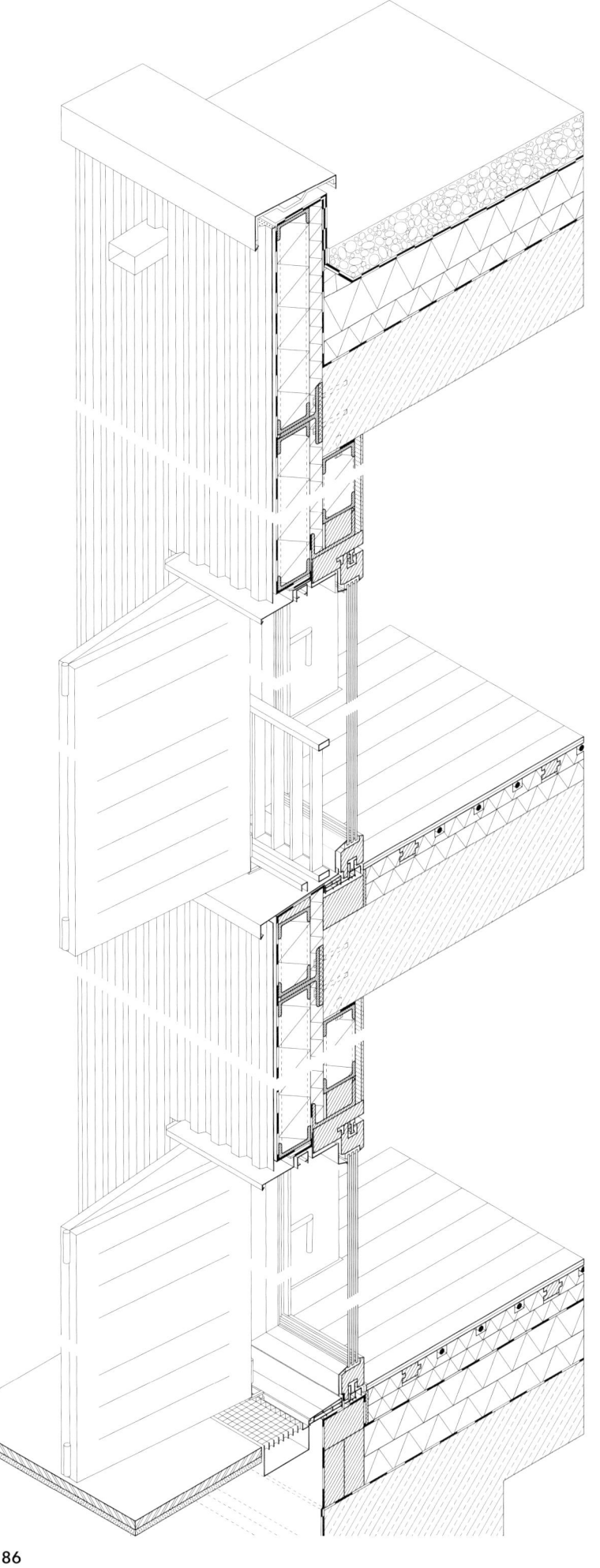

Concrete: Existing Conditions

Protect
parapet coping, gravel fill
corrugated sheet metal

Insulate
hemp infill insulation
foam glass panels
foam glass gravel

Support
reinforced concrete frame (existing)

Seal
EPDM sealant membrane
PE film

Cover
reclaimed wood floorboards
loam render on loam construction panel

Isometric illustration   scale 1:20
Vertical section   scale 1:20

1 Parapet coping, canted, clamped to anchor clip
   anchor clip, screw connections to top stud wall
   steel channel, corrosion-proof
2 Roof construction:
   Gravel fill
   2-ply EPDM sealant, membrane-to-membrane weld, attached to parapet, clamped by anchor clip
   foam glass panel insulation, min. 2 % to falls, loose layout
   PE film, diffusion-resistant, attached to parapet, clamped by anchor clip
   reinforced concrete slab (existing)
3 Steel T-section, corrosion-proof, as anchor for curtain wall construction with perforations to reduce thermal bridges, thermal separation, bolt connections to reinforced concrete slab (existing)
4 Steel folding / sliding shutter, corrosion proof
5 Sliding window:
   Triple glazing in wood frame, reused
   plastic dry sealant, sorted by type
6 Hand-crank-operated folding / sliding shutters
7 Steel RHS fall protection with flat steel balusters, corrosion-proof, vertical screw connections to reveal

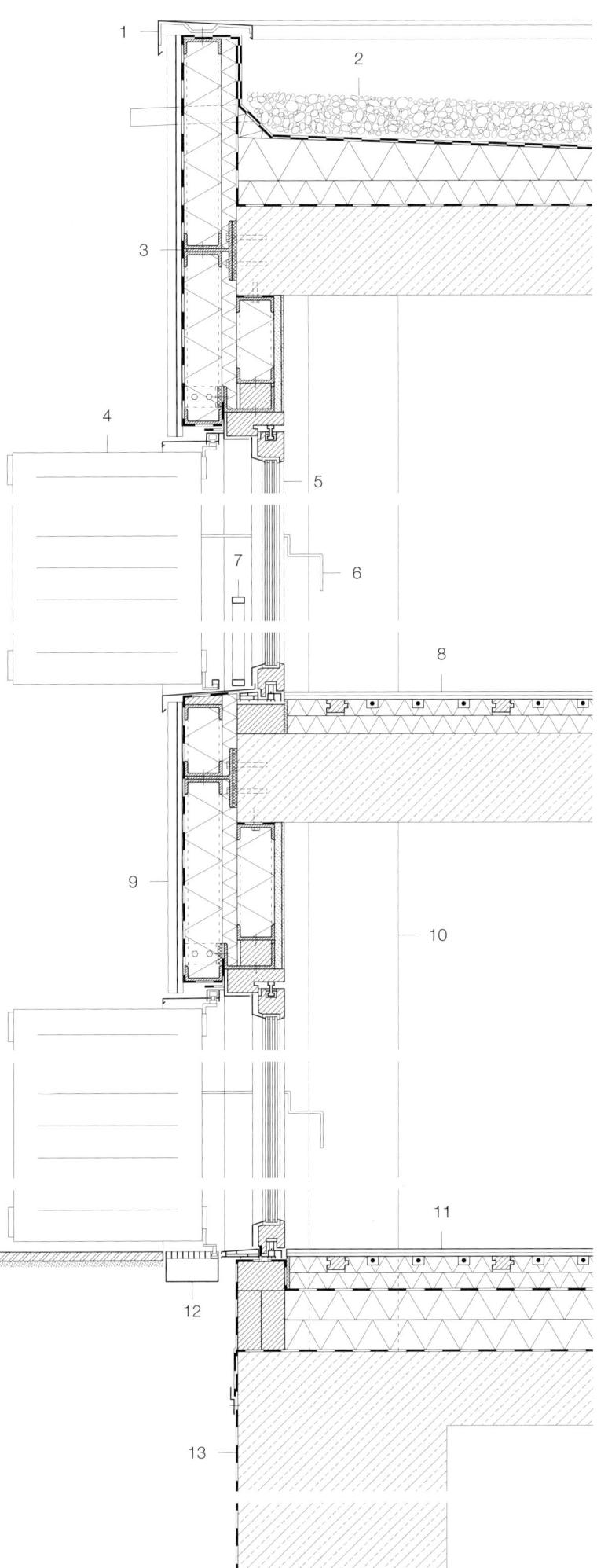

8 Floor construction:
Reclaimed wood floorboards, concealed tongue-and-groove screw connections
solid timber battens, screw connections
inlaid softwood fibre panel, lignin-bonded, as installation panel with grooves and heat diffusor plates to install underfloor heating pipes
softwood fibre panel levelling insulation, lignin-bonded
reinforced concrete ceiling (existing)

9 Exterior wall construction:
Corrugated sheet metal, galvanised, exposed screw connections
steel channel framing, corrosion-proof, bolt connections, thermal separation
sarking membrane, windproof, breathable, min. 30 cm overlap
stud wall construction:
Steel channel, corrosion-proof, with perforations to reduce thermal bridges
hemp insulation infill
steel channel, corrosion-proof, with perforations to reduce thermal bridges
loam construction panel with double loam render layer, reinforcement, screw connections

10 Reinforced concrete columns (existing)

11 Floor construction:
Reclaimed wood floorboards, concealed tongue-and-groove screw connections
solid timber battens, screw connections
inlaid softwood fibre panel, lignin-bonded, as installation panel with grooves and heat diffusor plates to install underfloor heating pipes
softwood fibre panel levelling insulation, lignin-bonded
PE film, diffusion-resistant, min. 30 cm overlap
double foam glass panel insulation
EPDM sealant against ground moisture, membrane-to-membrane weld, min. 30 cm overlap, loose layout
reinforced concrete slab (existing)

12 Stainless steel channel drain

13 EPDM sealant against ground moisture, membrane-to-membrane weld, clamped
reinforced concrete foundation (existing)

Isometric illustration, roof
Isometric illustration, plinth/floor
not to scale

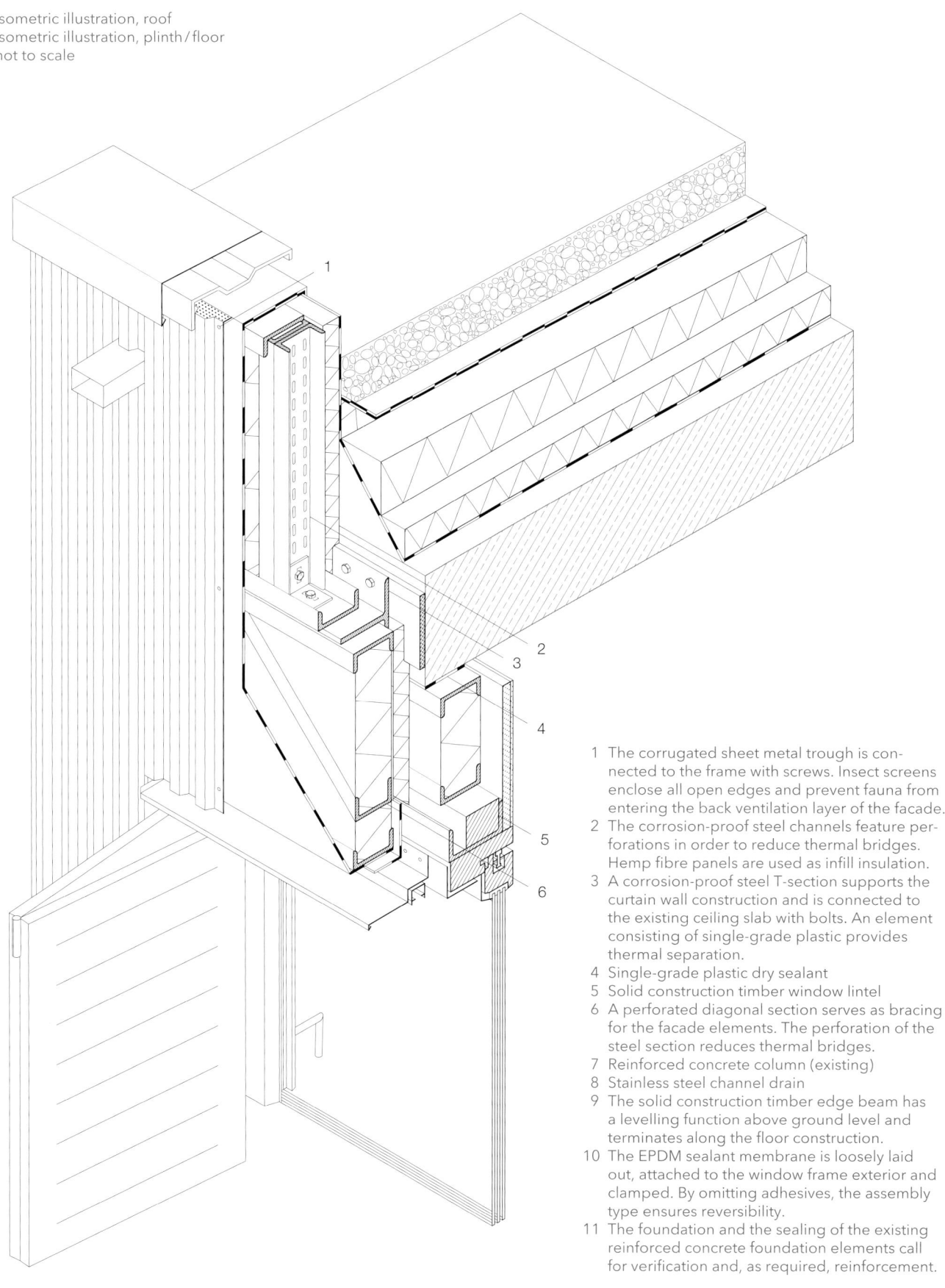

1. The corrugated sheet metal trough is connected to the frame with screws. Insect screens enclose all open edges and prevent fauna from entering the back ventilation layer of the facade.
2. The corrosion-proof steel channels feature perforations in order to reduce thermal bridges. Hemp fibre panels are used as infill insulation.
3. A corrosion-proof steel T-section supports the curtain wall construction and is connected to the existing ceiling slab with bolts. An element consisting of single-grade plastic provides thermal separation.
4. Single-grade plastic dry sealant
5. Solid construction timber window lintel
6. A perforated diagonal section serves as bracing for the facade elements. The perforation of the steel section reduces thermal bridges.
7. Reinforced concrete column (existing)
8. Stainless steel channel drain
9. The solid construction timber edge beam has a levelling function above ground level and terminates along the floor construction.
10. The EPDM sealant membrane is loosely laid out, attached to the window frame exterior and clamped. By omitting adhesives, the assembly type ensures reversibility.
11. The foundation and the sealing of the existing reinforced concrete foundation elements call for verification and, as required, reinforcement.

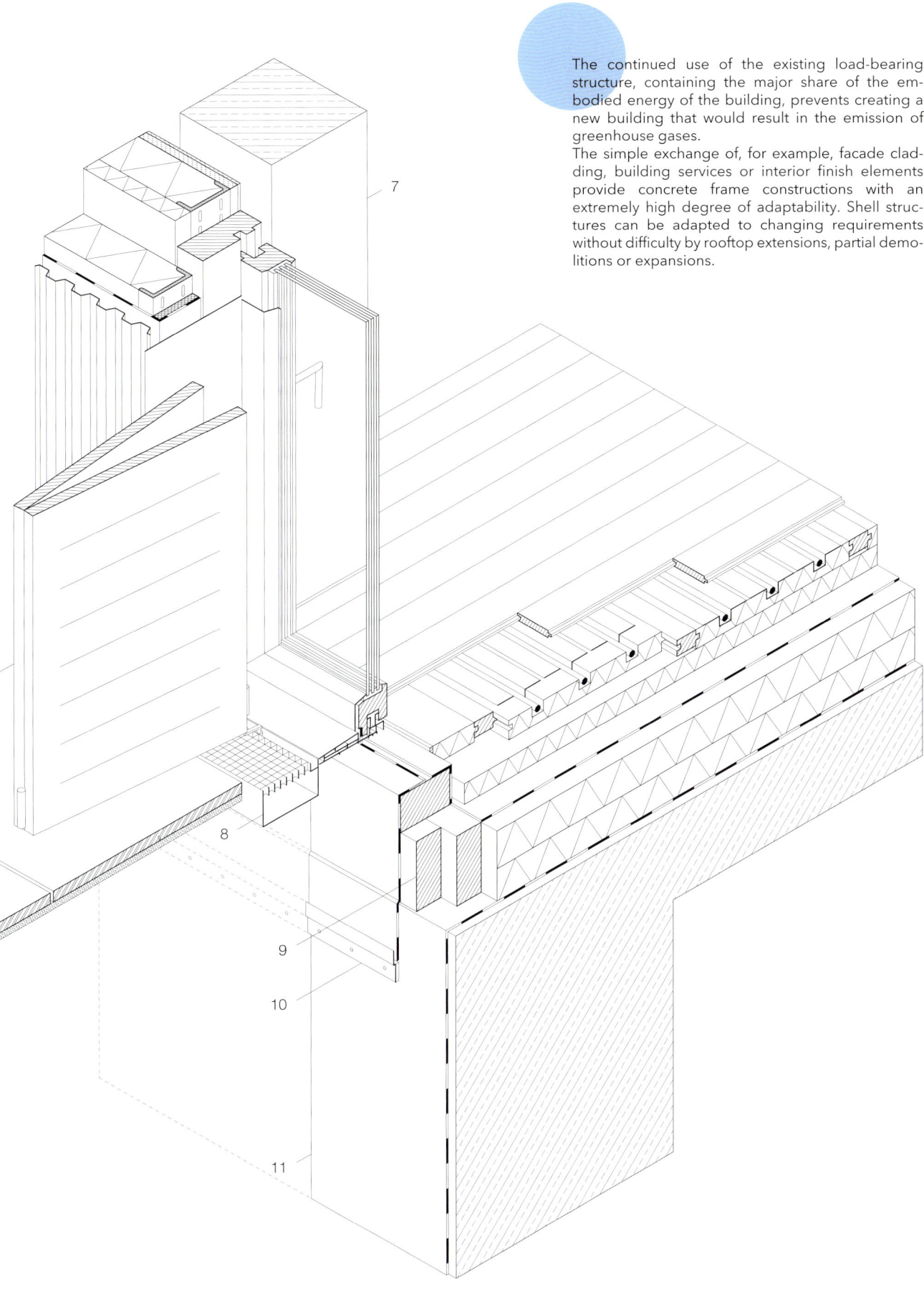

The continued use of the existing load-bearing structure, containing the major share of the embodied energy of the building, prevents creating a new building that would result in the emission of greenhouse gases.
The simple exchange of, for example, facade cladding, building services or interior finish elements provide concrete frame constructions with an extremely high degree of adaptability. Shell structures can be adapted to changing requirements without difficulty by rooftop extensions, partial demolitions or expansions.

Focus on Concrete

Focus on Steel

Steel is an alloy consisting of iron and carbon with a share of the latter of below 2%. By adding other metals such as silicon, manganese, molybdenum, chromium, nickel or copper, alloyed steel can be created with specific characteristics that are determined by the particular material composition [1]. For construction purposes, most of all non-alloyed or low-alloyed steel types (mild steel) find use and, to a lesser degree, high-alloy (stainless) steel types [2].

The production of steel follows either the blast furnace route or the electric arc furnace route. The first method comprises smelting iron ore, additives such as coke and reducing agents such as coal, oil or gas in order to produce molten pig iron in a blast furnace. In the following step, the pig iron is transformed into so-called oxygen steel in an oxygen converter. This process allows using up to 20% of scrap steel for steel production. The electric arc furnace route uses an electric arc furnace to melt scrap steel and produce crude steel. This process is suited to using exclusively scrap steel as the basic material. It is also significantly less energy-intensive than the blast furnace route. This is why the share of electric steel in the overall production of steel has steadily increased in the past years [3]. In the building sector, at about 84%, steel is the most frequently used metal construction material. Other metals such as aluminium, copper or zinc reach a combined share of less than 16% [4].

Raw material deposits
As a construction material defined by its industrial production method, steel is most of all available in the form of semi-finished products and rolled steel products, such as sheet metal, sections or cables. Construction-grade steel sections are available on the market in standardised forms as bars, structural sections or hollow sections at different quality grades [5]. Construction-grade steel finds use as reinforcement for concrete construction and as raw material for steel construction.

The most frequently used sections are so-called I-beams (rolled steel joists, universal beams or wide flange steel beams as per DIN 1025), as well as channels, angles, Z-shaped or T-shaped sections. Beyond that, many kinds of hollow or smaller solid circular or rectangular cross sections are available on the market (Fig. 1).

Sheet metal commonly refers to steel with a thickness of less than 5 mm. Processed into semi-finished products, sheet metal exists in various profiles. The most widely known are corrugated sheet metal types that are highly dimensionally stable despite their thinness, due to their three-dimensional profiling. Perforated sheet metal and expanded metal are special types: Perforated sheet metal is perforated by punching holes, yet remains planar. Expanded metal is pulled apart following the mechanical punching process. As a result, sheet metal assumes a three-dimensional grid mesh form with rhomboid perforations, leading to a high degree of surface stability as well as light and air permeability. For exterior applications, special titanium-zinc sheet metal types are available. Beyond that, a diverse range of steel products and semi-finished products exists, such as steel cable and wire, grating, nets, mesh, panels or connectors.

Characteristics
Steel is characterised by its ductility, hardness and strength. It permits hot or cold forming, bending, rolling, drawing or forging. Its ability to absorb compressive, tensile, bending, shear and torsional forces makes it the highest-performing common building material [6]. Steel is also very durable and easy to maintain. Unlike many other construction materials,

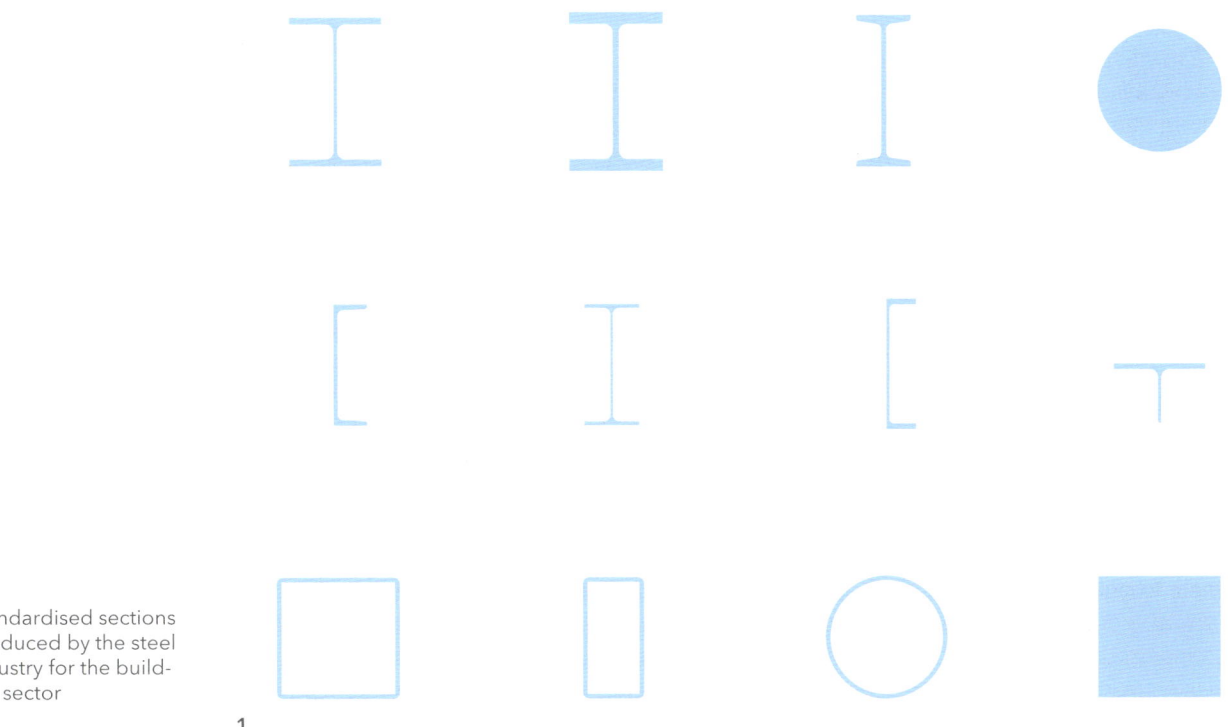

1
Standardised sections produced by the steel industry for the building sector

it can be reclaimed sorted by type during demolition work and its material value can be retained or even increased. Steel displays very good thermal conductivity. Thus, steel construction elements that penetrate the thermal building envelope should be avoided or very diligently planned.

In building construction terms, the disadvantages of steel include its behaviour in the case of fire and its tendency to corrode. Steel construction components are non-combustible, but they lose their ductility, hardness and strength when subjected to very high temperatures. This also results in the loss of load-bearing capacity following relatively short exposure to fire [7]. Depending on application type, additional fireproofing measures often become necessary, such as in the form of coatings or encapsulation. They tend to counteract the positive characteristics of the material when sorting by type. Above a humidity of about 65–70% [8] construction grade steel begins to corrode. (Hot dip) galvanisation increases corrosion resistance.

Stainless steel is a possible alternative: It describes steel with a chromium content of more than 12%. This also includes chromium-nickel steel or chromium-nickel-molybdenum steel of the V2A or V4A type. They develop a chromium-oxide film on their surface, acting as a protective layer against rust [9].

Due to its high cost, stainless steel is used only very sparingly for building construction purposes. For corrosion proofing sorted by type and most of all for exterior applications, so-called weathering steel types exist, for instance, the steel commonly known under the brand name corten. When exposing this particular type of steel to the impacts of the weather, it soon develops a dense protective layer resembling rust, due to its alloy composition containing a higher share of copper and phosphorus. The layer protects the material from further corrosion [10].

Black steel finds use particularly for interior applications. This non-alloyed steel type comprises an oxide coating that develops during production. Black steel can be protected against corrosion by a thin, reversible oil layer.

Aside from corrosion effects related to weather and humidity, the direct contact between metal types with different electrical potential leads to risks of so-called galvanic corrosion. In such cases, metal parts begin to corrode when they come into contact [11]. For this reason, a diligently executed separation layer is required when integrating different metal types into a building.

Typical construction methods

Steel can be applied in a highly flexible manner based on very different semi-finished product types. In the form of corrugated sheet steel, the material can be used for secondary structures such as ceilings, or as steel cable tension bracing for load-bearing constructions. Steel elements can also serve to clad the exterior of a building or to create gutters and leaders for roof drainage. Further, steel can be manufactured into window frames, door elements and many other components. Due to the broad range of standardised linear construction elements, steel is most of all suited to creating filigree constructions, such as frame structures for buildings. Screws, bolts, rivets or welded connections can combine individual parts into very large, high-performance load-bearing structures. Many bridges and halls consist of steel. However, the use of steel load-bearing structures for building types with limited spans is rare, primarily for economic reasons. Early steel and iron constructions were mostly assembled using riveted connections, a fact that can still be seen along older bridges. Steel building elements today are more commonly bolted or welded, depending on application type.

Rivets, in physical terms, serve to create a form-fit connection between two building elements by inserting a rivet into a hole and fixing it in place, thus preventing the building elements from moving apart. The contact pressure exerted by the rivet on the construction element results in a force-fit connection. Screws and bolts also establish force-fit connections by fixing in place multiple perforated construction elements, similar to rivets. Typically, bolts are fastened with locknuts. By tightening the connection, pressure exerted on the construction element is increased. The clamping effect results in a force-fit connection. Screws and bolts can also be inserted into pre-threaded or pre-tapped holes without the need for locknuts.

Welding creates material-locking connections between different steel construction elements. For this purpose, filler materials are applied directly to or between the construction elements to be connected, by use of a welding device under application of very high temperatures and by melting them. A connection to the molten material is created, which solidifies, forming a seam or spot weld. Screw connections enable good separability. Rivet connections can also be separated well, while requiring a certain effort and leading to a degree of loss. Welded connections, however, are permanent and cannot be separated without them being destroyed. Sheet steel and semi-finished steel products are suitable for different types of processing and reworking, including milling, cutting, drilling and lasering. As a result, the material can be used in a flexible manner and also offers an exceptionally broad range of design opportunities.

Circular potential

Metallic raw materials can remain in a technological cycle with almost no loss of quality [12]. In principle, this is also the case for steel. Unless steel elements have been coated with varnish or similar, steel is highly recyclable, without any substantial loss in quality.

Standardised semi-finished products, such as used for beams or columns, also permit reuse and repurposing very well. The opinion issued by Manfred Helmus and Anne Randel in the "Status Report on Steel Recycling in the Building Industry" indicates that about 88% of all steel components used in buildings are recycled and even up to 11% are reused as complete construction elements [13].

In the course of pre-processing for purposes of material recycling, scrap steel is first shredded and sorted according to steel quality. Then, melting in an electric arc furnace takes place once more. This process allows the alloy composition to be changed and, thus, also the steel grade as required. Melting in an electric arc furnace requires only 25% of the large amount of energy needed for steel production in a blast furnace. This still constitutes a significant expenditure of energy [14]. The reuse of steel building elements, thus, remains preferable to recycling.

Notes
[1] Moro 2021, p. 294
[2] Eggen/Sandaker 1996, p. 30–37
[3] Helmus/Randel 2014
[4] ibid., p. 1
[5] Hestermann/ Rongen 2015, p. 262
[6] ibid.
[7] Belz 1999, p. 54
[8] ibid., p. 55 and cf. note 2, p. 35
[9] cf. note. 2, p. 99 and note 5, p. 264
[10] ibid., p. 264
[11] ibid., p. 264
[12] Hillebrandt et al. 2018, p. 63
[13] cf. note. 3, p. 4
[14] Kuhnhenne 2018, p. 215

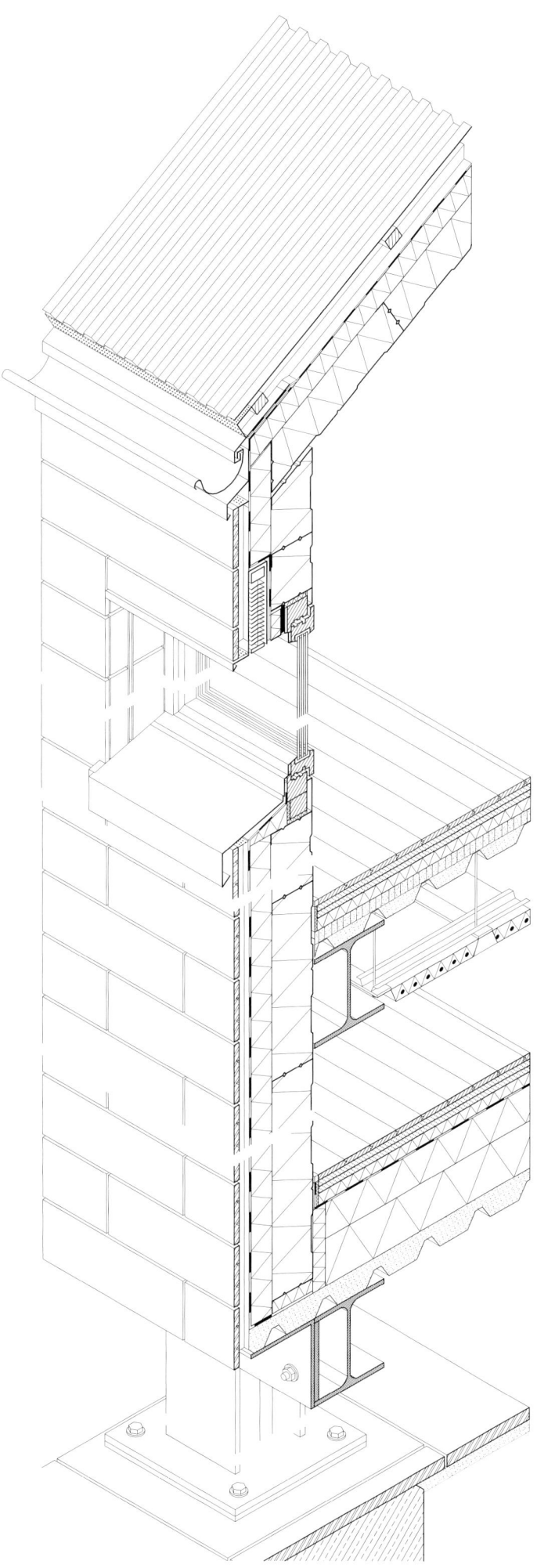

Steel: Frame Construction I

Protect
corrugated sheet steel, corrosion-proof
natural stone tile

Insulate
hemp insulation
foam glass panels

Support
wide flange steel beam, corrosion-proof
prefabricated reinforced concrete foundation element, reused

Seal
PE film
sarking membrane

Cover
sheet steel coffer system, corrosion-proof

Isometric illustration   scale 1:20
Vertical section   scale 1:20

1 Roof construction
  corrugated sheet steel, corrosion-proof, reversed, screw connections
  solid timber battens, screw connections
  counterbattens, screw connections/back ventilation
  sarking membrane, windproof, breathable, min. 30 cm overlap
  hemp insulation, anchored to coffer system with insulation dowels
  sheet steel coffer system with foam glass insulation, canted, corrosion-proof, screw connections to structural steel frame
  wide flange steel beam, corrosion-proof, interior
2 Sheet steel insect screen, corrosion-proof, screw connections to sheathing and counterbattens
3 Window: Triple glazing in wood/aluminium frame, reused
  plastic dry sealant, single grade
4 Steel bracket, corrosion proof, thermal separation, to support facade substructure
5 Floor construction:
  Reclaimed wood floorboards, concealed tongue-and-groove screw connections
  solid timber battens, screw connections
  inlaid double loam construction panel
  cork impact soundproofing/installation layer

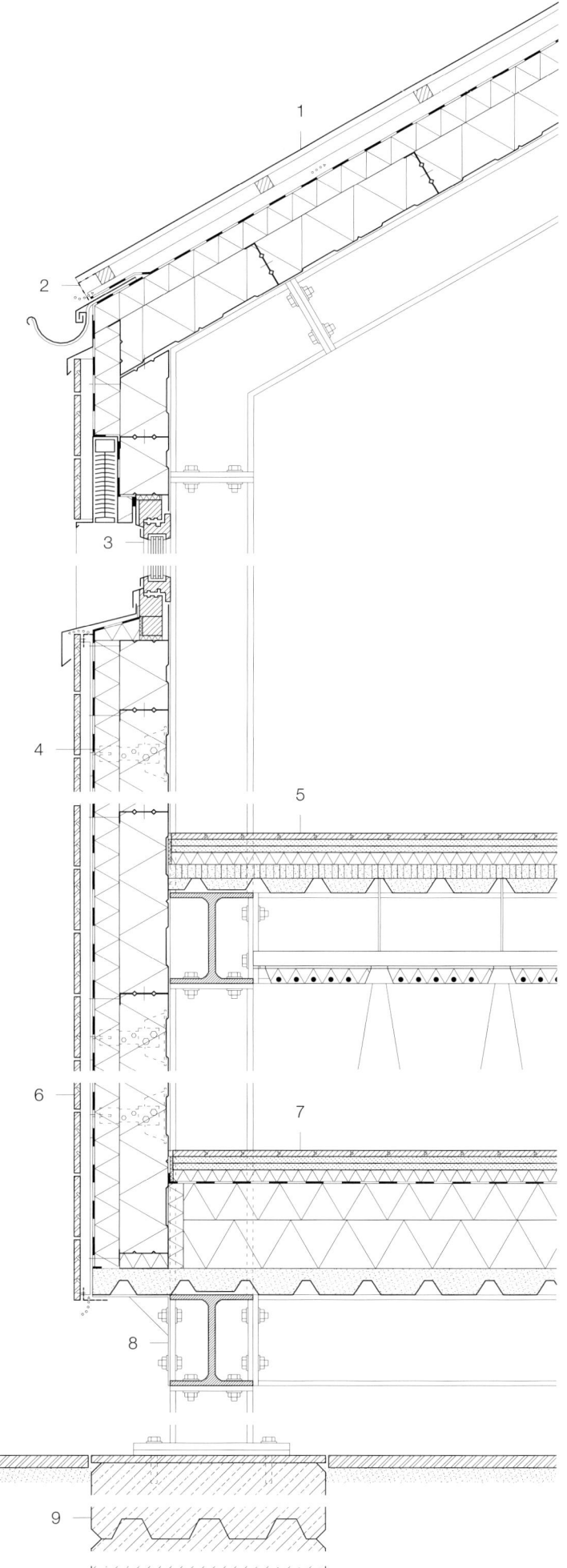

1 cardboard honeycomb panel
corrugated sheet steel, corrosion-proof, point-by-point screw connections to wide flange steel beam, corrosion-proof
mass infill, sand, to improve soundproofing, in honeycomb panels and between corrugated sheet metal ribs
wide flange steel beam, corrosion-proof, bolt connections to column
hung ceiling system:
Threaded steel bracket, screw connections
steel channel framing, screw connections
heating/cooling ceiling, steel panel with heating pipes, inlaid hemp insulation, integrated lighting, screw connections to framing
6 Exterior wall construction:
Natural stone tile, wild bond, interlocked
steel frame and clamp fasteners, corrosion-proof, screw connections penetrating insulation in sheet steel coffer system
sarking membrane, windproof, breathable, min. 30 cm overlap, clamped
hemp insulation, anchored to coffer system with insulation dowels
coffer system, canted sheet steel with foam glass fill, corrosion-proof, screw connections to steel frame
wide flange steel beam, corrosion-proof, interior
7 Floor construction:
Reclaimed wood floorboards, concealed tongue-and-groove screw connections
solid timber battens, screw connections
inlaid double loam construction panel
cork impact soundproofing / installation layer
PE film, diffusion-resistant, min. 30 cm overlap
double foam glass panel thermal insulation, loose layout
corrugated sheet steel, corrosion-proof, perforated, point-by-point screw connections to wide flange steel beam, corrosion-proof
mass fill, crushed rock, to improve soundproofing, loose fill between corrugated sheet metal ribs
wide flange steel beam, corrosion-proof, screw connections to column
8 Wide flange steel beam load-bearing structure, corrosion-proof, exterior, screw connections
thermal separation between interior and exterior structure
9 Prefabricated reinforced concrete foundation element, reused

Focus on Steel 195

Isometric illustration, roof
Isometric illustration, ceiling / wall
not to scale

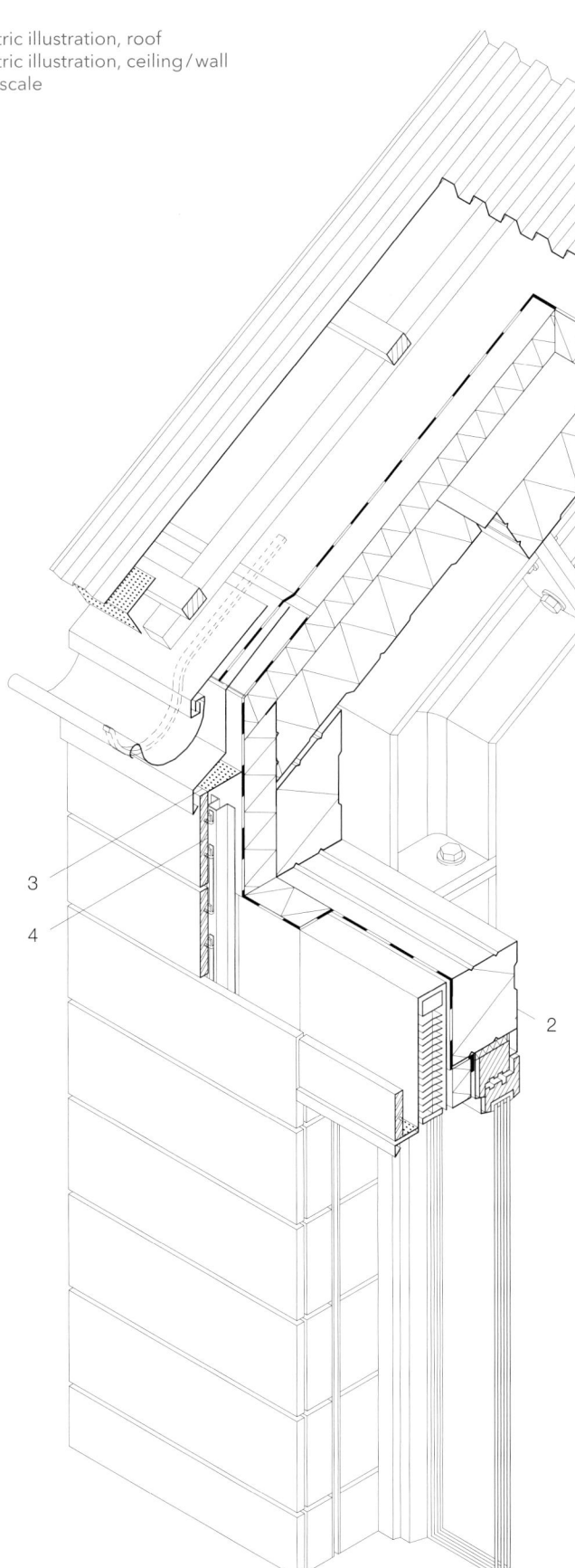

1   A continuous layer of insulation thermally separates the corrugated sheet metal facade surface and the corrosion-proof interior sheet steel coffer system.
2   The coffer system is predominantly assembled horizontally and reversibly. It serves as a load-bearing interior wall system component.
3   Perforated, corrosion-proof sheet steel elements cover all open edges and prevent insects from entering the facade back ventilation layer.
4   The facade structure consists of corrosion-proof steel channels that are connected to the coffer system elements by screws, penetrating their insulation layer. Hooks allow reversibly hanging facade panels into the structure.
5   Window sill with corrosion-proof sheet steel coping.
6   The sheet steel is connected to the wide flange steel beam structure by bolts. To improve soundproofing, a mass fill consisting of sand is introduced. Cardboard honeycomb panels laid out on top of the corrugated sheet metal secure the fill in place and simplify the infill process.
7   The hangers are connected to the corrugated sheet metal troughs by screws.
8   The ceiling system features steel panels including heating / cooling coils. The panels are infilled with loose organic hemp insulation. The ceiling system additionally contains integrated lighting.

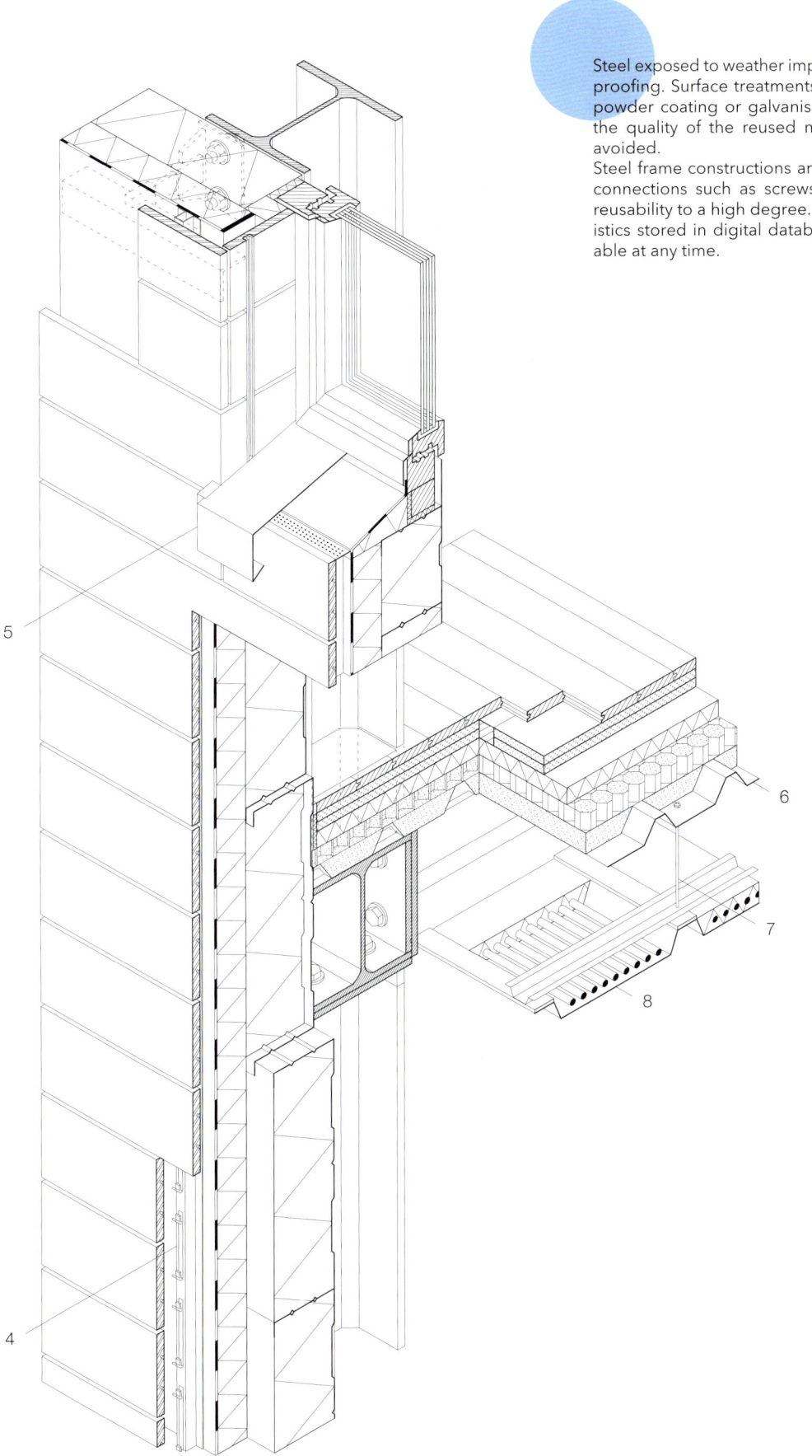

Steel exposed to weather impacts requires corrosion-proofing. Surface treatments such as paint finishes, powder coating or galvanisation, however, reduce the quality of the reused material and should be avoided.

Steel frame constructions are based on detachable connections such as screws and generally permit reusability to a high degree. Key material characteristics stored in digital databases should be retrievable at any time.

Focus on Steel

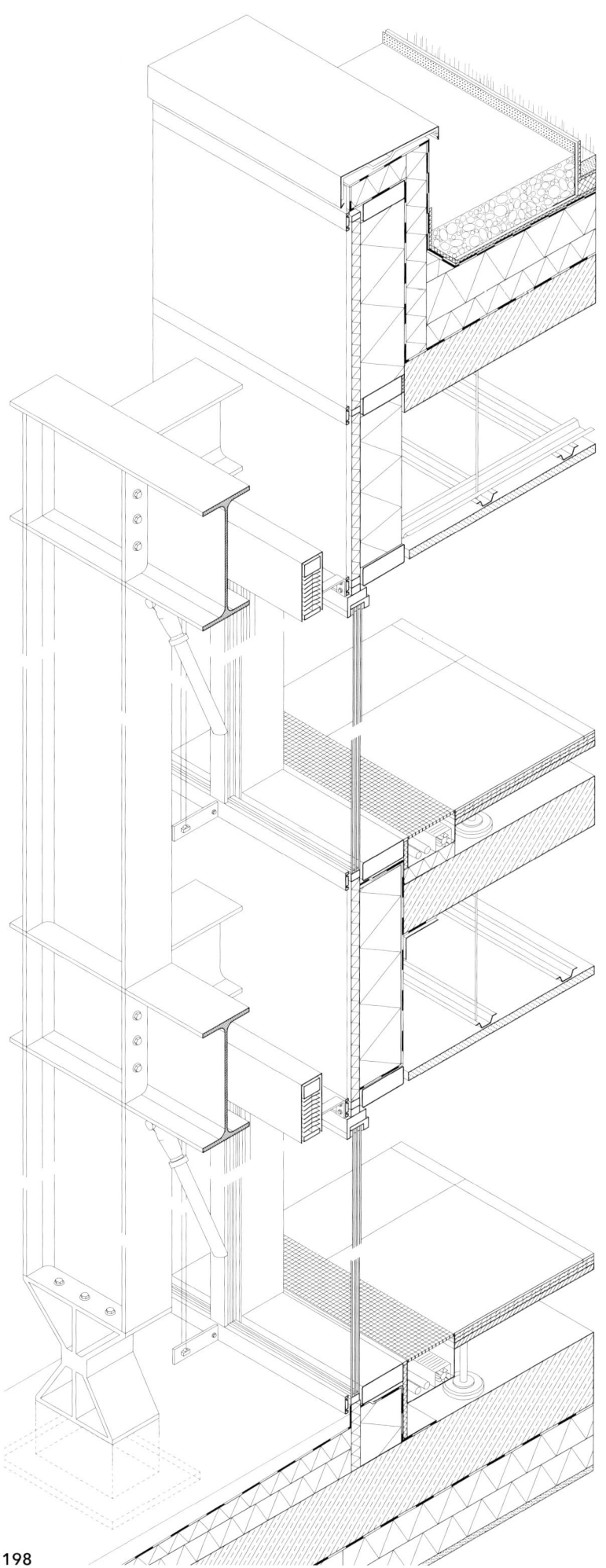

Steel: Frame Construction II

Protect
parapet coping, extensive greening
sheet steel coffer system element, galvanised
mullion transom structure

Insulate
foam glass panels, hemp fibre panels
mullion transom structure

Support
wide flange steel beam, corrosion-proof
prefabricated reinforced concrete foundation element

Seal
EPDM sealant, PE film
mullion transom structure

Cover
carpet tile, loam fine render

Isometric illustration   scale 1:20
Vertical section, facade   scale 1:20
Section, ceiling slab   scale 1:20

1 Roof edge:
 Parapet coping, canted, clamped to anchor clip
 anchor clip, screw connections to mullion transom structure
2 Roof construction:
 Extensive greening
 gravel perimeter strip
 vegetation layer, loose fill
 polypropylene filter fleece, loose layout
 HDPF or stainless steel drainage element
 single grade, loose layout, reused
 plastic storage protection mat, single-grade loose layout
 2-ply EPDM roof sealant, membrane-to-membrane weld, attached to parapet, clamped
 foam glass panel insulation, min 2 % to falls, loose layout
 foam glass panel insulation, loose layout in bond
 PE film, diffusion-resistant, min. 30 cm overlap, attached to parapet, clamped
 prefabricated reinforced concrete ceiling element
 wide flange steel beam, corrosion-proof,

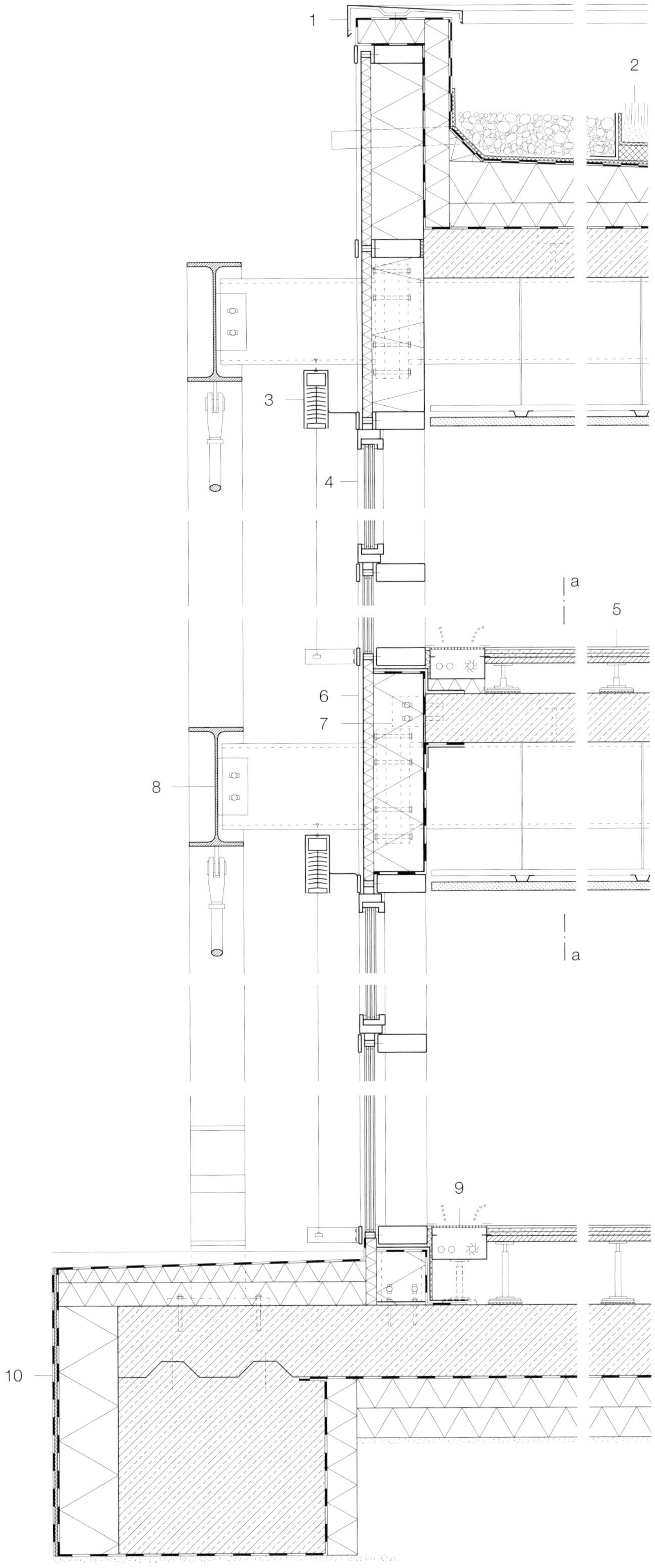

1 Roof construction:
   extensive green roof, loose layout
   filter fleece, loose layout
   drainage layer, loose layout
   protective fleece, loose layout
   EPDM sealant, membrane-to-membrane weld, loose layout
   foam glass panel insulation, loose layout
   vapour barrier, PE film, min. 30 cm overlap, loose layout
   prefabricated reinforced concrete ceiling element, reused
   wide flange steel beam, corrosion-proof, welded pre-threaded element, single grade plastic casing, thermal separation
2 Parapet construction:
   sheet steel cap, canted, corrosion-proof, screw connections
   EPDM sealant, membrane-to-membrane weld, clamped
   gravel, loose layout
   root-resistant protection fleece, loose layout
   EPDM sealant against humidity, membrane-to-membrane weld, loose layout
   wood framing, screw connections
   foam glass panel insulation, loose layout
   vapour barrier, PE film, min. 30 cm overlap, loose layout
   sheet steel parapet element, canted, corrosion-proof, screw connections to steel beam
   wide flange steel beam, corrosion-proof, welded pre-threaded element, thermal separation
   hung ceiling system:
   Sheet steel mounting channel, corrosion-proof, screw connections
   loam construction panel with loam render, screw connections
3 Sun protection, venetian blind
   screw connections to steel channel and mullion transom structure
4 Window:
   Triple glazing set into mullion transom structure, aluminium pressure bar and cap, reused
   plastic dry sealant, single grade
5 Floor construction:
   Carpet tile flooring, loose layout
   dry screed, double gypsum fibre panels, offset arrangement, screw connections
   sheathing, wood framing, height-adjustable steel levelling pedestals
   plastic bearing, single grade
   prefabricated reinforced concrete ceiling element, screw connections, reused
   wide flange steel beam, corrosion-proof, welded pre-threaded element, single grade plastic casing, thermal separation

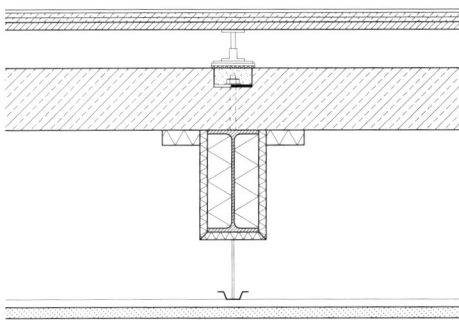

aa

   hung ceiling system:
   Sheet steel mounting channel, corrosion-proof, screw connections
   loam construction panel, double loam render layer, reinforcement, screw connections
6 Exterior wall construction:
   Sheet steel coffer system element, canted, galvanised
   inlaid double hemp insulation panel, coffer framing set between mullions, clamped
   PE film, diffusion-resistant, min. 30 cm overlap, clamped
7 Steel bracket support for mullion transom facade, bolt connections to ceiling slab
8 Wide flange steel beam, corrosion-proof, exterior
9 Underfloor convective heating system with access hatch
10 EPDM sealant against ground moisture, membrane-to-membrane weld, clamped by mullion cap
   dimpled sheet, min. 30 cm overlap
   EPDM sealant against ground moisture, membrane-to-membrane weld, loose layout
   foam glass panel perimeter insulation, loose layout
   prefabricated reinforced concrete floor slab element, reused

Focus on Steel

Isometric illustration / roof
Isometric illustration, plinth / floor
not to scale

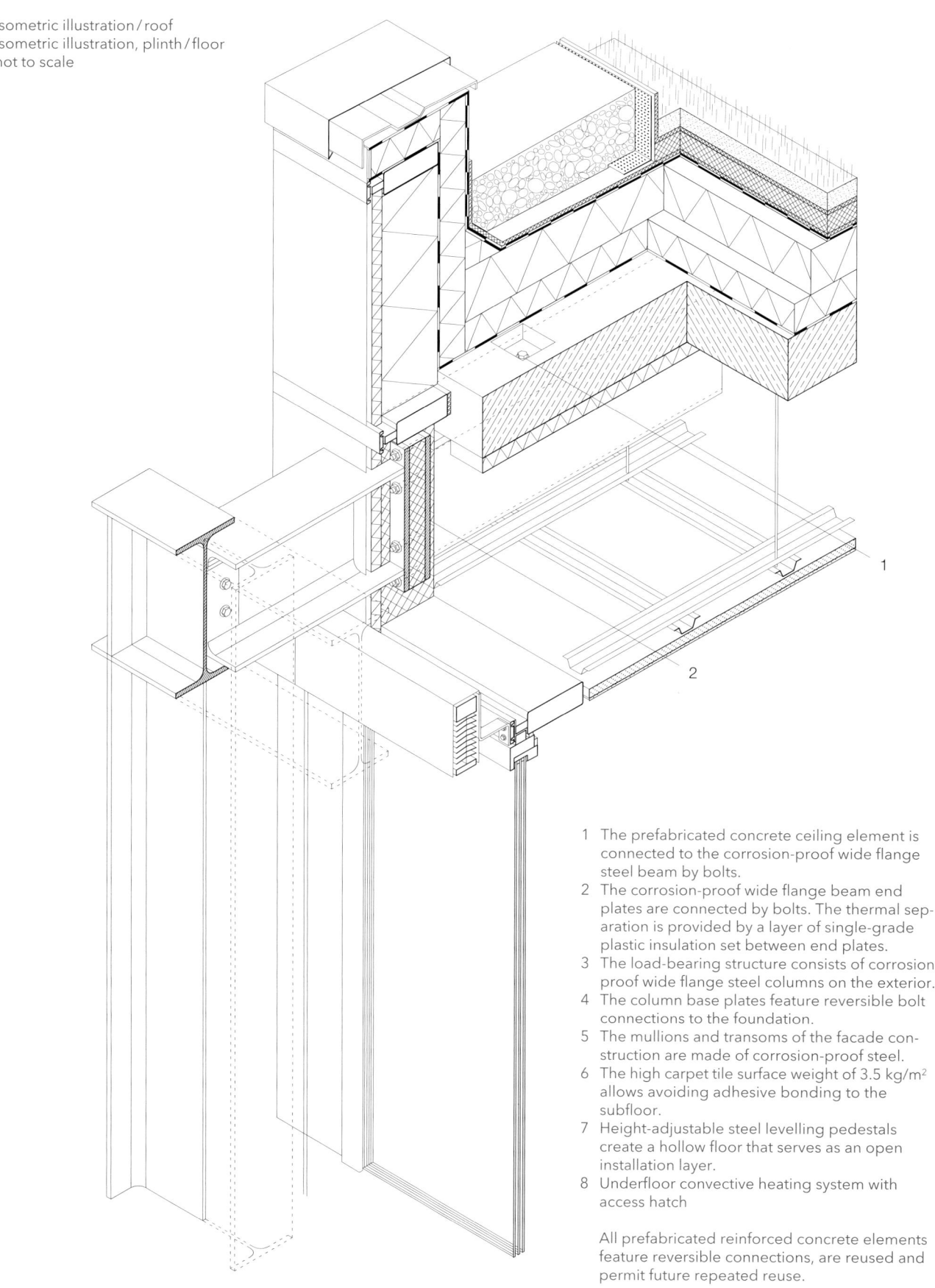

1. The prefabricated concrete ceiling element is connected to the corrosion-proof wide flange steel beam by bolts.
2. The corrosion-proof wide flange beam end plates are connected by bolts. The thermal separation is provided by a layer of single-grade plastic insulation set between end plates.
3. The load-bearing structure consists of corrosion-proof wide flange steel columns on the exterior.
4. The column base plates feature reversible bolt connections to the foundation.
5. The mullions and transoms of the facade construction are made of corrosion-proof steel.
6. The high carpet tile surface weight of 3.5 kg/m² allows avoiding adhesive bonding to the subfloor.
7. Height-adjustable steel levelling pedestals create a hollow floor that serves as an open installation layer.
8. Underfloor convective heating system with access hatch

All prefabricated reinforced concrete elements feature reversible connections, are reused and permit future repeated reuse.

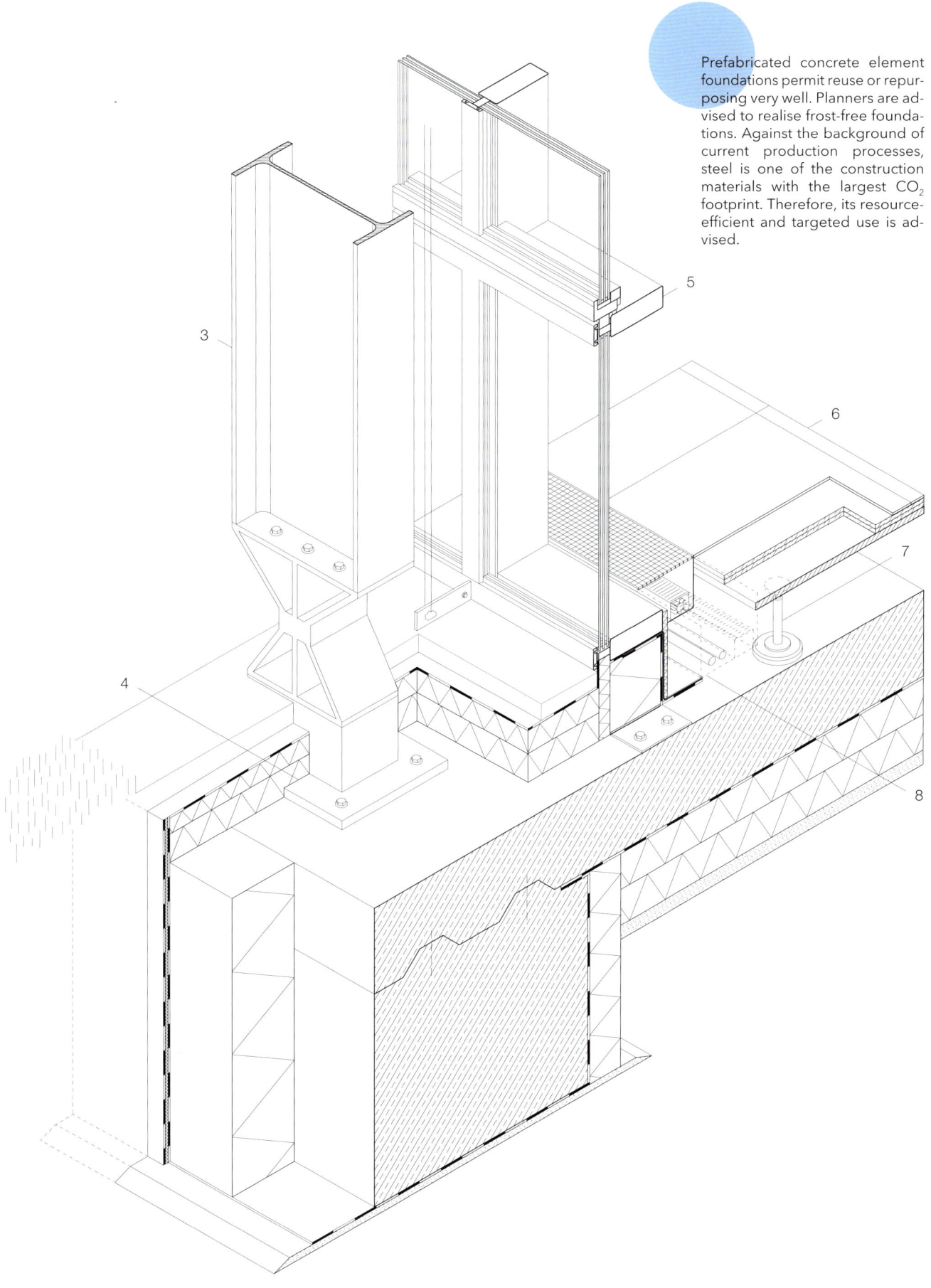

Prefabricated concrete element foundations permit reuse or repurposing very well. Planners are advised to realise frost-free foundations. Against the background of current production processes, steel is one of the construction materials with the largest CO_2 footprint. Therefore, its resource-efficient and targeted use is advised.

Focus on Steel

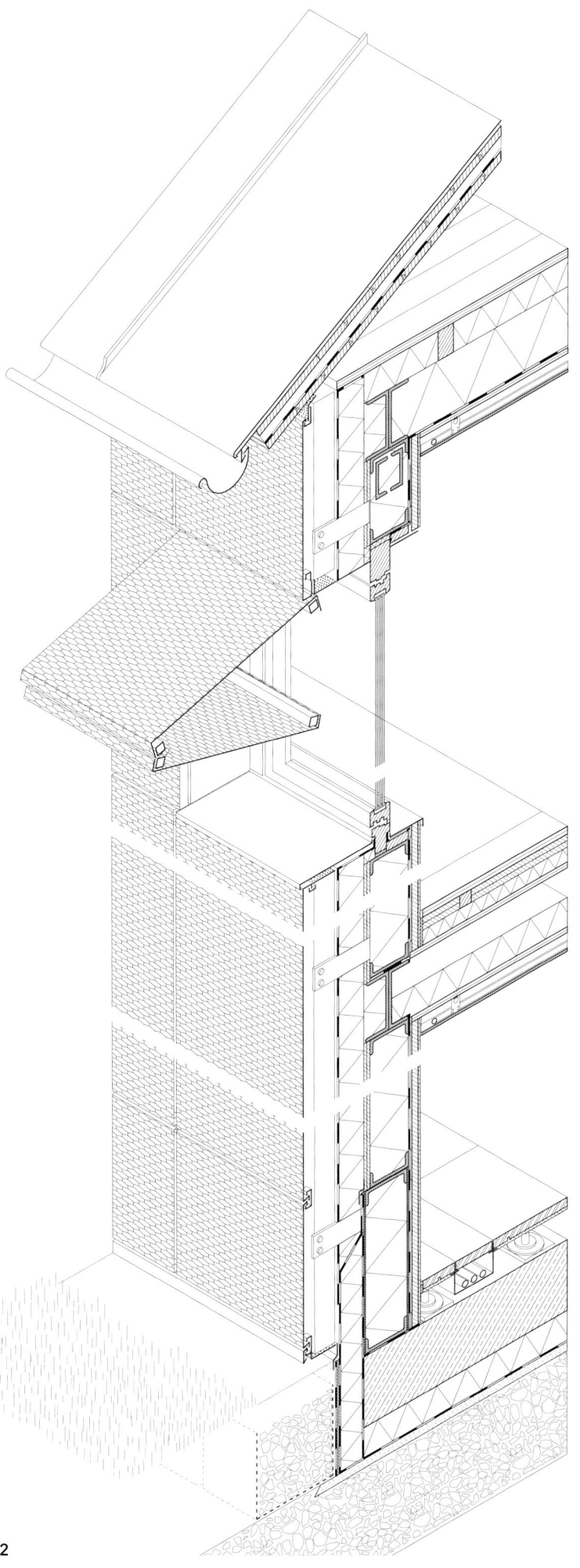

Steel: Lightweight Construction

Protect
standing seam roofing
expanded metal panels

Insulate
hemp insulation, reed cane insulation
foam glass panels

Support
lightweight steel frame wall and ceiling
prefabricated reinforced concrete floor slab element

Seal
EPDM sealant membrane
PE film, sarking membrane

Cover
reclaimed wood floorboards
natural stone tile
loam render on loam construction panel
heating/cooling ceiling

Isometric illustration   scale 1:20
Vertical section, facade   scale 1:20
Section, ceiling slab   scale 1:20

1 Cold roof construction:
   Standing seam roofing, clamped to anchor clip
   solid timber sheathing, screw connections
   min. 40 mm counterbattens, screw connections/
   back ventilation
   sarking membrane, windproof, diffusion-open,
   min. 30 cm overlap
   solid timber panel, diagonal sheathing with
   dovetail joints, screw connections
   steel channel rafter, corrosion-proof, bolt
   connections to bracket and lightweight steel
   wall construction
2 Ceiling construction beneath cold roof:
   Reclaimed wood floorboards, concealed
   tongue-and-groove screw connections
   solid timber sheathing, screw connections
   inlaid reed cane insulation
   2× steel channel primary beam, bolt connections
   to load-bearing steel stud wall construction
   steel channel secondary beam, set into primary
   beam, bolt connections
   inlaid reed cane insulation
   PE film, diffusion-resistant, min. 30 cm overlap,
   clamped
   solid timber sheathing, screw connections

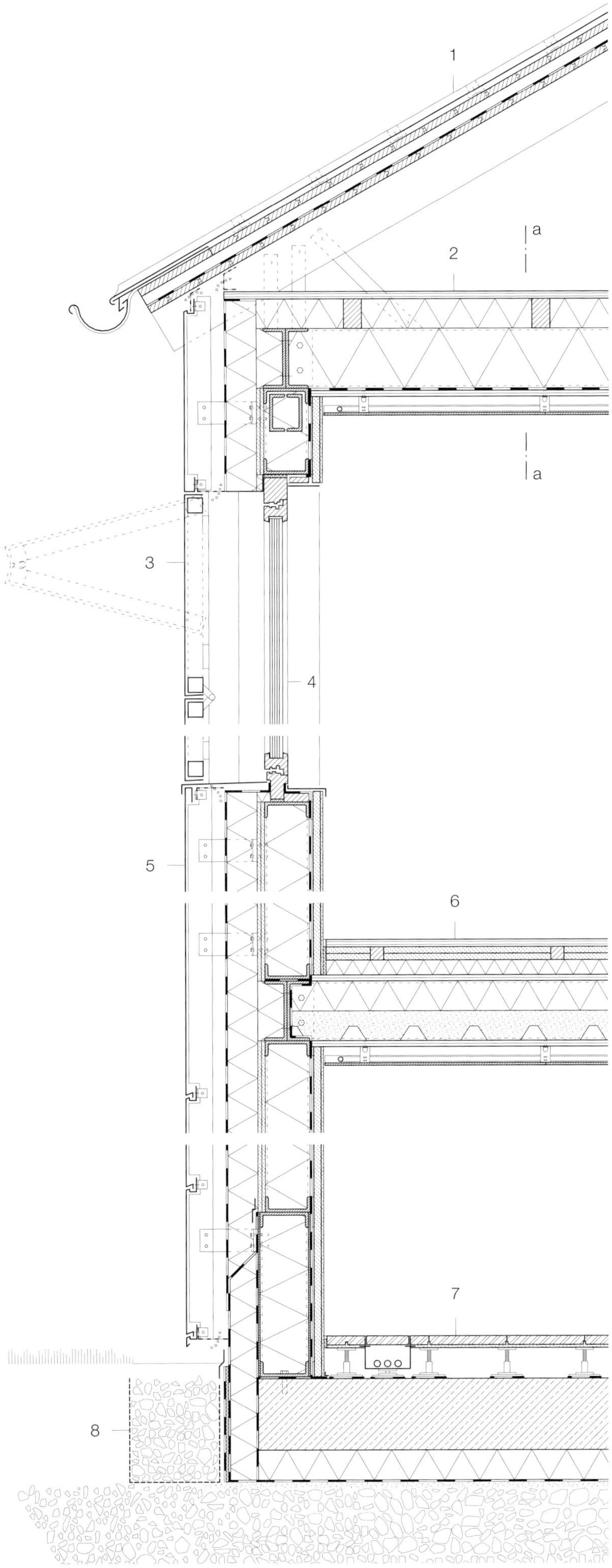

hung ceiling system:
Steel channel framing, screw connections
corrosion-proof steel ceiling system with
integrated heating / cooling coils
3 lifting folding shutter: Steel RHS, corrosion-proof, rail mounted, with gas spring
expanded metal cladding

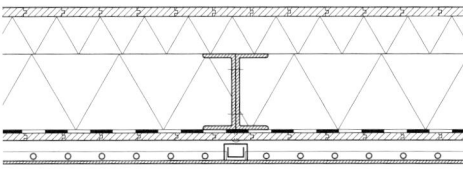

aa

4 Window: Triple glazing in wood frame, reused
plastic dry sealant, single grade
5 Expanded metal panel, concealed screw connections to steel angle framing, corrosion-proof
sarking membrane, single grade
hemp insulation, anchored to load-bearing steel
stud wall construction with insulation dowels
double gypsum fibre panel, screw connections
lightweight steel channel load-bearing stud wall
construction, corrosion-proof
inlaid hemp insulation, PE film, diffusion-resistant,
min. 30 cm overlap, clamped
loam construction panel with double loam render
layer, reinforcement, screw connections
6 Reclaimed wood floorboards, concealed tongue-and-groove screw connections
solid timber battens, screw connections
inlaid double loam construction panel
cork impact soundproofing / installation layer
solid timber sheathing, screw connections to
primary beam
2× steel channel primary beam, screw connections to steel stud wall construction
steel channel secondary beam, set into primary
beam, screw connections, reed cane insulation
mass fill, sand, to improve soundproofing, infilled
between corrugated sheet metal ribs
solid timber sheathing, screw connections
hung ceiling system:
Steel channel framing, screw connections
ceiling system with integrated heating / cooling
coils, steel, corrosion-proof
7 Natural stone tile flooring
height-adjustable steel levelling pedestals
plastic bearing, single grade
prefabricated reinforced concrete floor slab
element, reused
foam glass panel perimeter insulation, loose layout
EPDM sealant layer against ground moisture,
membrane-to-membrane weld, loose layout
sand base layer
foam glass gravel fill perimeter insulation
8 Foam glass gravel perimeter insulation in textile
fabric bags as construction support, mechanical
connections to reinforced concrete
EPDM sealant against ground moisture,
membrane-to-membrane weld, clamped by
mounting track
dimpled sheet, min. 30 cm overlap
EPDM sealant layer, membrane-to-membrane
weld, attached to foam glass panel
prefabricated reinforced concrete floor slab
element, reused

Focus on Steel

Isometric illustration, roof
Isometric illustration, plinth / floor
not to scale

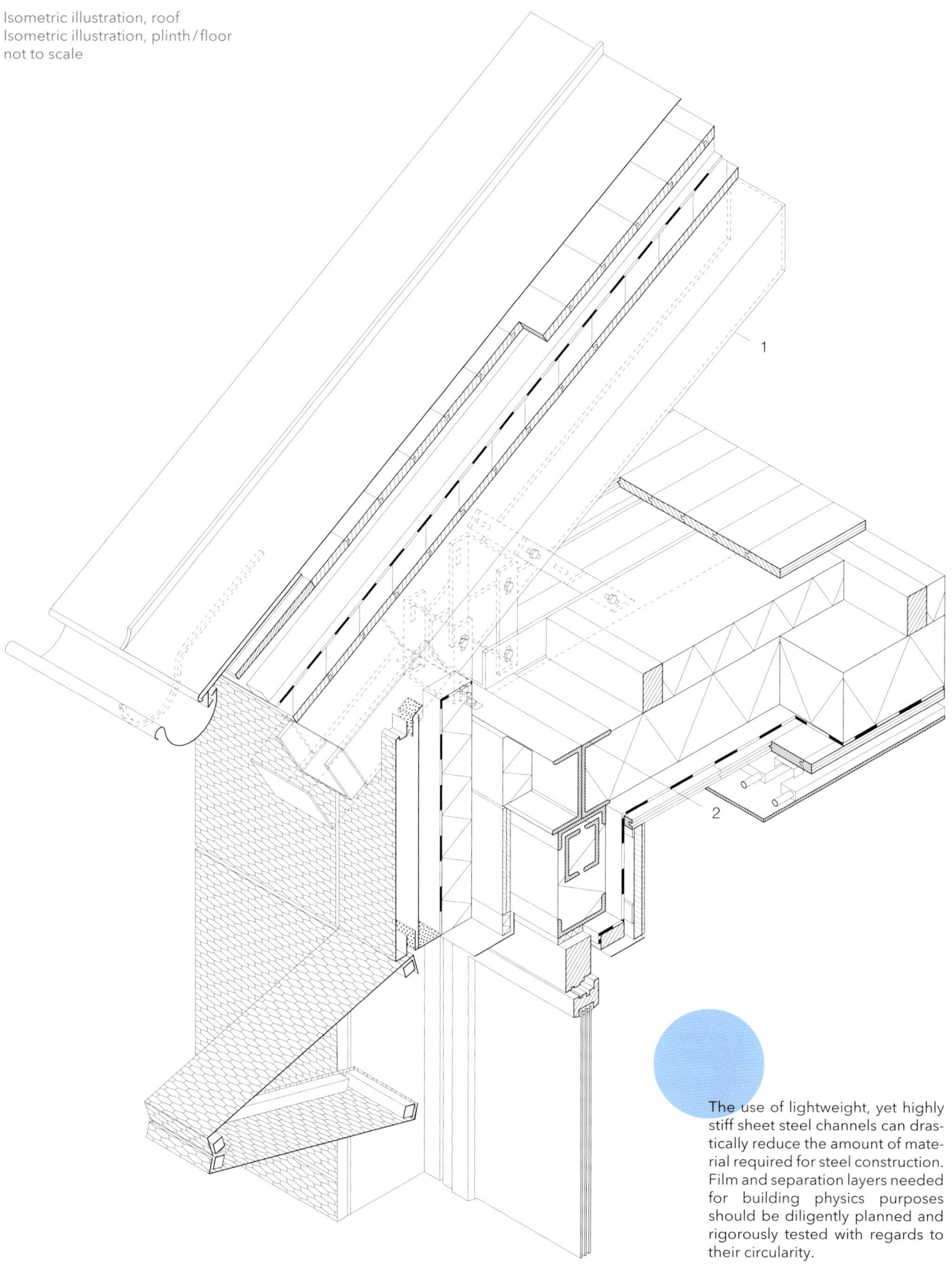

The use of lightweight, yet highly stiff sheet steel channels can drastically reduce the amount of material required for steel construction. Film and separation layers needed for building physics purposes should be diligently planned and rigorously tested with regards to their circularity.

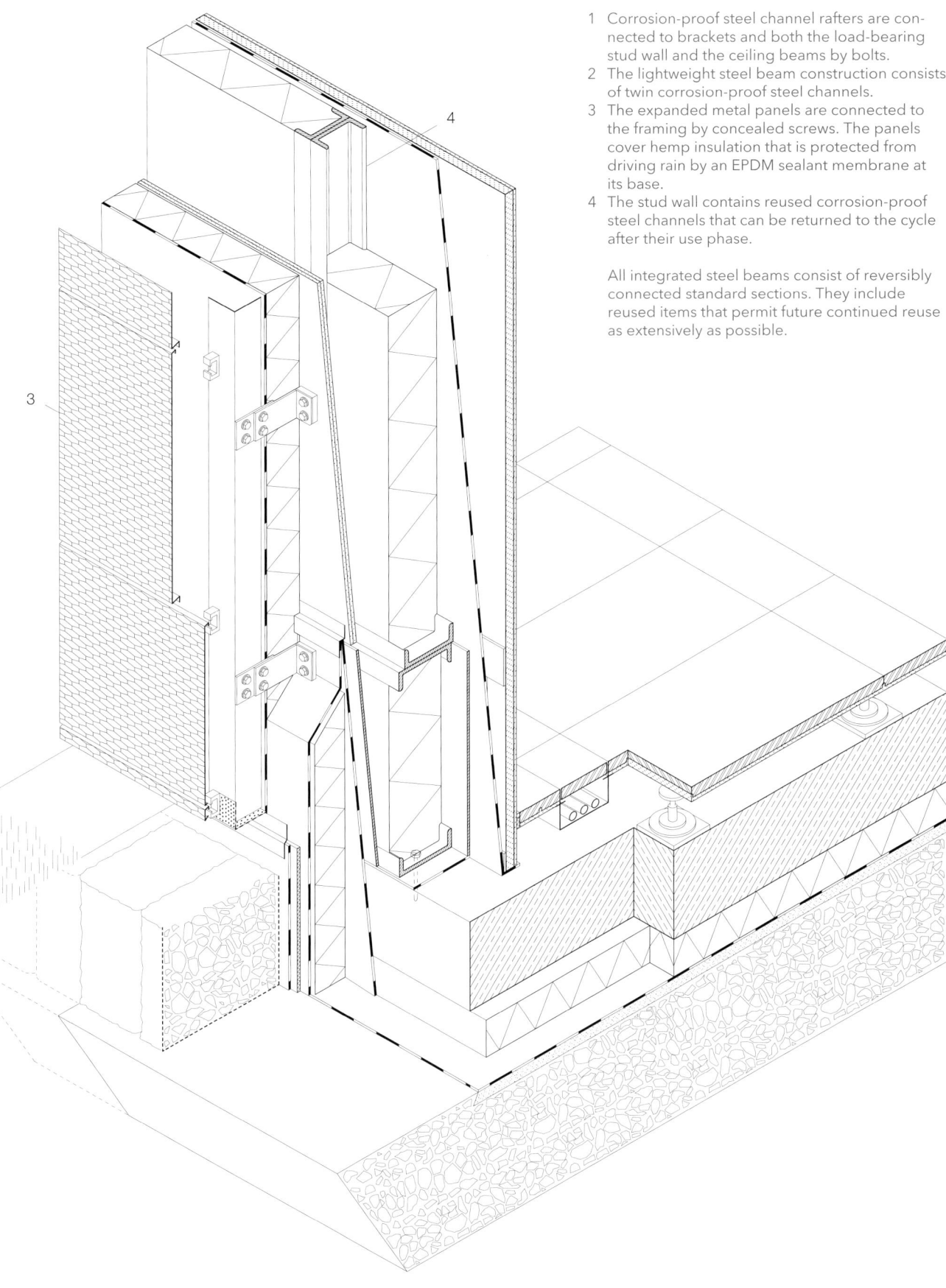

1 Corrosion-proof steel channel rafters are connected to brackets and both the load-bearing stud wall and the ceiling beams by bolts.
2 The lightweight steel beam construction consists of twin corrosion-proof steel channels.
3 The expanded metal panels are connected to the framing by concealed screws. The panels cover hemp insulation that is protected from driving rain by an EPDM sealant membrane at its base.
4 The stud wall contains reused corrosion-proof steel channels that can be returned to the cycle after their use phase.

All integrated steel beams consist of reversibly connected standard sections. They include reused items that permit future continued reuse as extensively as possible.

Focus on Steel

Focus on Loam

Both traditional and modern loam constructions have long been a viable method for building affordable housing worldwide. Today, about one-fifth of the world's population lives in houses that were partially or completely constructed with loam. In developing countries, the share is about one-fourth [1]. The construction material is popular in Germany as well, while the quota of existing buildings containing loam only reaches about 2–3 %. Most of all, this refers to traditional half-timbered houses, where the infill between timber elements consists of loam or features loam render. Germany's tallest loam building is the six-storey socalled Pisé-Haus (pisé French for rammed earth). It is located in Weilburg an der Lahn in the state of Hesse and consists of solid rammed earth walls. It was built in the late 19th century.

Raw material deposits
The reason for the extensive and global distribution of loam construction and its related methods is that the material is available in almost all countries of the world in large quantities and in a composition suitable for construction, often locally and free of charge (Fig. 1). Loam as a construction material for traditional building methods can also be fully recycled and reprocessed. Long transport routes are not required and fossil energy is hardly or not at all necessary, if loam is processed in its original form. As a result, loam as a construction material displays an extremely small carbon footprint.

Traditional means of building with loam are still currently practiced in many regions, having been passed on from one generation to the next. They are deeply rooted in cultural terms, expressed in different typological manifestations and applied according to a broad range of methods [2]. However, the dominance of imported building materials in developing regions of the world has grown massively and, as a result, traditional loam building is on the retreat. In such cases, building with concrete plays an increasing role, since concrete is industrially standardised, permits simple handling and appears to be cost-efficient. Often, negative consequences are disregarded, such as economic dependability, resource depletion, high carbon emissions as outcome of the production of cement, or lacking capacities for recycling the original raw resources of the processed material.

Following the onset of industrialisation and the advent of newly developed materials in Europe in the 19th century, loam was increasingly considered an inferior building material of less privileged social groups. Only during times of crisis, for instance in the aftermath of wars,did loam construction experience a short-lived renaissance, due to the previously mentioned reasons, such as easy accessibility and simple handling. Time and again, the large amount of labour required for traditional handling curtailed the use of loam in the years following the wars. A digital and industrial production process could change this. And, as a result, the application of an easily accessible and circular building material with very good mechanical, physical and health-related characteristics could once more become appealing, also from an economic point of view.

Characteristics
In principle, loam is a mix of sand, silt and clay. Water is added to create a moist mix that can be shaped and then subsequently dries out. Loam deposits differ based on the mixing ratio of the three components, the grain size and form of the included minerals, as well as the type of clay. The latter serves as binder for the loam mix. Loam with a large share of clay is described as "fat" while loam containing a low amount of clay is termed as "lean". As a result, only a limited range of loam types is suitable as a basis for loam construction. The proper composition of adequate types of clay, silt and sand are key. A "fat" mix tends to crack during the drying process. "Lean" mixes, however, become crumbly and dimensionally unstable. Often, natural fibre such as straw or other grass types finds use as a natural form of reinforcement within the loam mix. Loam displays good adhe-

1
Overview, highly suitable soil types for loam construction

sion to these fibres. The finer the admixture, the better the handling of the mix. This is also the case for visibly exposed render finishes. Traditionally, cow or horse dung was also used, due to the fact that the remaining fibre within the manure tends to be very fine and short. Further, lime can be added. It acts as a disinfectant and supports the chemical dissolution of fibre components in order to achieve an even better and more intricate adhesion between mineral and biological elements.

For a long time, uniform standards for loam construction in Europe were lacking. It was too difficult to standardise the different composition types of the material due to differences between deposit locations. In 2013 binding standards were established in Germany. However, they are predominantly oriented on factory-produced loam blocks, loam masonry mortar and mortar-based render – not for rammed earth applications or building with cob. In 2018 the standards were revised and amended. Also, a new standard for loam panels and masonry construction was introduced. Preceding these developments, a new awareness of loam as a building material had already been on the rise in Europe – even though the lack of standards or larger-scale applications, due to non-industrial building processes and their dependence on manual labour, led to an impression of economic non-competitiveness. In recent years, pioneers of loam construction, including Martin Rauch, Anna Heringer, Gernot Minke, Christof Ziegert, Eike Roswag-Klinge, Uwe Seiler and Andrea Klinge have developed new methods that permit industrial-scale application. Many of their buildings are considered ecological and technological milestones. They contributed to liberating the material from prejudice concerning its traditional and often declared "backward" character. As a result, it is perceived today as a modern building material with excellent building physics characteristics.

Loam can regulate indoor humidity both adequately and naturally: It absorbs water particles and discharges them into ambient air when conditions are arid. As a result, it can also absorb pollutants. Loam as a construction material is capable of diffusion and acts as an excellent thermal storage with regulating effect. Its inertia allows balancing temperatures during daily or annual cycles. Interiors of loam buildings in summer are perceived as pleasantly cool, due to moisture discharge. During winter, on the other hand, the moisture regulating characteristics of loam prevent indoor air from becoming dry. It is free of pollutants and is resilient against pests and bacterial infestation. These characteristics, however, can only be maintained if the loam mix does not include synthetic products aimed at improved handling. They obstruct the natural processes described above, because they cause clay particles, which enable the absorption and discharge of water on microscopic levels, to stick together. As a result, they can no longer actively contribute to the process of water regulation. This also occurs when cement is added to the mix in order to enhance technical characteristics (compressive strength). Here as well, clay particles stick together while, at the same time, active processes of water absorption and discharge are obstructed. Further, one of the greatest advantages of the natural material is eliminated: being fully recyclable.

Typical construction methods

In the case of half-timbered constructions, wood and loam comprise a building system that harmonises well. The two materials show similar swelling and shrinkage behaviour. Further, loam can preserve and protect wood components, if it is applied and handled properly. For this purpose, a number of methods exist for wall systems: One, the spaces between timber posts, transoms and bracing can be amended by battens, stakes or wattle as substrate for throwing and trowelling straw-reinforced loam on both sides. Further, the interstices of a half-timbered or timber frame construction can be filled with loam masonry brick and mortar. In such cases, the gaps between loam blocks and the timber structure may require filling. Beyond that, loam block infill subject to weathering should be protected by a double layer of lime render.

Existing walls can also be amended by loam blocks, lightweight loam wet construction, or stacked dry loam construction systems as facing shells on the interior or exterior. This approach increasingly finds use if thermal insulation or soundproofing measures are required. Loam, in combination with a timber frame system, can also serve to build ceiling constructions. This can be supported by placing battens, slats or stakes between primary beams. Moist straw loam can be applied to the wood structure and trowelled either below or above. Contemporary methods comprise a planar wood substrate that is connected either to the top or bottom of primary beams. Loam components can be attached to the substrate by use of dry construction methods. This includes air-dried loam blocks, panels or elements that feature grooves in order to install heating coils. Aside from the described hybrid timber loam constructions, so-called monolithic applications exist that can be found across the globe in easily distinguishable variants. One example is rammed earth construction, which entails compacting loam by ramming it into formwork, resulting in an homogeneous, monolithic edifice. By use of shifting or sliding formwork, this construction method can be applied to building tasks very efficiently and relatively quickly, once the material has been prepared accordingly. Lintels, ceilings and ring beams, as well as roof load-bearing structures are often made of other materials, such as timber or concrete. Building with cob is a further and comparable example. Here, loam is mixed with grass types, moistened and applied. Then the surface is treated with a spade to correct irregularities. The result is a monolithic structure.

Loam brick masonry is a further monolithic application. Here, loam is pressed into formwork to create bricks or also applied after moistening and laid out to dry. Bricks can be assembled into masonry wall systems or vaulted structures. The latter can be distinguished into barrel vaults (Nubian) and dome vaults (Catalan) [3]. In such cases, significant construction height is required at dimensions that are difficult or impossible to realise in contemporary multistorey construction.

Further application types include loam render or mortar methods, modern loam construction panels or large-scale prefabricated loam construction elements. They can be moved to the construction site easily and gaps between panels or elements can be filled with moist loam.

Circular potential

One of the greatest strengths of pure loam construction is its 100% circularity. However, this is also its greatest weakness: Loam should not come into contact with water when a building is in use. Precipitation or construction water can lead to mechanical deficiencies resulting in loss of compressive strength, swelling and shrinkage stress, or crack formation. In climate zones with strong seasonal precipitation, water impact is one of the greatest problems loam construction is confronted with. Yet, here as well, new application types display innovative results: Waterproof stones rammed into the wall or horizontal seals reduce the flow rate of rainwater along exposed wall elements and thus stop the erosion of the material within the construction component. As a result, loam buildings could then even be built without requiring a deep roof overhang or other protective measures [4]. The following examples refer to modern loam application types and indicate opportunities for rethinking how we use and apply this very traditional building material. The described advantages in terms of health and climate are too great to simply ignore them.

Notes
[1] Marsh/Kulshreshtha 2021
[2] Hebel/Moges/Gray 2015
[3] ibid.
[4] Sauer 2015

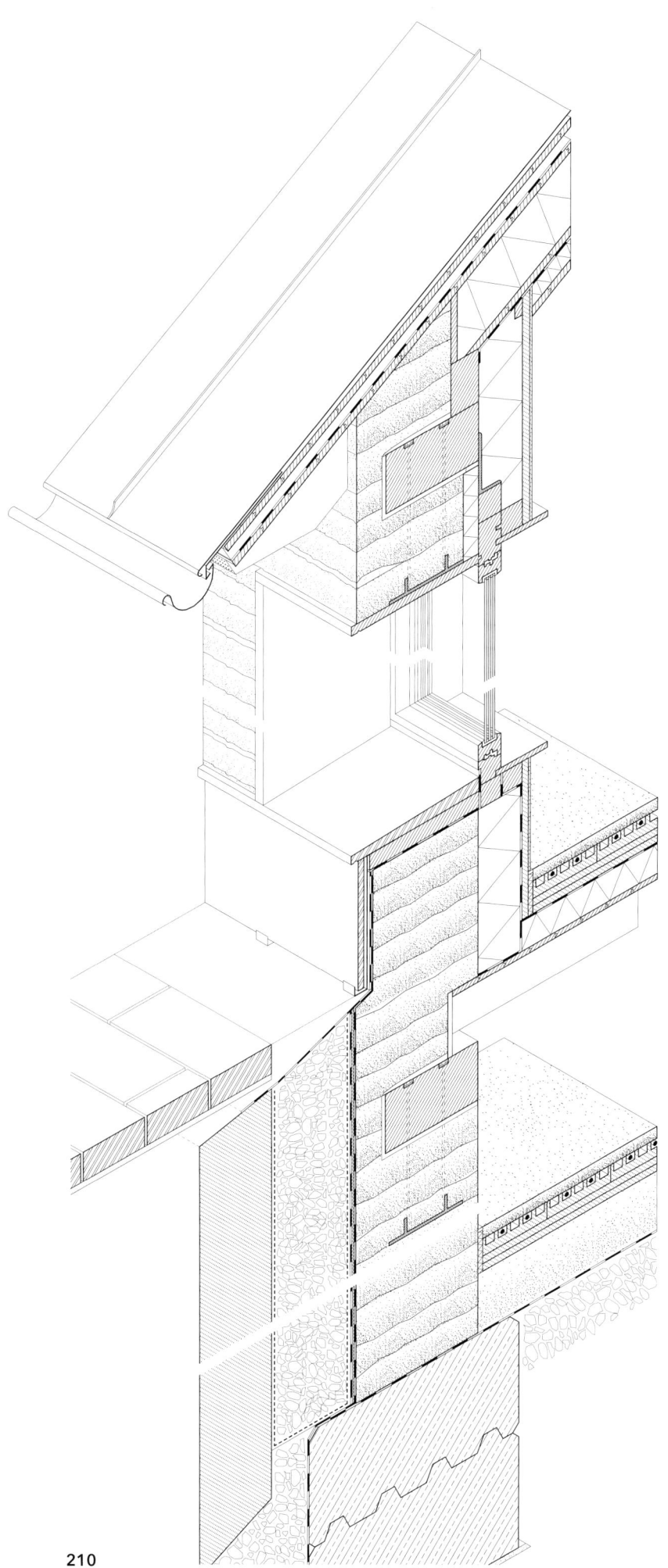

Rammed Earth: Solid Construction I

Protect
standing seam roofing
rammed earth with erosion proofing
natural stone tile

Insulate
reed cane insulation
foam glass gravel

Support
rammed earth
solid construction timber ring beam with steel plate anchor tie
timber beam, rafters
prefabricated reinforced concrete foundation element, reused

Seal
EPDM sealant layer
sarking layer

Cover
rammed earth floor, timber substrate
loam render on loam construction panel

Isometric illustration   scale 1:20
Vertical section   scale 1:20

1 Roof construction:
   Standing seam roofing, clamped to anchor clip
   min. 24 mm solid timber sheathing, screw connections
   min. 40 mm counterbattens, screw connections/ back ventilation
   sarking membrane, windproof, breathable, min. 30 cm overlap
   solid construction timber rafters, birdsmouth joint to wall plate/ridge beam, screw connections
   inlaid reed cane insulation, loose fill
   solid timber panel, diagonal sheathing with tongue-and-groove joints, screw connections
   PE film, diffusion-proof, min. 30 cm overlap, attached to wall plate
   solid timber battens, screw connections
   inlaid reed cane insulation/installation layer
   solid timber sheathing, screw connections
2 Sheet steel insect screen, corrosion-proof, screw connections to sheathing and counterbattens

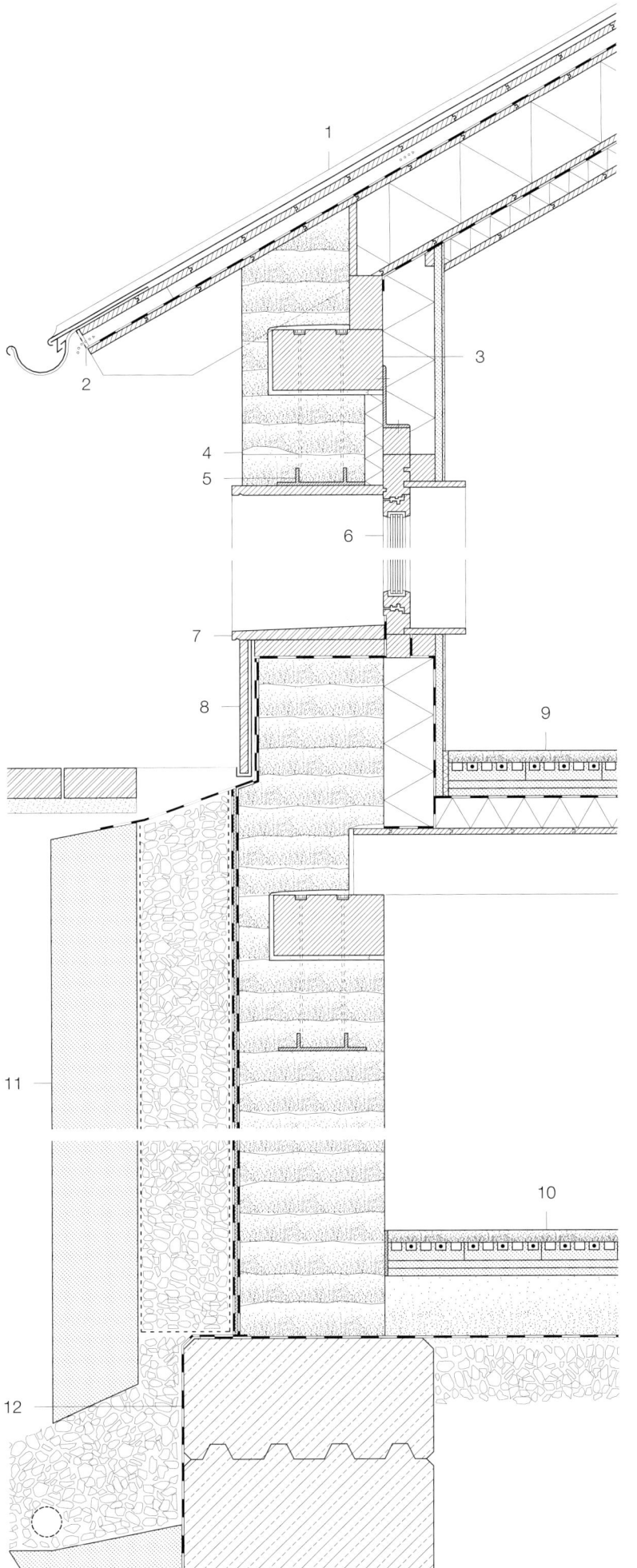

3 Solid construction timber ring beam, covered in loam to prevent wood moisture formation
flat steel anchor tie, screw connections
4 Trass lime mortar erosion proofing
5 Flat steel window lintel, steel bar ring beam suspension, rammed into loam, steel bar suspension, welded
6 Window: Triple glazing in wood frame, reused plastic dry sealant, single grade
7 Natural stone window sill
8 Exterior wall construction
natural stone tile plinth coping, steel bracket suspension, screw connections
EPDM sealant against ground moisture, membrane-to-membrane weld, 2-ply, clamped
rammed earth exterior wall
sill plate, bolt connections
inlaid reed cane thermal insulation / installation layer
loam render panel with double loam render layer, reinforcement, screw connections
9 Floor construction:
Rammed earth flooring, waxed finish
loam block, air-dried, with grooves and heat diffusor plates for installing underfloor heating pipes
double loam construction panel
PE film, diffusion-resistant, min. 30 cm overlap, clamped
reed cane impact soundproofing
solid timber sheathing, screw connections
solid construction timber ceiling beams, visibly exposed, bolt connections
10 Floor construction:
Rammed earth flooring, waxed finish
loam blocks, air-dried, with grooves and heat diffusor plates to install underfloor heating pipes
double loam construction panel
rammed earth floor slab
EPDM sealant layer against ground moisture, membrane-to-membrane weld, loose layout
sand base layer
foam glass gravel fill perimeter insulation
11 Fat loam
foam glass gravel perimeter insulation in textile fabric bags as construction support, mechanical connections to rammed earth
EPDM sealant layer against ground moisture, membrane-to-membrane weld, attached to window frame, clamped
dimpled sheet, min. 30 cm overlaps
EPDM sealant layer against ground moisture, membrane-to-membrane weld
rammed earth basement wall
12 EPDM sealant layer against ground moisture, membrane-to-membrane weld, clamped
prefabricated reinforced concrete foundation element, reused, frost-free foundation
sand base layer

Focus on Loam

Isometric illustration, roof
Isometric illustration, plinth / ceiling
not to scale

1. A layer of trass lime mortar serves as erosion proofing.
2. The solid construction timber ring beam is covered in loam in order to prevent moisture formation on wood surfaces. Flat steel hangers connect the ring beam to the rammed earth wall.
3. The flat steel window lintel is hung from the ring beam with steel bar connectors.
4. The solid timber window reveal with milled drip edge is connected to the window lintel with screws.
5. Along the plinth, natural stone tile is supported by steel bracket hangers in order to protect the rammed earth wall from water intrusion.

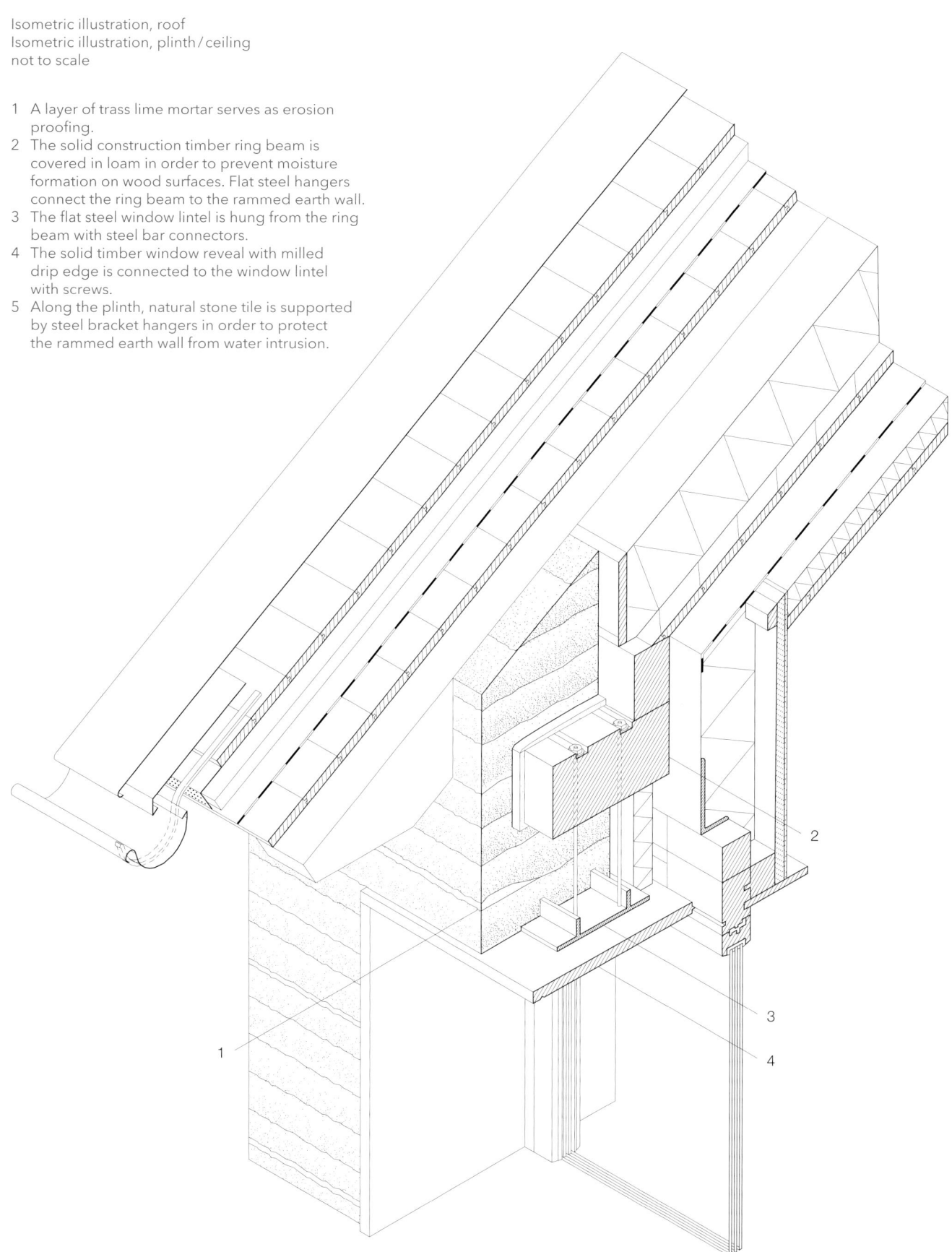

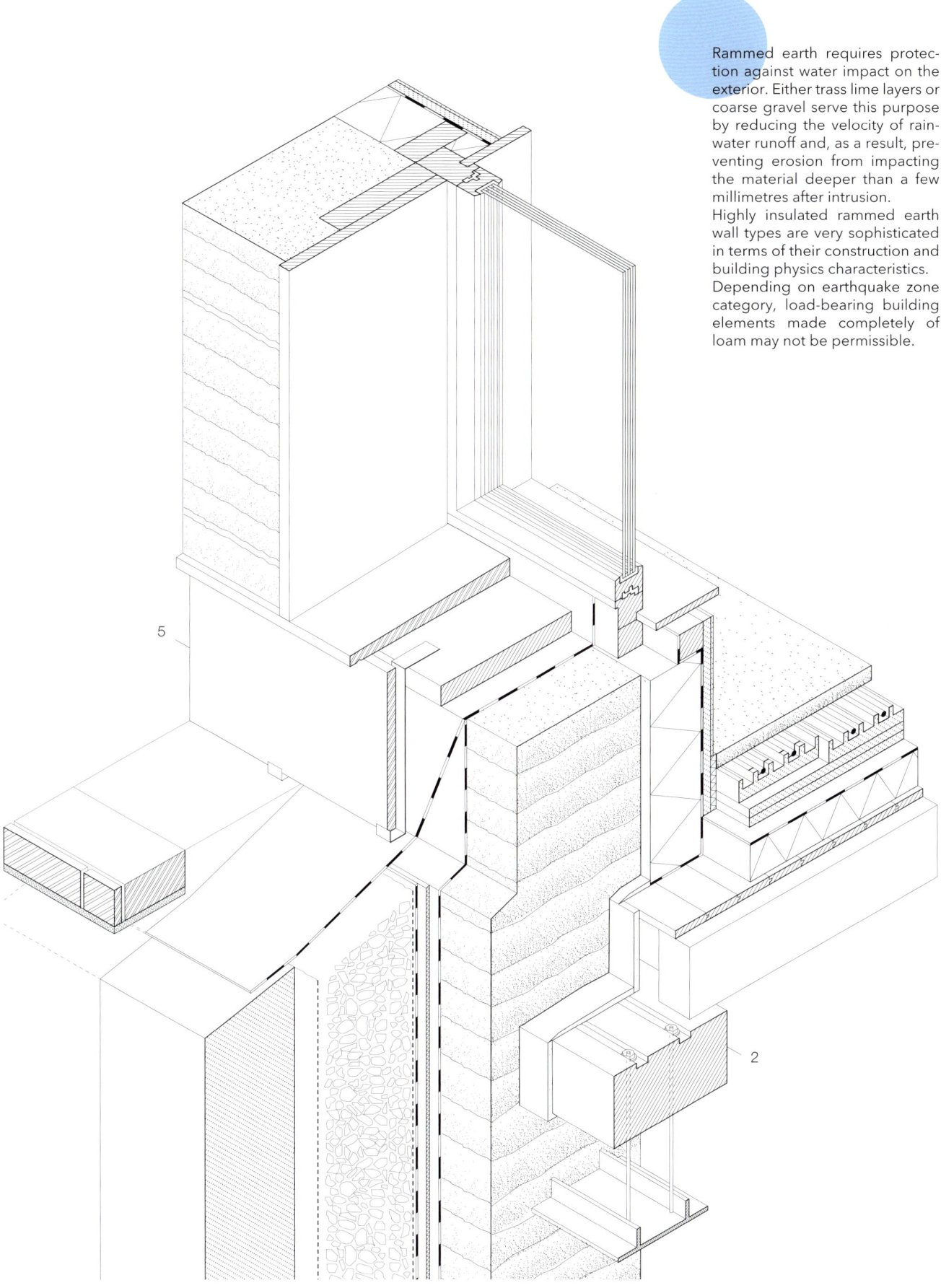

Rammed earth requires protection against water impact on the exterior. Either trass lime layers or coarse gravel serve this purpose by reducing the velocity of rainwater runoff and, as a result, preventing erosion from impacting the material deeper than a few millimetres after intrusion.
Highly insulated rammed earth wall types are very sophisticated in terms of their construction and building physics characteristics. Depending on earthquake zone category, load-bearing building elements made completely of loam may not be permissible.

Focus on Loam   213

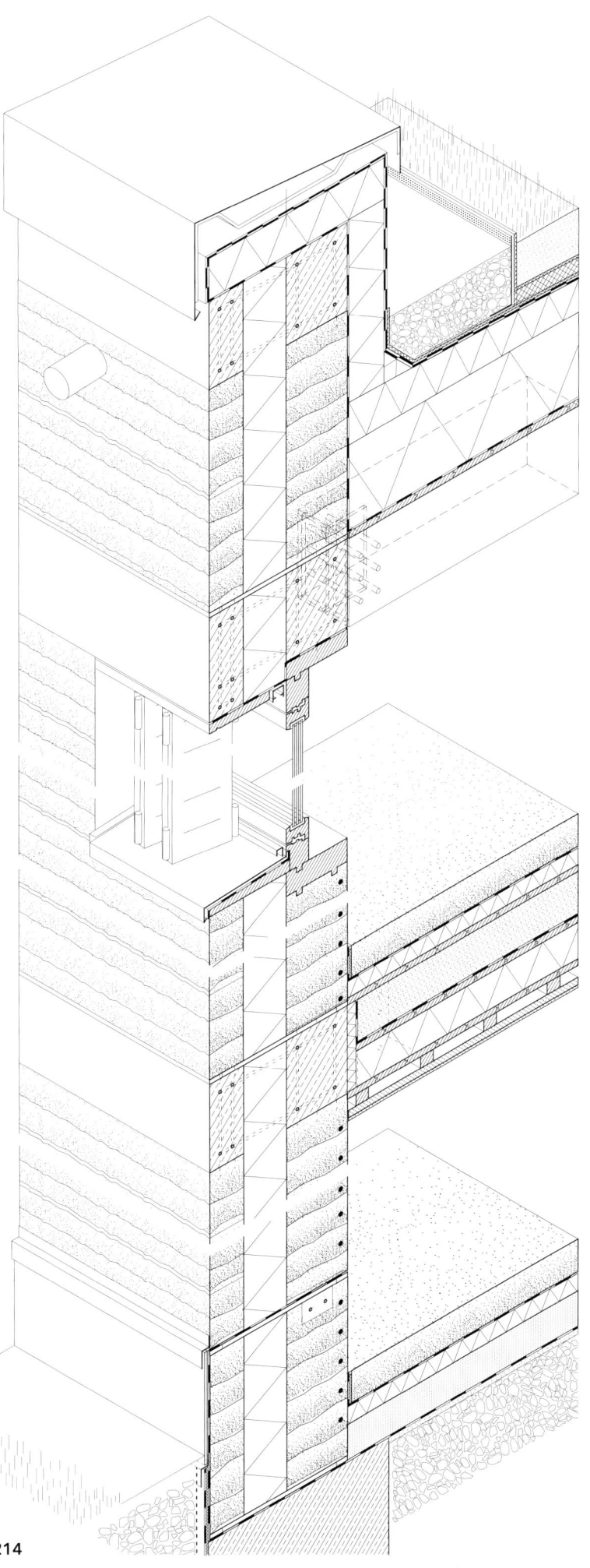

Rammed Earth: Solid Construction II

Protect
parapet coping, extensive greening
rammed earth with erosion proofing

Insulate
foam glass panels
foam glass gravel

Support
prefabricated rammed earth element
prefabricated reinforced concrete element, reused
prefabricated reinforced concrete foundation element, reused

Seal
EPDM sealant membrane
PE film

Cover
rammed earth
rammed earth floor, loam render

Vertical section scale 1:20
Isometric illustration scale 1:20

1 Roof edge:
   Parapet coping, canted, clamped to anchor clip
   anchor clip, screw connections to parapet
2 Roof construction:
   Extensive greening
   gravel perimeter strip
   vegetation layer, loose fill
   polypropylene filter fleece, loose layout
   HDPF or stainless steel drainage element
   single grade, loose layout, reused
   plastic storage protection mat, single grade, loose layout
   EPDM roof sealant, membrane-to-membrane weld, 2-ply, attached to parapet, clamped
   foam glass panel insulation, min. 2 % to falls, loose layout
   foam glass panel insulation, loose layout in bond
   PE film, diffusion-resistant, min. 30 cm overlap, attached to parapet, clamped
   solid timber panel, diagonal sheathing with dovetail joints, screw connections
   solid construction timber ceiling beam,

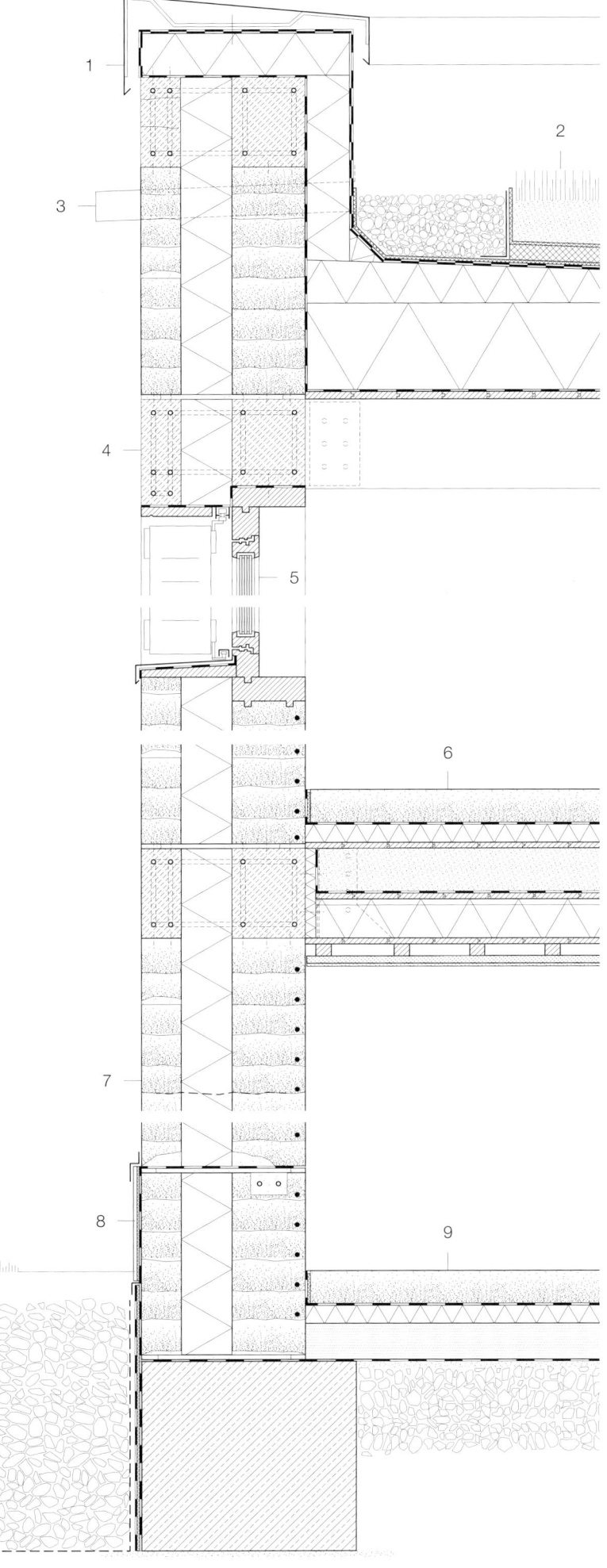

untreated, exposed bolt connections through head plate and sheet metal to ring beam
3 Zinc or copper emergency drainage, pressed into prefabricated rammed-earth element
4 Prefabricated reinforced-concrete ring beam, thermal separation, steel reinforcement as connection between both rammed earth shells
5 Window: triple glazing in wood frame, reused plastic dry sealant, single grade
6 Floor slab construction:
Rammed earth floor, waxed finish
PE film, diffusion-resistant, min. 30 cm overlap, clamped
cork impact soundproofing/installation layer
solid timber sheathing, screw connections
solid construction timber beam, untreated, exposed bolt connections through head plate and sheet metal to ring beam
loose mass fill, sand, to improve soundproofing, in honeycomb panel
PE film, diffusion-resistant, min. 30 cm overlap, clamped
insert panel, loose placement
foam glass panel infill insulation between beams, loose layout
solid timber sheathing, screw connections
solid timber battens for hung ceiling, screw connections to ceiling beam
loam construction panel with double loam render layer, reinforcement, screw connections
7 Exterior wall construction:
Min. 12 cm prefabricated rammed earth element, self-supporting
foam glass panel thermal insulation
min. 24 cm prefabricated rammed earth element, self-supporting, interior with integrated wall heating system
8 Lime render, plinth splash water protection on render substrate panel, plinth height min. 50 cm above ground level
9 Floor construction:
Rammed earth floor, waxed finish
PE film, diffusion-resistant, min. 30 cm overlap
cork impact soundproofing/installation layer
floor slab, cork trass lime mix
EPDM sealant against ground moisture, membrane-to-membrane weld, loose layout
sand base layer
foam glass gravel fill perimeter insulation
10 Foam glass gravel perimeter insulation, filled in textile fabric bags to support construction, mechanical connections to reinforced concrete
EPDM sealant layer against ground moisture, membrane-to-membrane weld, clamped by mounting track
dimpled sheet, 30 cm overlap
EPDM sealant layer against ground moisture, membrane-to-membrane weld, loose layout
prefabricated reinforced concrete foundation element, reused, frost-free foundation
sand base layer

Focus on Loam

Isometric illustration, roof
Isometric illustration, plinth / floor
not to scale

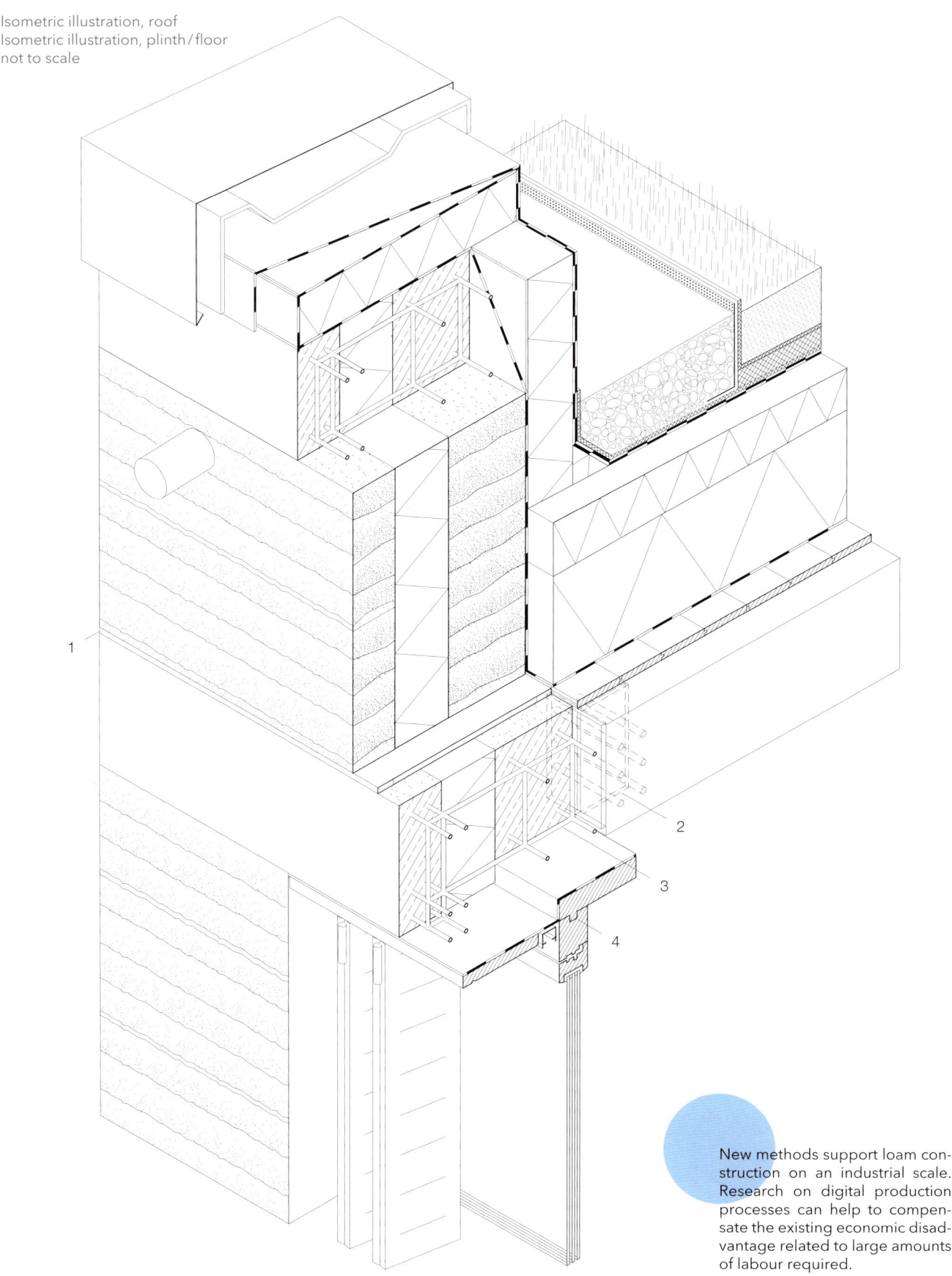

New methods support loam construction on an industrial scale. Research on digital production processes can help to compensate the existing economic disadvantage related to large amounts of labour required.

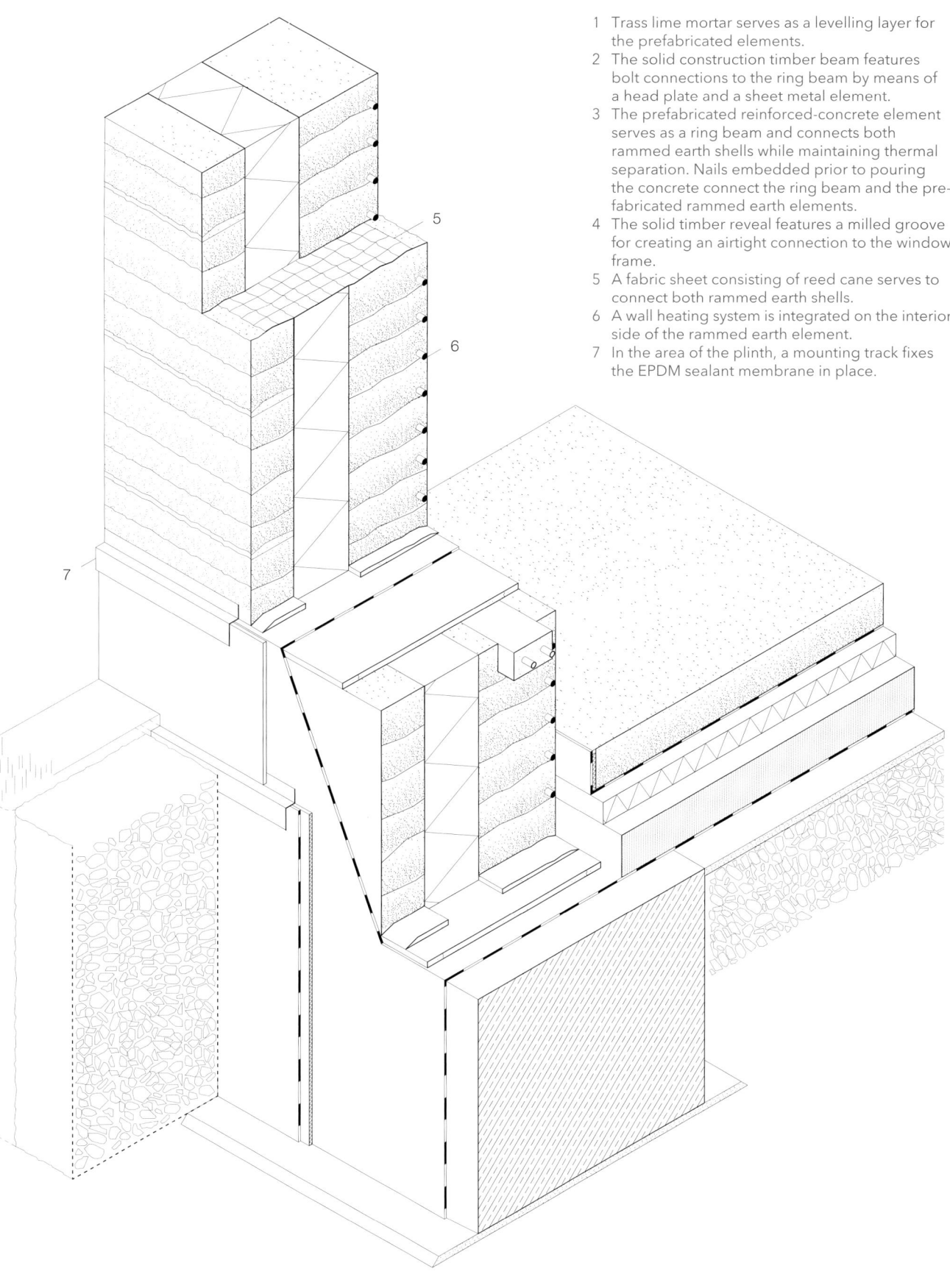

1 Trass lime mortar serves as a levelling layer for the prefabricated elements.
2 The solid construction timber beam features bolt connections to the ring beam by means of a head plate and a sheet metal element.
3 The prefabricated reinforced-concrete element serves as a ring beam and connects both rammed earth shells while maintaining thermal separation. Nails embedded prior to pouring the concrete connect the ring beam and the prefabricated rammed earth elements.
4 The solid timber reveal features a milled groove for creating an airtight connection to the window frame.
5 A fabric sheet consisting of reed cane serves to connect both rammed earth shells.
6 A wall heating system is integrated on the interior side of the rammed earth element.
7 In the area of the plinth, a mounting track fixes the EPDM sealant membrane in place.

Focus on Loam

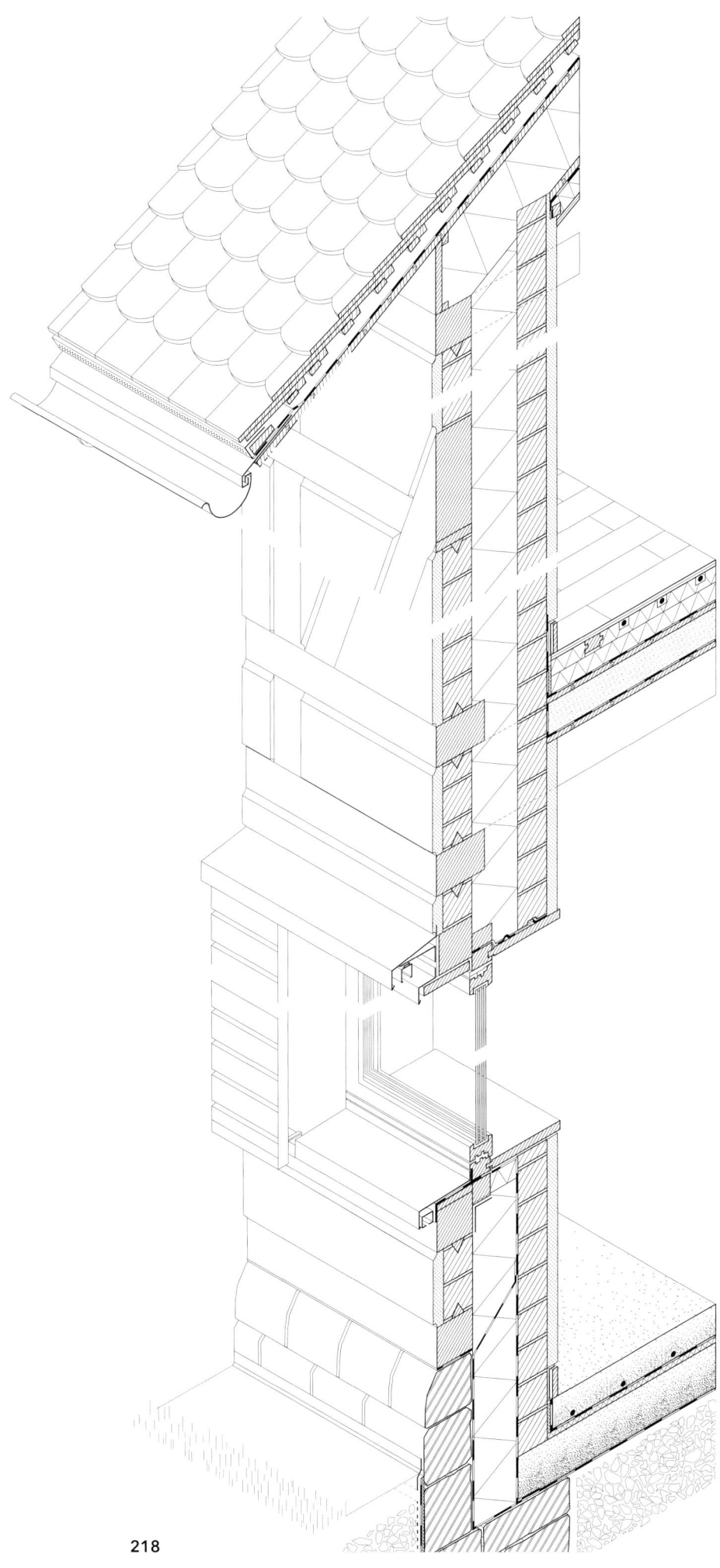

Wood and Loam: Half-timbered Construction

Protect
plain roofing tile, clay
solid construction timber,
half-timbered frame
loam blocks, lime render finish

Insulate
reed cane insulation
foam glass gravel

Support
solid construction timber,
half-timbered frame
natural stone foundations

Seal
EPDM sealant
PE film, sarking membrane

Cover
reclaimed wood floorboards
rammed earth floor
loam render

Isometric illustration   scale 1:20
Vertical section   scale 1:20

1 Roof construction:
   Plain tile roofing, clay, interlocking
      solid timber battens, screw connections
      min. 40 mm counterbattens, screw connections /
      back ventilation
      sarking membrane, windproof, breathable,
      min. 30 cm overlap
      solid timber sheathing, screw connections
      solid construction timber rafters, birdsmouth
      joint to wall plate / ridge beam, screw connections
      inlaid reed cane insulation
      solid timber panel, diagonal sheathing with
      tongue-and-groove joints, screw connections
      PE film, breathable, min. 30 cm overlap,
      attached to wall plate, clamped
      solid timber battens, screw connections to rafters
      inlaid reed cane insulation as installation layer

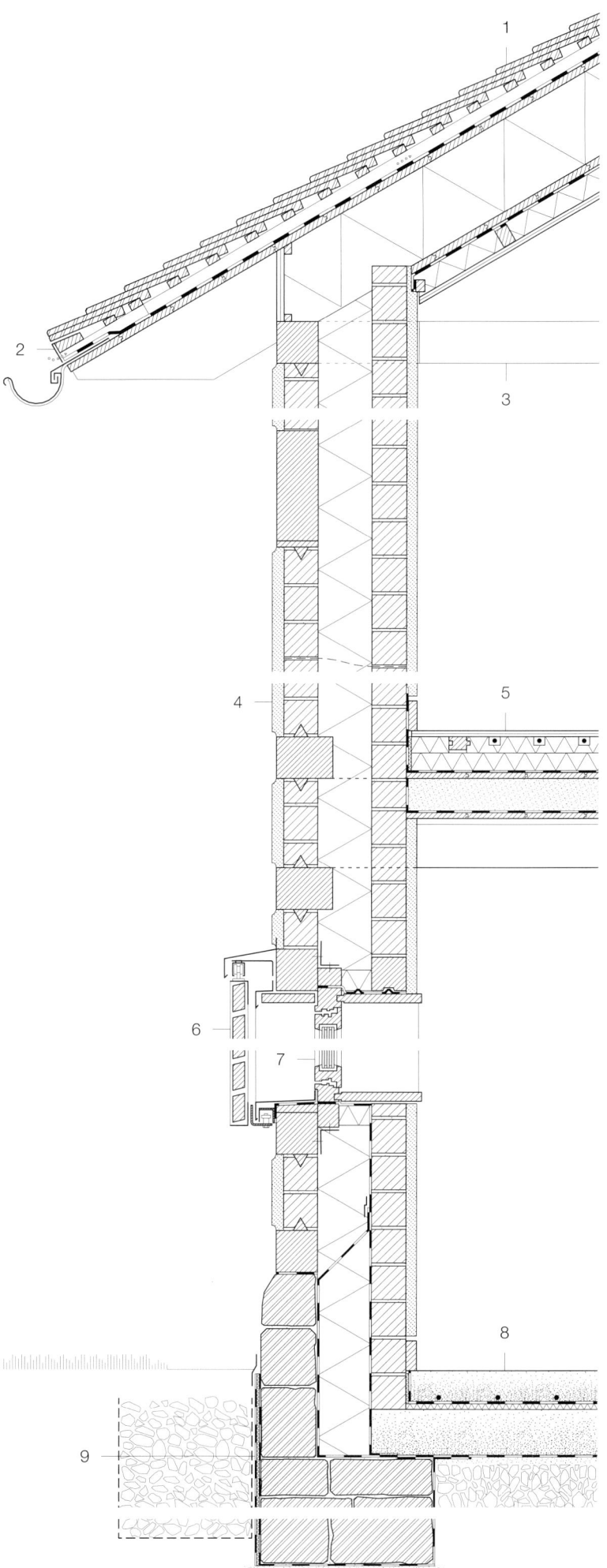

1. solid timber sheathing, visibly exposed, screw connections
2. Sheet steel insect screen, corrosion-proof, screw connections to sheathing and counterbattens
3. Solid construction timber tie beam, tensile member, interlocked, screw connections
4. Exterior wall construction:
   Lime render
   loam block masonry wall, trass lime mortar, as half-timbered frame infill
   solid construction timber, screw connections
   reed cane insulation, anchored to masonry wall with insulation dowels
   loam block masonry wall, non-load-bearing, trass lime mortar joints
   double render layer, grass fibre mat reinforcement
5. Floor construction
   reclaimed wood floorboards, concealed tongue-and-groove screw connections
   solid timber battens, screw connections
   inlaid soft wood fibre panel, lignin-bonded, as installation panel with grooves and heat diffusor plates to install underfloor heating pipes
   softwood fibre levelling insulation, lignin-bonded
   PE film, diffusion-resistant, min. 30 cm overlap, clamped
   solid timber sheathing, screw connections
   solid construction timber beam, visibly exposed, supported by top plate
   inlaid mass fill, sand, to improve soundproofing
   trickle protection mat, min. 30 cm overlap
   insert panel, set on top of battens, screw connections
   solid timber battens, screw connections
6. Wood slat sliding shutter
7. Window: Triple glazing in wood frame, reused plastic dry sealant, sorted by type
8. Floor construction:
   Rammed earth floor, waxed finish, pressed-in underfloor heating pipes
   PE film, diffusion-resistant, min. 30 mm overlap, clamped
   reed cane impact soundproofing
   rammed earth floor slab
   EPDM sealant layer against ground moisture, membrane-to-membrane weld, loose layout
   sand base layer
   foam glass gravel perimeter insulation
9. EPDM sealant layer against ground moisture, membrane-to-membrane weld, clamped by mounting track
   dimpled sheet, min. 30 cm overlap
   EPDM sealant layer against ground moisture, membrane-to-membrane weld, loose layout
   natural stone foundation, e.g. granite, as masonry wall, frost-free foundation
   sand base layer

Focus on Loam

Isometric illustration, roof
Isometric illustration, plinth/floor
not to scale

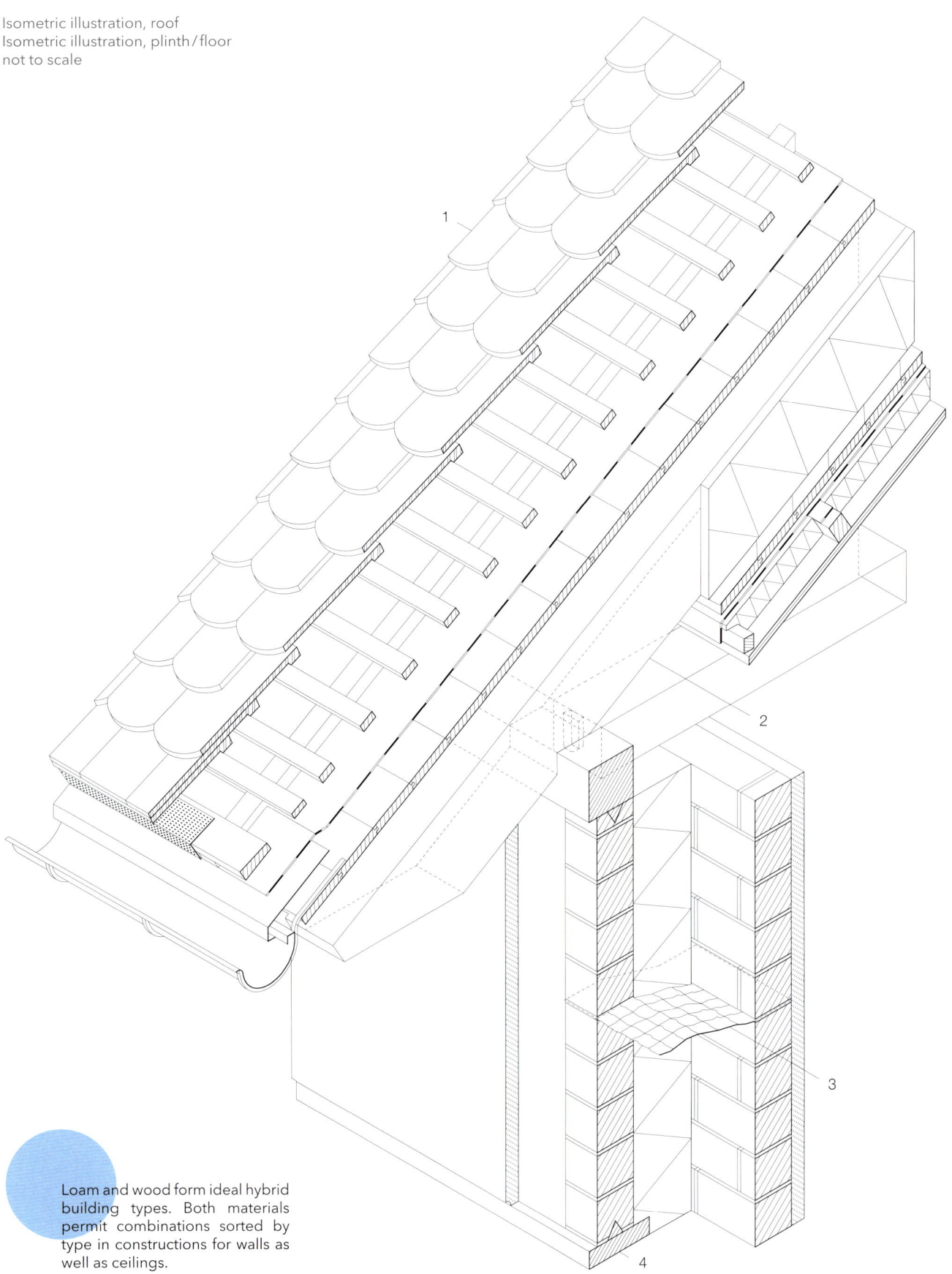

Loam and wood form ideal hybrid building types. Both materials permit combinations sorted by type in constructions for walls as well as ceilings.

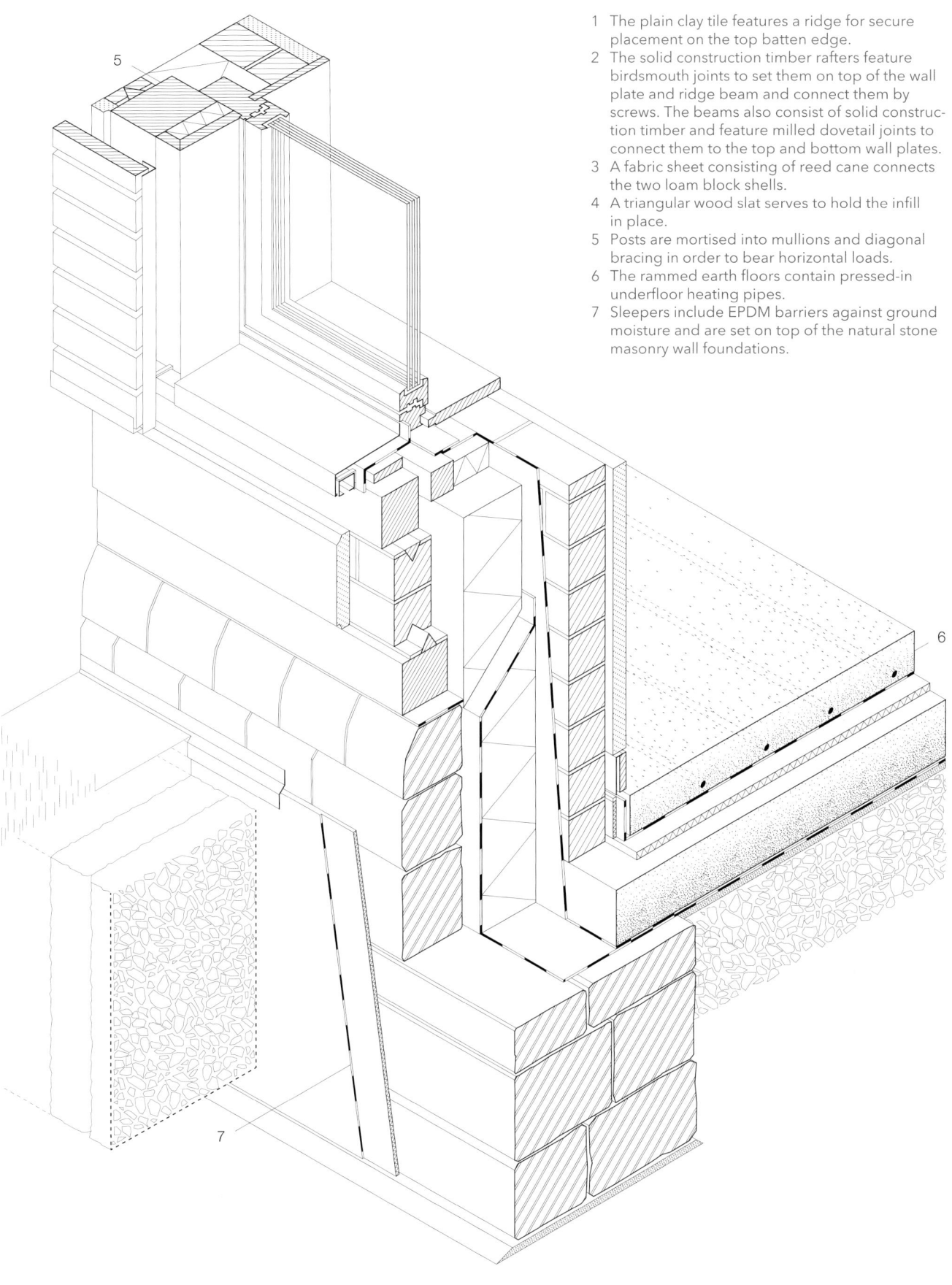

1. The plain clay tile features a ridge for secure placement on the top batten edge.
2. The solid construction timber rafters feature birdsmouth joints to set them on top of the wall plate and ridge beam and connect them by screws. The beams also consist of solid construction timber and feature milled dovetail joints to connect them to the top and bottom wall plates.
3. A fabric sheet consisting of reed cane connects the two loam block shells.
4. A triangular wood slat serves to hold the infill in place.
5. Posts are mortised into mullions and diagonal bracing in order to bear horizontal loads.
6. The rammed earth floors contain pressed-in underfloor heating pipes.
7. Sleepers include EPDM barriers against ground moisture and are set on top of the natural stone masonry wall foundations.

Biographies

Dirk E. Hebel
Dirk E. Hebel is Professor of Sustainable Construction and Dean of the Department of Architecture at the Karlsruhe Institute of Technology (KIT). He is the author of numerous publications, such as most recently, "Circular Construction and Circular Economy" (Birkhäuser, 2022, with Felix Heisel and Ken Webster) and "Urban Mining and Circular Construction – the City as Raw Material Storage" (Urban Mining und kreislaufgerechtes Bauen. Die Stadt als Rohstofflager, Fraunhofer IRB, 2021, with Felix Heisel). His work is exhibited globally, such as most recently in "Plastic: Remaking our World" (Vitra Design Museum, Weil am Rhein, 2022, with Nazanin Saeidi, Alireza Javadian, Sandra Böhm and Elena Boerman). As faculty representative, in cooperation with Prof. Andreas Wagner, he won the first Solar Decathlon competition held in Germany in 2022 in Wuppertal as a member of Team RoofKIT (Regina Gebauer and Nicolas Carbonare). He is a member of the BDA, practising architect and co-founder of the office 2hs Architekten und Ingenieur, Hebel/Heisel/Schlesier, with a focus on resource-oriented construction and circular use of materials. Together with Werner Sobek and Felix Heisel, he realised widely recognised construction projects, such as the Urban Mining and Recycling research unit within the NEST project of the EMPA in Dübendorf near Zurich, or the Added.VALUE.Pavilion (MehrWert Pavillon) for the State Institute for the Environment of Baden-Württemberg on occasion of the German Federal Garden Show in Heilbronn, together with Felix Heisel and Karsten Schlesier in cooperation with Lisa Krämer and Simon Sommer.

Ludwig Wappner
Ludwig Wappner is Professor of Building Construction at the Department of Architecture at KIT and Executive Director (Speaker) of the Institute of Building Design and Technology (IEB). He is a founding partner of Allmann Sattler Wappner Architekten in Munich. Since 2022, the firm has operated under the name allmannwappner. Ludwig Wappner was a guest professor at various national and international universities. Currently, he is chairperson of the design advisory boards of the cities of Mannheim and Pforzheim following related roles in Munich, Bamberg, Ingolstadt and Trier. He is member of the BDA and chairperson of the Schelling Architecture Foundation in Karlsruhe. His experience includes many years as a judge of competitions and author of expert reports. Lectures, publications, contributions as guest critic and within workshops both nationally and internationally are cornerstones of his activity in practice, teaching and research. In 2022 he initiated and realised the design build living lab project Tiny House in Karlsruhe-Durlach in cooperation with students, his research team and carpentry apprentices from Freiburg. The project received recognition in the Timber Construction University Award in 2023. It is considered a lighthouse project of circular construction sorted by type for a timber residential house. Aside from the significant knowledge generation for all involved participants in terms of underlying technological questions, it also offers an experience of beautiful architecture.

Werner Sobek
Werner Sobek is an architect and consulting engineer. He is founder of the Institute for Lightweight Structures and Conceptual Design (ILEK) at the University of Stuttgart and initiator of the Collaborative Research Centre 1244 for adaptive skins and structures of the built environment of tomorrow. Werner Sobek is founder and honorary president of numerous charitable initiatives, including the aed e.V. Association for the Promotion of Architecture, Engineering and Design. As the only architect and engineer, the magazine Cicero included him in their 2022 list of the 500 most important German speaking intellectuals. In 2022 he was awarded the Federal Cross of Merit for his exceptional social commitment in the field of architecture and engineering with a focus on sustainability.

Thomas Auer
Thomas Auer is Professor of Building Technology and Climate Responsive Design at the Technical University of Munich (TUM) and partner in Transsolar. He cooperates with renowned architectural offices across the world within award-winning projects that are characterised by innovative design and integral climate strategies. Thomas Auer taught at various universities, including Yale University in New Haven. In 2014 he was appointed as full professor at TUM. His research focus includes the decarbonisation of the building sector, climate adaptation and related impacts on quality of stay. He is member of the Academy of Arts and extraordinary member of the BDA.

Katharina Blümke
Katharina Blümke is a scientific associate in teaching, research and innovation at the Professorship of Sustainable Construction, KIT Department of Architecture. She studied at HTWG Konstanz, the Escola da Cidade São Paulo and KIT, where she received her master's degree with honours in 2019. Her master's thesis "Roter Ochse – die Resozialisierung der Architektur" deals with the conversion of a soon-to-be vacant prison in Halle an der Saale. It received the Friedrich-Weinbrenner Award and was also nominated for the wa award. In addition to her university responsibilities, she is cooperating with Falk Schneemann on the realisation of the garage rooftop extension Heilbronner Strasse in Karlsruhe. The lighthouse project for innovative and circular residential construction received funding from the State of Baden-Württemberg.

Elena Boerman
Elena Boerman is a scientific associate in teaching, research and innovation at the Professorship of Sustainable Construction, KIT Department of Architecture. She studied architecture at KIT and received her master's degree in 2021. For her master's thesis, she deliberated on strategies of urban design and architecture aimed at existing building structures from the post-WWII era. The thesis received a Friedrich-Weinbrenner Award honourable mention. Together with Sandra Böhm, she is responsible for the creation of an innovative materials library at the KIT Department of Architecture with a focus on circular architecture. She is active as a volunteer in the Karlsruhe section of Architects for Future, which she founded in 2020 with Alisa Schneider.

Steffen Bytomski
Steffen Bytomski is a scientific associate at the Professorship of Sustainable Construction, KIT Department of Architecture. He studied at the TH Nürnberg and at KIT, where he received his master's degree in 2022. In his master's thesis, he dealt with strategies of urban design and architecture for the revitalisation of vacant industrial sites. As a student member of Team RoofKIT, he won the first Solar Decathlon Competition held in Germany in Wuppertal. Aside from his university responsibilities, he is a contributor in the office of Kühnl + Schmidt Architekten in Karlsruhe.

Valerio Calavetta
Valerio Calavetta is an architect and scientific associate in teaching, research and innovation at the Professorship of Building Construction, KIT Department of Architecture. He studied architecture at KIT and received his master's degree with honours in 2017. His master's thesis received numerous national and international awards. From 2017 to 2021 he worked for allmannwappner in Munich, where he was responsible for competitions, multiple commissions and other competitive processes. Since 2022 he has held a teaching position for design and illustration at the Hochschule für Technik Stuttgart (HfT). Aside from his teaching assignments, he is active as a freelance architect with Studio Sozia.

Lisa Häberle
Lisa Häberle is an architect and scientific associate in teaching, research and innovation at the Professorship of Building Construction, KIT Department of Architecture. She studied architecture at KIT, the Università degli Studi di Napoli Federico II and TUM, where she received her master's

degree with honours in 2020. Her master's thesis and theoretical study on the topic of social sustainability and future-proof redensification strategies in subsidised residential construction were published numerous times. As a freelancer, she collaborated in various projects across all service phases in the offices of Studio Knack and FAM Architekten in Munich. Aside from her teaching activity, she is a freelance architect with Studio Sozia.

Andreas Hild
Andreas Hild is Associate Professor of Architectural Design, Rebuilding and Conservation at TUM. He studied architecture at ETH Zurich and TUM, where he received his degree in 1989. Together with Tillman Kaltwasser, he co-founded the office of Hild und Kaltwasser Architekten in Munich in 1992. The office, currently led by Andreas Hild, Dionys Ottl and Matthias Haber, has operated under the name Hild und K Architekten since 1998. Hild's teaching roles, including substitute and guest professorships, have led him to the Technical University of Kaiserslautern, the Munich University of Applied Sciences, the University of Fine Arts of Hamburg, the Graz University of Technology and the Technical University of Darmstadt. Further, he was a member of the design advisory boards of the cities of Bregenz, Munich and Regensburg. Since 2014, he has been member of the design advisory board of the city of Zurich.

Peter Hoffmann
Peter Hoffmann is an architect and scientific associate in teaching, research and innovation at the Professorship of Building Construction, KIT Department of Architecture. He studied architecture at KIT, where he received his degree with honours in 2014. He then worked in the office of Prof. Peter Krebs, Büro für Architektur in Karlsruhe. Aside from his scientific activities, he is a founding partner of Schneider Hoffmann Architekten. As project manager in cooperation with Falk Schneemann and Helge Hörmann, he was responsible for the living lab project Tiny House at the Professorship of Building Construction. The project was built sorted by type and recognised in the Timber Construction University Award in 2023.

Christian Holl
Christian Holl is a freelance author and journalist. He is also a curator and member of the exhibitions committee of the architekturgalerie am weißenhof in Stuttgart and managing director of BDA Hessen. He studied architecture at RWTH Aachen, the Università degli Studi di Firenze and the University of Stuttgart. Next, he was editor of the architectural magazine db deutsche bauzeitung. In 2004 he co-founded frei04 publizistik together with Ursula Baus and Claudia Siegele. From 2005 to 2010 he was scientific associate at the Institute of Urban Planning and Design at the University of Stuttgart and held various subsequent teaching positions. Since 2017 he is publisher of the online magazine Marlowes in cooperation with Ursula Baus and Claudia Siegele. In 2021 Marlowes was awarded a BDA special prize for architectural criticism.

Paula Holtmann
Paula Holtmann is a scientific associate in teaching, research and innovation at the Professorship of Sustainable Construction, KIT Department of Architecture. She studied architecture at the University of Siegen, the Universidad Ramon Llull in Barcelona and KIT, where she received her master's degree in 2022. In her master's thesis, she deliberated on existing vacant buildings from the post-WWII era and developed innovative strategies for reuse in order to address the current housing shortage. She received a young talent grant for housing strategies from the Ikea Foundation. She was also a working student and intern in the offices of Nidus Studio in Düsseldorf, Caruso St John in Zurich and Malo Architekten in Karlsruhe.

Hauke Horn
Hauke Horn is Professor of Architectural Theory and Science at the Technical University of Darmstadt. As an architect, energy consultant and cultural scientist, his research and teaching topics focus on the intersection of architectural history, sustainability and communication theory. His numerous publications include two award-winning dissertations in architecture (Dr.-Ing.) and art history (Dr. phil.), which deal with transformation processes in the built environment and the formation of identity. For his habilitation, he received a postdoctoral funding grant from the Gerda Henkel Foundation. Aside from his academic activities, he directs the office energiewaende with a focus on energy consultancy and sustainable renovation in the existing context.

Hanna Hoss
Hanna Hoss is an architect and scientific associate in teaching, research and innovation at the Professorship of Sustainable Construction, KIT Department of Architecture. She studied architecture at KIT, where she received her master's degree in 2020. Her master's thesis "Pro Humla – Geburtshaus in Westnepal" was nominated for the 2020 BDA-SARP Award and received a Friedrich-Weinbrenner Award honorary mention in 2020. From 2020 to 2023 she worked in the office of Löffler_Schmeling Architekten in Karlsruhe with a focus on climate-friendly architecture. In 2022 she was also scientific associate at the STO Foundation Guest Professorship for "Sustainable Materials for a new Architectural Practice – Entering a circular economy", held by Peter van Assche and Katja Hogenboom.

Daniel Lenz
Daniel Lenz is an architect and scientific associate at the Department of Architecture and Civil Engineering at the Technical University of Applied Sciences Lübeck. He studied architecture at the Technical University of Darmstadt and at the Politecnico di Milano. He was lead architect for Pfeifer Kuhn Architekten and Kuhn und Lehmann Architekten in Freiburg. From 2018 to 2023 he was a scientific associate in teaching, research and innovation at the Professorship of Sustainable Construction, KIT Department of Architecture. From 2021 to 2022 he worked for Prof. Dr. Anupama Kundoo, STO Foundation Guest Professor for "Sustainable Materials for a new Architectural Practice – Entering a circular economy". Aside from his academic responsibilities, he has contributed to various cooperative projects as a freelance architect.

Falk Schneemann
Falk Schneemann is an architect and scientific associate in teaching, research and innovation at the Professorship of Building Construction, KIT Department of Architecture. Following his carpentry apprenticeship, he studied architecture at the HfT Stuttgart and TU Delft, where he received his degree with honours. For many years, he worked in leading positions for Foster + Partners in London and Herzog & de Meuron in Basel. In 2019 at KIT he successfully presented his dissertation on "skyscrapers as fabric of design and technology" (Das Hochhaus als Gewebe von Gestaltung und Technik, Jovis Verlag 2021). He co-founded the architectural office FSA and in 2023 completed the garage rooftop extension Heilbronner Strasse in Karlsruhe. As an example of innovative and circular housing construction, the lighthouse project received funding from the State of Baden-Württemberg.

Daniela Schneider
Daniela Schneider is an architect and senior consultant for cradle-to-cradle projects at EPEA – Part of Drees & Sommer in Stuttgart. She studied architecture at the HfT Stuttgart, where she received her degree in 2008. For a number of years, she worked as a construction and project manager in the field of sustainable building. In 2012 she received her master's degree in the study programme "Architecture and Environment" from Wismar University of Applied Sciences. She is currently active in practice and also a doctoral candidate at the Professorship of Sustainable Construction, KIT Department of Architecture, with a focus on circular construction. Her field of activity comprises the development of comprehensive circular concepts and consulting for investors, clients and planners in relation to hazard-free construction materials and recyclable constructions. She is a DNGB auditor and member of the DNGB committee for life cycles and circular building.

Photo credits

The editors, authors and the publisher would like to express their sincere gratitude to everyone who supported the realisation of this book by providing photos and images, giving permission for their reproduction and supplying related information. All featured drawings were specifically created for this book. Despite our most intense efforts, we could not identify the originators of some images. Nevertheless, their copyright remains protected. We kindly request related information.

INTRODUCTION
Building Sorted by Type
1 NASA, USA
2 Drawing: Sebastian Kreiter, as per Richard Rogers: Cities for a small planet. New York, 1993.
3 Drawing: Sebastian Kreiter, as per Annie Leonard: The Story of Stuff: How Our Obsession with Stuff is Trashing the Planet, Our Communities, and Our Health – and a Vision for Change. London, 2010.
4 Drawing: Sebastian Kreiter, as per Rhine-Westphalia Institute for Economic Research (RWI Essen), Fraunhofer Institute for Systems and Innovation Research (ISI), Federal Institute for Geosciences and Natural Resources (BGR): Trends der Angebots- und Nachfragesituation bei mineralischen Rohstoffen. Final report, research project no. 09/05, Federal Ministry of Economics and Technology (BMWi). Berlin, 2005, p. 17
5 shutterstock.com
6 Drawing: Sebastian Kreiter, as per Müller, Felix; Lehmann, Christian; Kosmol, Jan; Keßler, Hermann; Bolland, Til: Urban Mining – Ressourcenschonung im Anthropozän. Dessau-Roßlau 2017, p. 32, www.umweltbundesamt.de/publikationen/urban-mining-ressourcenschonung-im-anthropozaen. Original source: Skinner, B. J.: Exploring the resource base. In: Unpublished notes for a presentation to the Conference on Depletion and the Long-Run Availability of Mineral Commodities held in Washington, DC. April 2001
7 Drawing: Sebastian Kreiter, as per Richard Rogers: Cities for a small planet. New York, 1993.
8 Drawing: Sebastian Kreiter, as per Ellen MacArthur Foundation: Circular economy systems diagram, 2019, ellenmacarthurfoundation.org

HISTORY AND STATUS QUO
History of the Building Culture of Reuse
1 Hauke Horn
2 Biller, Thomas; Wendt, Achim: Die Burgen im Welterbegebiet Oberes Mittelrheintal. Ein Führer zu Architektur und Geschichte. Regensburg 2013, p. 140
3 Wikimedia Commons, CC BY-SA 4.0, photo: Carlos Teixidor Cadenas, 2022
4 Hauke Horn
5 Vio, Ettore (ed.): San Marco. Geschichte, Kunst und Kultur. Munich 2001, p. 99
6 Bosman, Lex: The Power of Tradition. Spolia in the architecture of St. Peter's in the Vatican. Hilversum 2004, p. 42
7a Puhle, Matthias (ed.): Aufbruch in die Gotik. Der Magdeburger Dom und die späte Stauferzeit und Europa, Vol. 1. Mainz 2009, p. 38
7b ibid., p. 370
8 Hauke Horn
9 Kappel, Kai: Memento 1945? Kirchenbau aus Kriegsruinen und Trümmersteinen in den Westzonen und der Bundesrepublik Deutschland. Kunstwissenschaftliche Studien, Vol. 145. Munich 2008, fig. XXV
10 ibid., fig. XXIII

More than a Mine – Existing Buildings as a Material and Cultural Resource
1-6 Christian Holl

Vernacular Architecture
1 Katiekk/shutterstock.com
2 Konstantin Litvinov/shutterstock.com
3 Anton_Ivanov/shutterstock.com
4 Peter Hoffmann
5 Drawing: Sebastian Kreiter, as per Krauth, Theodor; Meyer, Franz Sales: Das Zimmermannsbuch: Die Bau- und Kunstzimmerei. Leipzig 1895, p. 89
6 Viollet-le-Duc, Eugène-Emmanuel: Dictionnaire raisonné de l'architecture française du XIe au XVIe siècle. Vol 7. Paris 1875, p. 43 and p. 47
7 Robert Schneider/stock.adobe.com
8 Ruedi Walti
9 Ricardo Canino/shutterstock.com
10 Hebel, Dirk; Moges, Melakeselam; Gray, Zara (eds.) in collaboration with Something Fantastic: SUDU – the Sustainable Urban Dwelling Unit. Berlin 2015
11 Minke, Gernot: Building with Bamboo, Design and Technology of a Sustainable Architecture. Basel 2012

Learning from Temporary Buildings
1 Vintage illustrations, Brockhaus Konversations-Lexikon 1908 / vector illustration by Hein Nouvens/shutterstock.com
2 Arnold Newman/Getty Images, Lustron Corporation, Columbus, Ohio
3 Fonds Prouvé/Bildkunst.de/bpk-Bildagentur
4 courtesy of Shigeru Ban Architects
5a-b Filip Dujardin
6 Jeroen van der Wielen

Contemporary Examples of Circular Construction
1 Rasmus Hjortshøj & Lendager
2-3 Lendager Group
4-5 in situ, photo: Martin Zeller
6-8 Zooey Braun

MATERIALS – CONNECTIONS – LAYERS
Materials of the Circular Economy
p. 70 left KME SE, Department of Architecture, KIT
p. 70, right materials library, Department of Architecture, KIT
p. 71 left MAGNA Glaskeramik GmbH, Department of Architecture, KIT
p. 71 right StoneCycling, Department of Architecture, KIT
p. 72 left Naporo Klima Dämmstoff GmbH, Department of Architecture, KIT
p. 72 right Neptu GmbH, Department of Architecture, KIT
p. 73 left Liapor GmbH & Co. KG, Department of Architecture, KIT
p. 73 right GLAPOR Werk Mitterteich GmbH, Department of Architecture, KIT
p. 74 left AMANN Die Dachmarke, Department of Architecture, KIT
p. 74 right p. 75 left Ampack Bautechnik GmbH, Department of Architecture, KIT
p. 75 right Hanno-Werk GmbH & Co. KG, Department of Architecture, KIT
p. 76 left Hirsch & Sohn Holzhandel GmbH, Department of Architecture, KIT
p. 76 right metal workshop, Department of Architecture, KIT
p. 77 materials library, Department of Architecture, KIT
p. 78 left CLAYTEC GmbH & Co. KG, Department of Architecture, KIT
p. 78 right M & K Filze, Department of Architecture, KIT
p. 79 left Smile Plastics, Department of Architecture, KIT
p. 79 right Tarkett Holding GmbH, Department of Architecture, KIT

Pollutants in the Cycle
1 Daniela Schneider, as per Gesamtverband Schadstoffsanierung e.V.: Schadstoffe in Innenräumen und an Gebäuden. Hamburg 2010, p. 67, 171–172, 179, 241, 310, 319–321
2 Daniela Schneider, as per Bavarian Chamber of Architects: Nachhaltigkeit gestalten. Munich 2018, p. 146, byak.de/data/Nachhaltigkeit_gestalten/Nachhaltigkeit_gestalten_Download.pdf (accessed 22.6.2022)
3 EPEA GmbH – Part of Drees & Sommer
4 Linden, Wolfgang; Marquardt, Iris (eds.): Ökologisches Baustofflexikon. Bauprodukte, Chemikalien, Schadstoffe, Ökologie, Innenraum. Berlin 2018, p. 622
5,6 Daniela Schneider, adapted and amended as per EPEA GmbH – Part of Drees & Sommer

Sources

Digitalisation in the Circular Economy
1. Iwan Baan
2. Gramazio Kohler Research, ETH Zurich
3. Matthias Rippmann
4. Zooey Braun
5. Drawing: Sebastian Kreiter, as per Madaster
6. Felix Heisel, Circular Construction Lab, Cornell University

(Re)Building Simply
1. TU Munich, Chair of Building Technology and Climate Responsive Design
2. Hild und K Architekten
3-5. Transsolar
6. Comm AG

Reversible Assembly and Connection Methods
1-2. Daniela Schneider, drawing: Sebastian Kreiter
3. Drawing: Sebastian Kreiter, as per Durmisevic, Elma: Transformable building structures, design for disassembly as a way to introduce sustainable engineering to building design & construction. Dissertation. Delft University of Technology, 2006, p. 178f.

Layering as a Circular Principle
1. Drawing: Luca Diefenbacher, as per Stewart Brand, Donald Ryan
2. Drawing: Luca Diefenbacher

DETAIL CATALOGUE
Focus on Wood
1. Drawing: Luca Diefenbacher

Focus on Concrete
1. Drawing: Luca Diefenbacher, as per Weber, Robert: Guter Beton. 24th edition. Verlag Bau+Technik, Düsseldorf 2014

Focus on Steel
1. Drawing: Luca Diefenbacher

Focus on Loam
1. Drawing: Luca Diefenbacher, as per Gandreau, David; Delboy, Leticia; Joffroy, Thierry: Patrimoine mondial, Inventaire et situation des biens construits en terre. UNESCO/CH/CPM. Paris 2010

INTRODUCTION
Building Sorted by Type
[1] United Nations: Rio Declaration on Environment and Development. The United Nations Conference on Environment and Development. Rio de Janeiro 3.–14.6.1992
[2] Winkels, Rebecca: Wer bestimmt die Erdzeitalter? In: Helmholtz-Gemeinschaft, 29.8.2016. www.helmholtz.de/newsroom/artikel/wer-bestimmt-die-erdzeitalter/ (accessed 22.8.2022)
[3] Crutzen, Paul J.; Stoermer, Eugene F.: The "Anthropocene". In: IGBP Global Change Newsletter, No. 41, May 2000, p.17f. http://www.igbp.net/download/18.316f18321323470177580001401/1376383088452/NL41.pdf (accessed 21.8.2022)
[4] Anthropocene Working Group: Subcommission on Quarternary Stratigraphy: Results of Binding Vote by the Anthropocene Working Group. 21.5.2019. http://quaternary.stratigraphy.org/working-groups/anthropocene/ (accessed 21.8.2022)
[5] Hauff, Volker (ed.): Unsere gemeinsame Zukunft: Der Brundtland-Bericht der Weltkommission für Umwelt und Entwicklung. Greven 1987, p. 46
[6] Milliman, John D.; Syvitski, James PM.: Geomorphic/Tectonic Control of Sediment Discharge to the Ocean: The Importance of Small Mountainous Rivers. In: The Journal of Geology, 5/1992, p. 525–544
[7] ECO Spezial: Sand – ein Milliardengeschäft. SFR. TV documentary. Zurich 2014. www.srf.ch/sendungen/eco/eco-spezial-sand-ein-milliardengeschaeft-2 (accessed 22.8.2022)
[8] Peduzzi, Pascal: Sand, Rarer than One Thinks. In: Global Environmental Alert Service. UNEP. March 2014. www.sciencedirect.com/science/article/pii/S2211464514000396?via%3Dihub, (accessed 22.8.2022)
[9] Delestrac, Denis: Sand Wars. Documentary. 2014
[10] Hassler, Uta; Topalovic, Milica; Grün, Armin: Constructed Land – Singapore 1924–2012. ETH Zurich 2014
[11] see note 8 and United Nations Commodity Trade Statistics Database (UNcomtrade). http://comtrade.un.org/db/help/ureadMeFirst.aspx?returnPath=%2fdb%2fce%2fceSnapshot.aspx%3fpx%3dS1%26cc%3d273 (accessed 22.8.2022)
[12] see note 7
[13] Boden, George; Courtney, Oliver: Shifting Sand Report – Supporting Documents. Global Witness Report. London/Brussels/Washington 2010
[14] Leonard, Annie: The Story of Stuff: How Our Obsession with Stuff is Trashing the Planet, Our Communities, and Our Health – and a Vision for Change. London 2010
[15] U.S. Geological Survey: Mineral Commodity Summaries 2021. https://pubs.usgs.gov/periodicals/mcs2021/ (accessed 23.8.2022)
[16] Institute of rare earths elements and strategic metals (ISE): Kupfergewinnung und -produktion. https://institut-seltene-erden.de/kupfergewinnung-und-produktion/ (accessed 24.8.2022)
[17] Deutsches Kupferinstitut Berufsverband e.V.: Ressourcenschonung dank Recycling, Düsseldorf. www.kupferinstitut.de/kupferwerkstoffe/nachhaltigkeit/recycling/ (accessed 23.8.2022)
[18] Gäth, Stefan; Eck, Francis: Zur falschen Zeit am falschen Ort. Müll als Ressource. Aus Politik und Zeitgeschichte (APuZ). Federal Agency for Civic Education. Bonn 2018. www.bpb.de/shop/zeitschriften/apuz/281502/zur-falschen-zeit-am-falschen-ort/#footnote-target-4 (accessed 23.8.2022)
[19] ibid.
[20] see note 17
[21] ibid.
[22] Umweltbundesamt: Verwertung von Bau- und Abbruchabfällen, Dessau-Roßlau 2021. www.umweltbundesamt.de/daten/ressourcen-abfall/verwertung-entsorgung-ausgewaehlter-abfallarten/bauabfaelle#verwertung-von-bau-und-abbruchabfallen (accessed 24.8.2022)
[23] Bundesverband Glasindustrie e.V. (ed.), BDE Bundesverband der deutschen Entsorgungs-, Wasser- und Rohstoffwirtschaft e.V.; bvse-Bundesverband Sekundärrohstoffe und Entsorgung e.V.: Qualitätsanforderungen an Glasscherben zum Einsatz in der Behälterglasindustrie. Specification T 120. Düsseldorf 2014
[24] Neroth, Günther; Vollenschaar, Dieter: Glas. In: Neroth, Günther; Vollenschaar, Dieter (eds.): Wendehorst Baustoffkunde. Wiesbaden 2011, p. 555–635. doi: 10.1007/978-3-8348-9919-4_9 (accessed 25.8.2022)
[25] European Commission (ed.): Establishing Criteria Determining When Glass Cullet Ceases to Be Waste under Directive 2008/98/EC of the European Parliament and of the Council, Commission Regulation (EU) 1179/2012. Brussels 2012. http://data.europa.eu/eli/reg/2012/1179/oj. (accessed 25.8.2022)
[26] Heisel, Felix; Hebel, Dirk E.: Urban Mining und kreislaufgerechtes Bauen. Die Stadt als Rohstofflager. Stuttgart 2021
[27] ibid.
[28] Kreislaufwirtschaft Bau, Bundesverband Baustoffe – Steine und Erden e.V.: Mineralische Bau- und Abbruchabfälle.

Monitoring report, https://kreislaufwirtschaft-bau.de/#services, (accessed 3.10.2022)
[29] see note 22
[30] Braungart, Michael; McDonough, William: Cradle to Cradle: Remaking the Way We Make Things. New York 2010
[31] Bavarian Chamber of Architects, executive board staff group on "social questions": Gebäudeklasse "E"xperiment, Rückkehr zu den wesentlichen Grundregeln der Architektur, Berlin: In: Deutsches Architektenblatt DAB, 1/2021

HISTORY AND STATUS QUO
History of the Building Culture of Reuse
[1] Froschauer, Eva et al. (eds.): Vom Wert des Weiterbauens. Konstruktive Lösungen und kulturgeschichtliche Zusammenhänge. Basel 2020; Horn, Hauke: Die Tradition des Ortes. Ein formbestimmendes Moment in der deutschen Sakralarchitektur des Mittelalters. Berlin 2015; Horn, Hauke: Erinnerungen, geschrieben in Stein. Spuren der Vergangenheit in der mittelalterlichen Kirchenbaukultur. Berlin 2017; Kappel, Kai; Müller, Matthias (eds.): Geschichtsbilder und Erinnerungskultur in der Architektur des 20. und 21. Jahrhunderts. Regensburg 2014; Hoffmann, Volker (eds.): Die „Denkmalpflege" vor der Denkmalpflege. Akten des Berner Kongresses 30. Juni – 3. Juli 1999. Conference volume. Bern 2005; Albrecht, Stephan: Die Inszenierung der Vergangenheit im Mittelalter. Die Klöster von Glastonbury und Saint-Denis. Munich 2003; Meier, Hans-Rudolf; Wohlleben, Marion (eds.): Bauten und Orte als Träger von Erinnerung. Die Erinnerungsdebatte und die Denkmalpflege. Zurich 2000
[2] Meier, Hans-Rudolf: Spolien. Phänomene der Wiederverwendung in der Architektur. Berlin 2020, p. 9.
Basic literature on the topic of spolia:
Meier 2020; Altekamp, Stefan; Marcks-Jacobs, Carmen; Seiler, Peter (eds.): Perspektiven der Spolienforschung 2. Zentren und Konjunkturen der Spoliierung. Berlin 2017;
Altekamp, Stefan; Marcks-Jacobs, Carmen; Seiler, Peter (eds.): Perspektiven der Spolienforschung 1. Spoliierung und Transposition. Berlin 2013;
Klein, Ulrich; Untermann, Matthias (eds.): Vom Schicksal der Dinge. Spolie – Wiederverwendung - Recycling. Paderborn 2014;
Brillant, Richard; Kinney, Dale (eds.): Reuse Value. Spolia and Appropriation in Art and Architecture. From Constantine to Sherrie Levine. Farnham 2011; Poeschke, Joachim (ed.): Antike Spolien in der Architektur des Mittelalters und der Renaissance. Munich 1996
[3] Horn, Hauke: Erinnerungen, geschrieben in Stein. Spuren der Vergangenheit in der mittelalterlichen Kirchenbaukultur. Berlin 2017, p. 61–77
[4] Bosman, Lex: The Power of Tradition. Spolia in the Architecture of St. Peter's in the Vatican. Hilversum 2004
[5] Gruber, Karl: Die Gestalt der deutschen Stadt. Leipzig 1937

More than a Mine – Existing Buildings as a Material and Cultural Resource
[1] Environmental Action Germany: Fördermittelcheck der Deutschen Umwelthilfe deckt auf. Milliardenschwere Fehlinvestitionen im Gebäudebereich gehen am Klimaschutz vorbei. Press release. 17.2.2022. www.duh.de/presse/pressemitteilungen/pressemitteilung/foerdermittelcheck-der-deutschen-umwelthilfe-deckt-auf-milliardenschwere-fehlinvestitionen-im-gebaeud/ (accessed 22.4.2022)
[2] Schröer, Thomas: Neubauten sollen die Ausnahme werden. In: Frankfurter Allgemeine Zeitung, 30.9.2022
[3] Architects for Future e. V. (ed.): Klimaneutrales bzw. klimapositives Bauen: Vorschläge für eine Muster(um)bauordnung. Bremen 2021, p. 11. https://drive.google.com/drive/folders/1F1FECQCFndKnYe4QmxrCDmmjTBN2fPZo (accessed 23.10.2022)
[4] Berkemann, Karin: Das Studentenwohnheim an der Billwiese muss gehen. In: moderneREGIONAL, 12.2.2022. www.moderne-regional.de/das-studentenwohnheim-an-der-billwiese-muss-gehen/ (accessed 22.04.2022)
[5] Berkemann, Karin: Stadthalle Braunschweig. Abriss oder Sanierung? In: moderneREGIONAL, 13.2.2022. www.moderne-regional.de/stadthalle-braunschweig-abriss-oder-sanierung/ (accessed 22.4.2022)
[6] Dittrich, Philipp: Zitat, Reminiszenz, Abriss. In: marlowes, 5.4.2022. www.marlowes.de/zitate-reminszenzen-abriss/ (accessed 22.4.2022)
[7] lernort garnisonkirche: Statt Wiederaufbau Kirchenschiff ein Haus der Demokratie und Erhalt des Rechenzentrums. 28.1.2022. http://lernort-garnisonkirche.de/?p=1770 (accessed 22.4.2022)
[8] see note 3, p. 11
[9] BDA-Denklabor, Der Architektur-Podcast: Die Kreislaufwirtschaft zum Laufen bringen. Episode 23, 24.2.2022, from min. 13:47 onward. https://bda-denklabor-dont-waste-the-crisis.stationista.com/bda-denklabor-23-die-kreislaufwirtschaft-zum-laufen-bringen_621762eaf625fe5edc3eef0d (accessed 22.4.2022)
[10] European Commission: Europäischer Grüner Deal: Neue Vorschläge zur Energieeffizienz von Gebäuden. Press release, 15.12.2021. https://germany.representation.ec.europa.eu/news/europaischer-gruner-deal-neue-vorschlage-zur-energieeffizienz-von-gebauden-2021-12-15_de (accessed 22.4.2022)
[11] see note 5
[12] Tajeri, Niloufar: Fast unsichtbar. In: Holl, Christian et al. (eds.): Die Region leben, Tübingen 2018, p. 206
[13] Groys, Boris: Über das Neue. Versuch einer Kulturökonomie. Munich 1992
[14] see e.g. Deutsche Stiftung Denkmalschutz: Es kommen schwere Zeiten für Denkmale in NRW. 6.4.2022. www.denkmalschutz.de/presse/archiv/artikel/es-kommen-schwere-zeiten-fuer-denkmale-in-nrw.html (accessed 11.5.2022)
[15] Kasparek, David: Die Villa zum Beispiel. In: marlowes, 30.11.2021. www.marlowes.de/die-villa-zum-beispiel/ (accessed 11.5.2022)
[16] as quoted in Sellner, Jan: Der Abriss der Schmitthenner-Villa hat begonnen, In: Stuttgarter Nachrichten, 27.12.2021. www.stuttgarter-nachrichten.de/inhalt.architektur-in-stuttgart-das-abriss-der-schmitthenner-villa-hat-begonnen.b8730094-d380-4874-a541-2cec1d-c2ba9d.html (accessed 11.5.2022)
[17] Confurius, Gerrit: Architektur und Baugeschichte. Der intellektuelle Ort der europäischen Baukunst. Bielefeld 2017, p. 102

Vernacular Architecture
[1] Schittich, Christian: Traditionelle Bauweisen. Ein Atlas zum Wohnen auf fünf Kontinenten. Basel 2019, p. 18
[2] Umweltbundesamt: Abfallaufkommen: in Auftrag des Bundesministerium für Umwelt, Naturschutz, nukleare Sicherheit und Verbraucherschutz. Dessau-Roßlau 2022. www.umweltbundesamt.de/daten/ressourcen-abfall/abfallaufkommen#geanderte-statistische-erfassung (accessed 24.10.2022)
[3] Statista: Wohnfläche je Einwohner in Wohnungen in Deutschland von 1991 bis 2021. Published 5.5.2023. https://de.statista.com/statistik/daten/studie/36495/umfrage/wohnflaeche-je-einwohner-in-deutschland-von-1989-bis-2004/ (accessed 30.10.2022)
[4] Marsh, Alastair T.M.; Kulshreshtha, Yask: The State of Earthen Housing Worldwide: How Development Affects Attitudes and Adoption. Building Research & Information, 5/2021. doi: 10.1080/09613218.2021.1953369
[5] see note 1, p. 16
[6] Statista: Anteil sekundärer Rohstoffe an der Produktion von Kupfer, Aluminium und Rohstahl in Deutschland im Jahr 2021. Published 14.4.2023. https://de.statista.com/statistik/daten/studie/259779/umfrage/recyclinganteil-bei-der-produktion-ausgewaehlter-metalle-in-deutschland/ (accessed 1.1.2022)

[7] Habenicht, Gerd: Kleben: Grundlagen, Technologien, Anwendungen. Berlin/Heidelberg 2009, p. 142
[8] ibid., p. 147

Learning from Temporary Buildings
[1] Schneider, Jens: Joseph Paxton, Erbauer des Londoner Kristallpalasts. db deutsche bauzeitung, 12/2005. www.db-bauzeitung.de/wissen/technik/ingenieurportraet-joseph-paxton (accessed 09.11.2022)
[2] Balzer, Willi: Das Hochhaus in den USA. Dissertation. Herten 1973, p. 46–49
[3] Huber, Benedikt; Steinegger, Jean-Claude: Jean Prouvé. Zurich 1971; Vitra Design Museum (ed.): Jean Prouvé. Die Poetik des technischen Objekts. Weil am Rhein 2006; Sulzer, Peter: Jean Prouvé: Œuvre complète. Basel/Berlin 2005
[4] Bureau SLA: People's Pavilion, Amsterdam 2017. www.bureausla.nl/project/peoples-pavilion (accessed 9.11.2022); ArchDaily: People's Pavilion/bureau SLA + Overtreders W 2017, www.archdaily.com/915977/peoples-pavilion-bureau-sla-plus-overtreders-w (accessed 9.11.2022)
[5] www.mero-tsk.de (accessed 10.11.2022)
[6] also see Konrad Wachsmann's construction systems
[7] www.neptunus.de (accessed 10.11.2022)

Contemporary Examples of Circular Construction
[1] Emke, Verena: Treffpunkt Toilette: 100 Jahre Marcel Duchamp. 6.4.2017. www.weltkunst.de/ausstellungen/2017/04/treffpunkt-toilette-100-jahre-marcel-duchamp (accessed 3.10.2022)
[2] Lendager Group: Wasteland Exhibition. https://lendager.com/project/wasteland-exhibition/ (accessed 29.10.2022)
[3] Lendager Group: Wasteland Exhibition. https://lendager.com/project/wasteland-exhibition/ (accessed 29.10.2022)
[4] Federal Institute for Geosciences and Natural Resources – Geologie der mineralischen Baustoffe, Arbeitsbereich Bergbau und Nachhaltigkeit: Aluminium – Informationen zur Nachhaltigkeit. Hanover 2020. www.deutsche-rohstoffagentur.de/DE/Gemeinsames/Produkte/Downloads/Informationen_Nachhaltigkeit/aluminium.pdf (accessed 30.10.2022)
[5] ArchDaily: "Wasteland" Provides a Tactile Insight into the World of Upcycling in Architecture. 6.4.2017. www.archdaily.com/868533/wasteland-provides-a-tactile-insight-into-the-world-of-upcycling-in-architecture (accessed 3.10.2022)

[6] Lendager Group: Upcycle House, https://lendager.com/project/upcycle-house (accessed 29.10.2022)
[7] Sörgel, Christian; Mantau, Udo; Weimar, Holger: Standorte der Holzwirtschaft – Aufkommen von Sägenebenprodukten und Hobelspänen. Universität Hamburg, Zentrum Holzwirtschaft. Arbeitsbereich Ökonomie der Holz- und Forstwirtschaft. Hamburg 2006, https://literatur.thuenen.de/digbib_extern/dn056893.pdf (accessed 29.10.2022)
[8] Umweltbundesamt: Holzindustrie, 24.9.2014, www.umweltbundesamt.de/themen/wirtschaft-konsum/industriebranchen/holz-zellstoff-papier-industrie/holzindustrie (accessed 29.10.2022)
[9] ArchDaily: Upcycle House/Lendager Arkitekter, www.archdaily.com/458245/upcycle-house-lendager-arkitekter (accessed 3.10.2022)
[10] baubüro in situ: K.118 – Kopfbau Halle 118, 2021, www.insitu.ch/projekte/196-k-118 (accessed 3.10.2022)
[11] Feil, Tanja: Aus Resten erbaut. In: Deutsche Bauzeitung, 7.11.2021, www.db-bauzeitung.de/bauen-im-bestand/projekte/halle-118-winterthur-insitu (accessed 3.10.2022)
[12] BauNetz: Erfolgreiche Bauteiljagd. Ateliergebäude in Winterthur von baubüro in situ, 25.11.2021, www.baunetz.de/meldungen/Meldungen-Atelierge-baeude_in_Winterthur_von_baubuero_in_situ_7780504.html (accessed 1.10.2022)
[13] ibid.
[14] see note 9

MATERIALS – CONNECTIONS – LAYERS
Materials Selection for Circular Construction
[1] German Cement Works Association (ed.): Zementindustrie im Überblick 2021/2022 (status August 2021). Berlin 2021, https://vdz.info/ziue21
[2] Umweltbundesamt: Rohstoffkonsum steigt wieder an – auf 16,1 Tonnen pro Kopf und Jahr. Press release no. 39/2018, 29.11.2018, www.umweltbundesamt.de/presse/pressemitteilungen/rohstoffkonsum-steigt-wieder-an-auf-161-tonnen-pro, (accessed 7.10.2022)
[3] Federal Statistical Office of Germany: Umwelt. Abfallbilanz 2020. Published 30.6.2022, www.destatis.de/DE/Themen/Gesellschaft-Umwelt/Umwelt/Abfallwirtschaft/Publikationen/Downloads-Abfallwirtschaft/abfall-bilanz-pdf-5321001.html (accessed 7.10.2022)
[4] see note 4
[5] Joachim, Mitchell. BBC Online. May 2013

Materials of the Circular Economy
[1] US Geological Survey (USGS). 2020. www.usgs.gov/centers/national-minerals-information-center/copper-statistics-and-information (accessed 31.3.2023)
[2] Kupferverband e.V.: Kupfer und Circular Economy. https://kupfer.de/kupferwerkstoffe/nachhaltigkeit/kreislaufwirtschaft/ (accessed 31.3.2023)
[3] Federal Ministry for the Environment, Nature Conservation, Nuclear Safety and Consumer Protection: Altholz (status: 26.7.2021). www.bmuv.de/themen/wasser-ressourcen-abfall/kreislaufwirtschaft/abfall-arten-abfallstroeme/altholz (accessed 31.3.2023)
[4] Umweltbundesamt: Glas und Altglas. 22.9.2022. www.umweltbundesamt.de/daten/ressourcen-abfall/verwertung-entsorgung-ausgewaehlter-abfallarten/glas-altglas (accessed 31.3.2023)
[5] ibid.
[6] Federal Statistical Office of Germany: Abfallaufkommen in Deutschland im Jahr 2019 weiter auf hohem Niveau. Press release no. 261, 4.6.2021. www.destatis.de/DE/Presse/Pressemitteilungen/2021/06/PD21_261_321.html (accessed 31.3.2023)
[7] Linden, Wolfgang; Marquardt, Iris (eds.): Ökologisches Baustofflexikon. Bauprodukte, Chemikalien, Schadstoffe, Ökologie, Innenraum. 4th completely revised and expanded edition. Berlin 2018, p. 97
[8] Hillebrandt, Annette et al.: Atlas Recycling: Gebäude als Materialressource. Munich 2018, p. 90
[9] AMANN die DachMarke GmbH: data sheet, PE Dampfsperre. Hard 2016. www.amann-dachmarke.at/fileadmin/medien/produkte/abdichtung/Ecovap_blue_beschreibung_datenblatt_leistungserklaerung.pdf (accessed 9.4.2023)
[10] see note 7, p. 568
[11] Fraunhofer IMWS: Einsatz von Schrott in der Stahlherstellung mindert CO_2-Ausstoß erheblich. Press release, 14.11.2019. www.imws.fraunhofer.de/de/presse/pressemitteilungen/stahl-schrott-kreislaufwirtschaft-co2.html (accessed 31.3.2023)
[12] see note 7, p. 361
[13] see note 7, p. 321
[14] Federal Ministry for Housing, Urban Development and Building: Ökobaudat. Informationsportal Nachhaltiges Bauen. Datenbanksuche. www.oekobaudat.de/no_cache/datenbank/suche.html (status: 27.03.23)
[15] see note 7, p. 364

Pollutants in the Cycle
[1] Schäfer, Siegfried: Die Dosis macht das Gift. Wiebelsheim 2021

[2] Kaub, Siegmund: Schadstoffe im Bauwesen. Wiesbaden 2021, p. 120
[3] Bavarian Chamber of Architects: Nachhaltigkeit gestalten. Munich 2018, p. 146. www.byak.de/data/Nachhaltigkeit_gestalten/Nachhaltigkeit_gestalten_Download.pdf (accessed 22.6.2022)
[4] Federal Institute for Materials Research and Testing (BAM), Hermann Rietschel Institute (HRI) of the Technical University Berlin, Umweltbundesamt (UBA): Bauprodukte: Schadstoffe und Gerüche bestimmen und vermeiden. Berlin 2007, p. 8. www.nachhaltigesbauen.de/fileadmin/pdf/PDF_weitere_leitfaeden/bauprodukte-schadstoffe-gerueche.pdf (accessed 7.10.2022)
[5] see note 3
[6] ibid.
[7] REACH – Regulation (EC) 1907/2006
[8] Gesamtverband Schadstoffsanierung e.V.: Schadstoffe in Innenräumen und an Gebäuden. Hamburg 2010, p. 170 and p. 179
[9] ibid., p. 66
[10] Glücklich, Detlef: Ökologisches Bauen. Munich 2005, p. 121
[11] Zwiener, Gerd; Lange, Frank-Michael: Handbuch Gebäude-Schadstoffe und gesunde Innenraumluft. Berlin 2015, p. 5
[12] Linden, Wolfgang; Marquardt, Iris: Ökologisches Baustofflexikon. Berlin 2018, p. 247
[13] ibid., p. 248
[14] Umweltbundesamt: Hexabromcyclododecan (HBCD): Antworten auf häufig gestellte Fragen. Dessau-Roßlau 2017, p. 8–12. www.umweltbundesamt.de/sites/default/files/medien/421/publikationen/faq_hbcd_de_17.pdf (accessed 26.8.2022)
[15] see note 12
[16] see note 14
[17] American Chemical Society: CAS Content. www.cas.org/about/cas-content (accessed 31.10.2022)
[18] Scinexx das Wissensmagazin: Zahl der Chemikalien verdreifacht: Neues Register erfasst mehr als 350 000 Industriechemikalien. 17.2.2020. www.scinexx.de/news/technik/zahl-der-chemikalien-verdreifacht (accessed 1.8.2022)
[19] see note 12, p. 522
[20] European Chemicals Agency (ECHA): REACH verstehen. https://echa.europa.eu/de/regulations/reach/understanding-reach (accessed 23.6.2022)
[21] European Chemicals Agency (ECHA): Liste der für eine Zulassung in Frage kommenden besonders besorgniserregenden Stoffe. https://echa.europa.eu/de/candidate-list-table (accessed 23.6.2022)
[22] Bund für Umwelt und Naturschutz Deutschland e.V. (BUND) – Friends of the Earth Germany: EU braucht dreizehn Jahre, um gefährliche Chemikalien zu verbieten. 28.7.2022. www.bund.net/themen/aktuelles/detail-aktuelles/news/eu-braucht-dreizehn-jahre-um-gefaehrliche-chemikalien-zu-verbieten/ (accessed 14.7.2022)
[23] BfGA – Beratungsgesellschaft für Arbeits- und Gesundheitsschutz: CMR (cancerogen mutagen reprotoxic) – Definition. www.bfga.de/arbeitsschutz-lexikon-von-a-bis-z/fachbegriffe-c-i/cmr/ (accessed 6.11.2022)
[24] Schneider, Daniela: Einfach intelligent konstruieren. In: Heisel, Felix; Hebel, Dirk E.: Urban Mining und kreislaufgerechtes Bauen. Stuttgart 2021, p. 127
[25] C2C Products Innovation Institute: Banned List of Chemicals, V3. http://www.c2c-centre.com/library-item/banned-lists-chemicals (accessed 14.08.2023)
[26] Label-Online, Die Verbraucherinitiative e.V.: Blauer Engel, Label-Suche. https://label-online.de/suche/sp/1/f0/sector%253ABauen%2Bund%2BWohnen/ (accessed 30.10.2022)
[27] Blauer Engel: Vergabekriterien. RAL gGmbH. www.blauer-engel.de/de/zertifizierung/vergabekriterien (accessed 21.7.2022)
[28] Emicode: Grenzwerte, www.emicode.com/grenzwerte/ (accessed 21.7.2022)
[29] Umweltbundesamt: Ausschuss zur gesundheitlichen Bewertung von Bauprodukten. Dessau-Roßlau, 4.11.2022. www.umweltbundesamt.de/themen/gesundheit/kommissionen-arbeitsgruppen/ausschuss-zur-gesundheitlichen-bewertung-von#ausschuss-zur-gesundheitlichen-bewertung-von-bauprodukten-agbb (accessed 20.11.2022)
[30] GUT e.V.: Die GUT-Produktprüfung. https://gut-prodis.eu/produktpruefung (accessed 4.11.2022)
[31] C2C Products Innovation Institute: Certified Products. www.c2ccertified.org/products/registry (accessed 31.10.2022)
[32] C2C Products Innovation Institute: Cradle to Cradle Certified Restricted Substances List. Rotterdam, www.c2ccertified.org/resources/detail/cradle-to-cradle-certified-restricted-substances-list-rsl (accessed 11.12.2022)
[33] EPEA GmbH: Cradle to Cradle. https://epea.com/ueber-uns/cradle-to-cradle (accessed 11.12.2022)
[34] Spritzendorfer, Josef: EGGBI Bewertungen von über 100 Gütezeichen und „Kennzeichnungen", Datenbanken, Zertifikate für Baustoffe, Gebäude, Hotels „Produkte für das Wohnumfeld" und „Berater" für Verbrauche mit „erhöhten" Anforderungen an die „Wohngesundheit". EGGBI publication series. Abensberg 2022, p. 65. www.eggbi.eu/fileadmin/EGGBI/PDF/EGGBI_UEberblick_Guetezeichen_Baustoffe_Gesundheit.pdf (accessed 15.11.2022)
[35] ibid.
[36] ibid.
[37] European Society for Healthy Building and Indoor Air Quality e.V. (EGGBI): Holz: Emissionen aus Holz und Holzwerkstoffen. www.eggbi.eu/forschung-und-lehre/zudiesemthema/holz-emissionen-aus-holz-und-holzwerkstoffen/#c468, (accessed 1.11.2022)
[38] European Society for Healthy Building and Indoor Air Quality e.V. (EGGBI): Greenwashing mit „Wohngesundheit" und „Ökologie", www.eggbi.eu/beratung/produktinformationen-guetezeichen/greenwashing (accessed 22.6.2022)
[39] EPEA GmbH: Ressourcenschutz Zukunft Bauen, Planen und Bauen für die Circular Economy. Ein Leitfaden für die richtige Material- und Produktauswahl pro HOAI-Phase. https://norocketscience.earth/alles-im-kreislauf/ (accessed 1.11.2022)
[40] ibid.
[41] Material Building Scout GmbH: DGNB, LEED, BREEAM, WELL, …? https://building-material-scout.com (accessed 14.11.2022)
[42] German Sustainable Building Council – DGNB System: Risiken für die gebaute Umwelt. Stuttgart, https://static.dgnb.de/fileadmin/dgnb-system/de/gebaeude/neubau/kriterien/02_ENV1.2_Risiken-fuer-die-lokale-Umwelt.pdf (accessed 14.11.2022)

Digitalisation in the Circular Economy
[1] www.block.arch.ethz.ch/brg/content/project/armadillo-vault-venice-italy (accessed 27.10.2022)
[2] www.empa.ch/web/nest/hilo (accessed 27.10.2022)
[3] Block, Philippe; Rippmann, Matthias; Paulson, Noelle: Beyond Bending. Munich 2017, p. 33
[4] ibid., p. 17ff.
[5] Winter, Klaus; Rug, Wolfgang: Innovationen im Holzbau: Die Zollinger Bauweise. In: Bautechnik, 4/1992, p. 193
[6] Troxler, Irène: Im Labor für digitales Bauen. In: Neue Zürcher Zeitung, 22.9.2016. www.nzz.ch/zuerich/aktuell/neubau-der-eth-zuerich-im-labor-fuer-digitales-bauen-ld.118280 (accessed 22.09.2022)
[7] ibid.
[8] Hettinger, Pia: Tragwerksplaner können in Sachen Nachhaltigkeit viel bewirken. In: DGNB Blog, German Sustainable Building Council, 30.9.2020. https://blog.dgnb.de/tragwerksplaner-nachhaltigkeit/ (accessed 22.9.2022)

[9] Hillebrandt, Annette et al.: Atlas Recycling: Gebäude als Materialressource. Munich 2018, p. 65 and p. 72
[10] Heisel, Felix; Dirk E. Hebel: Urban Mining und kreislaufgerechtes Bauen. Stuttgart 2021, p. 99
[11] ibid., p. 157
[12] Presskit Concular 2021. https://concular.de/wp-content/uploads/2021/09/Presskit-Concular-2021.pdf (accessed 27.10.2022)
[13] Homepage Restado. https://restado.de (accessed 27.10.2022)
[14] Campanella, Dominik: DGNB Sustainability Award 2020. www.youtube.com/watch?v=FU28DEkX2VY (accessed 27.10.2022)
[15] ibid.
[16] see note 10, p. 158
[17] ibid.
[18] Sobek, Werner; Hebel, Dirk E.; Heisel, Felix. http://nest-umar.net/portfolio/umar/ (accessed 6.10.2022)
[19] Heisel, Felix; Rau-Oberhuber, Sabine: Calculation and Evaluation of Circularity Indicators for the Built Environment Using the Case Studies of UMAR and Madaster. In: Journal of Cleaner Production, 243/2020. doi: 10.1016/j.jclepro.2019.118482
[20] Heisel, Felix; McGranahan, Joseph: Enabling Design for Circularity with Computational Tools. In: De Wolf, Catherine; Cetin, Sultan; Bocken, Nancy M.P.: A Circular Built Environment in the Digital Age. Circular Economy and Sustainability. Cham 2023

(Re)Building Simply
[1] Federal Ministry for Economic Affairs and Energy – BMWi (ed.): Energieeffizienzstrategie Gebäude. Wege zu einem nahezu klimaneutralen Gebäudebestand. Brochure. Berlin 2015
[2] Pérez-Lombard, Luis et al.: A Review on Buildings Energy Consumption Information. In: Energy and Buildings, 40/2008, p. 394–398;
European Commission: Climate Action. 2050 Long Term Strategy. Brussels 2018. https://ec.europa.eu/clima/policies/strategies/2050_en (accessed 10.01.2023);
Federal Ministry for Economic Affairs and Energy (BMWi): Grünbuch Energieeffizienz. Discussion paper of the Federal Ministry for Economic Affairs and Energy. Berlin 2016;
FIZ Karlsruhe – Leibniz Institute for Information Infrastructure (ed.): BINE Informationsdienst: Nutzerverhalten bei Sanierungen berücksichtigen. Project organisation: Federal Ministry for Economic Affairs and Energy (BMWi). Project participants: RWTH Aachen University, E.ON Energy Research Centre. Project info 02/2015

[3] Delzendeh, Elham et al.: The Impact of Occupants' Behaviours on Building Energy Analysis: A Research Review. In: Renewable and Sustainable Energy Reviews, 80/2017, p. 1061–1071
[4] Einfach Bauen: Ganzheitliche Strategien für ein energieeffizientes, einfaches Bauen: Untersuchung der Wechselwirkungen von Raum, Technik, Material und Konstruktion. Final report. Zukunft Bau (BBSR). Grant number SWD-10.08.18.7-16.29. TU Munich 2018
[5] Rhein, Beate: Robuste Optimierung mit Quantilmaßen auf globalen Metamodellen. Dissertation. University of Cologne. Berlin 2014;
Maderspacher, Johannes: Robuste Optimierung in der Gebäudesimulation, Entwicklung einer Methode zur robusten Optimierung für die energetische Sanierung von Gebäuden unter unsicheren Randbedingungen. Dissertation. TU Munich 2017
[6] Brand, Stewart: How Buildings Learn – and What Happens After They Are Built. London 1997
[7] Heisel, Felix; Hebel, Dirk E. (eds.): Urban Mining und kreislaufgerechtes Bauen. Die Stadt als Rohstofflager. Stuttgart 2021
[8] www.einfach-bauen.net/ (last accessed 10.1.2023)
[9] Fisch, M. Norbert; Plesser, Stefan; Bremer, Carsten: "EVA – Evaluierung von Energiekonzepten für Bürogebäude." Final report. Institute for Building and Solar Technology – IGS. Braunschweig 2007

Reversible Assembly and Connection Methods
[1] Schneider, Daniela: Reuse of Traditional Joining Techniques for Single-Variety Construction. KIT ChangeLab. Karlsruhe 2021. https://changelab.exchange/reuse-of-traditional-joining-techniques-for-single-variety-construction/ (accessed 10.1.2023)
[2] DIN 8580:2003-08. Fertigungsverfahren Fügen – Teil 0: Allgemeines; Ordnung, Unterteilung, Begriffe, p. 3
[3] Bender, Beate; Gericke, Kilian: Pahl/Beitz Konstruktionslehre. Methoden und Anwendung erfolgreicher Produktentwicklung. Berlin/Heidelberg 2021, p. 729–730
[4] Hering, Ekbert: Taschenbuch für Wirtschaftsingenieure. Munich/Vienna 2009, p. 120–130
[5] Mayer, Bernd; Groß, Andreas (eds.): Kreislaufwirtschaft und Klebtechnik. Study. Fraunhofer Institute for Manufacturing Technology and Applied Materials Research IFAM. Bremen 2020, p. 35
[6] ibid.
[7] Rieg, Frank; Weidermann, Frank et al.: Decker Maschinenelemente. Munich 2018, p. 156–253

[8] DIN 8580:2003-08. Fertigungsverfahren Fügen – Einleitung, Begriffe, p. 12
[9] Steinhilper, Waldemar, Röper, Rudolf: Maschinen- und Konstruktionselemente 2: Verbindungselemente. Berlin/Heidelberg 2000, p. 1f.
[10] ibid., p. 36
[11] Doobe, Marlene: Deutsche Klebstoffindustrie auf Rekordniveau. In: Springer Professional, 6.4.2016. www.springer-professional.de/verbindungstechnik/deutsche-klebstoffproduktion-auf-rekordniveau/11102508 (accessed 10.1.2023)
[12] see note 1
[13] ibid.
[14] Durmisevic, Elma: Transformable Building Structures: Design for Disassembly as a Way to Introduce Sustainable Engineering to Building Design & Construction. Dissertation. TU Delft 2006, p. 178f.
[15] ibid.

Layering as a Circular Principle
[1] Brand, Stewart: How Buildings Learn – and what happens after they are built. London 1997, p. 178
[2] ibid., primarily p. 12–23 and p. 178–221
[3] Heisel, Felix; Hebel, Dirk E. (eds.): Urban Mining und kreislaufgerechtes Bauen. Die Stadt als Rohstofflager. Stuttgart 2021
[4] Institute of Constructive Design, ZHAW Department of Architecture, Design and Civil Engineering, Stricker, Eva et al. (eds.): Bauteile wiederverwenden: Ein Kompendium zum zirkulären Bauen. Zurich 2021, p. 129–142
[5] Heisel, Felix; Rau-Oberhuber, Sabine: Materialpässe und Materialkataster für die Dokumentation und Planung. In: Heisel, Felix; Hebel, Dirk E. (eds.): Urban Mining und kreislaufgerechtes Bauen. Die Stadt als Rohstofflager. Stuttgart 2021, p. 157–167
[6] Einfach Bauen: Ganzheitliche Strategien für ein energieeffizientes, einfaches Bauen. Untersuchung der Wechselwirkungen von Raum, Technik, Material und Konstruktion. Research project final report. TU Munich 2018. www.einfach-bauen.net/wp-content/uploads/2019/04/einfach-bauen-schlussbericht.pdf (accessed 7.9.2022)
[7] Francis Duffy, cited in: see note 1, p. 12f.
[8] ibid., p. 12f.
[9] Stahel, Walter R.: Wirtschaften in Kreisläufen: Eine Begriffserklärung für den Bausektor. In: Heisel, Felix; Hebel, Dirk E. (eds.): Urban Mining und kreislaufgerechtes Bauen. Die Stadt als Rohstofflager. Stuttgart 2021, p. 39f.
[10] see note 1, p. 13

DETAIL CATALOGUE

Focus on Wood

[1] Collins, Peter: Concrete. The Vision of a New Architecture. Montreal 2004, p. 19–56
[2] Gerbig, Chris; Greschat, Isabel; Timm, Christoph: Sie bauten eine neue Stadt. Der Neuaufbau Pforzheims nach 1945. Regensburg 2015
[3] Volz, Michael: Das Vollholz. In: Herzog, Thomas et al.: Holzbau Atlas. Munich 2003, p. 31ff.
[4] Stuttgarter Nachrichten, 27.2.2018. www.stuttgarter-nachrichten.de/inhalt.die-floesserei-im-schwarzwald-wie-schwarzwaldtannen-nach-holland-kamen.256745b4-bafe-4769-b869-90c04754a360.html (accessed 24.09.2022)
[5] Statista: Holzbau – Quote der genehmigten Wohngebäude in Deutschland bis 2021. https://de.statista.com/statistik/daten/studie/456639/umfrage/quote-der-genehmigten-wohngebaeude-in-holzbauweise-in-deutschland/ (accessed 05.10.2022)
[6] Umweltbundesamt: Potenziale von Bauen mit Holz. Berlin 2020. www.umweltbundesamt.de/sites/default/files/medien/5750/publikationen/2020_10_29_texte_192_2020_potenziale_von_bauen_mit_holz_aktualisiert.pdf (accessed 24.09.2022)
[7] Beck-O'Brien, Meghan et al.: Alles aus Holz: Rohstoff der Zukunft oder kommende Krise. Ansätze zu einer ausgewogenen Bioökonomie. Berlin 2020, p. 15 und p. 46. www.wwf.de/fileadmin/fm-wwf/Publikationen-PDF/Wald/WWF-Studie-Alles-aus-Holz.pdf (accessed 05.10.2022)
[8] see note 3
[9] Selberherr, Julia: Holzbau für institutionelle Investoren. Aktuelle Marktentwicklungen und zukünftige Chancen. In: Rinke, Mario; Martin Krammer (eds.): Architektur fertigen. Konstruktiver Holzelementbau. Zurich 2003, p. 15
[10] ibid.
[11] Meyer, Frederike: Reversibel Bauen. In: Baunetzwoche#531. Das Querformat für Architekten: Holz im Loop. 17.03.2023, p. 25. www.baunetz.de/baunetzwoche/baunetzwoche_ausgabe_6437916.html (accessed 05.10.2022)
[12] ibid.

Focus on Masonry

[1] Dierks, Klaus; Wormuth, Rüdiger: Baukonstruktion. Cologne 2012, p. 65
[2] Moro, José Luis: Baukonstruktion – vom Prinzip zum Detail, Volume 1 "Grundlagen", 3rd edition. Berlin 2021, p. 254–258
[3] ibid., p. 258
[4] Hillebrandt, Annette; Riegler-Floors, Petra; Rosen, Anja; Seggewies, Johanna-Katharina: Atlas Recycling. Gebäude als Materialressource. Munich 2018, p. 69
[5] Pfeifer, Günter; Ramcke, Rolf et al.: Mauerwerk Atlas. Basel 2001, p. 10
[6] see note 1
[7] see note 5, p. 92-95
[8] Federal Association of the Sand Lime Brick Industry (ed.): Kalksandstein – Maurerfibel. Düsseldorf 2019
[9] Belz, Walter: Zusammenhänge. Bemerkungen zur Baukonstruktion und dergleichen. Cologne 1999, p. 86
[10] Kreislaufwirtschaft Bau: Mineralische Bauabfälle Monitoring 2018. Report, 2021. https://kreislaufwirtschaft-bau.de/Download/Bericht-12.pdf (accessed 19.10.2022)
[11] Müller, Anette: Baustoffrecycling: Entstehung – Aufbereitung – Verwertung. Wiesbaden 2018, p. 234–235
[12] Martens, Hans; Goldmann, Daniel: Recyclingtechnik. Wiesbaden 2016, p. 358
[13] see note 4, p. 69
[14] see note 4, p. 50

Focus on Concrete

[1] Koenders, Eduardus; Weise, Kira; Vogt, Oliver: Werkstoffe im Bauwesen. Einführung für Bauingenieure und Architekten. Wiesbaden 2020
[2] Binder, Markus; Riegler-Floors, Petra: Einstoffliche Bauweisen. In: Hillebrandt, Annette u. a: Atlas Recycling: Gebäude als Materialressource. Munich 2018, p. 105
[3] Beton und Naturstein Babelsberg: Geopolymerbeton. 13.09.2019. https://bnb-potsdam.de/geopolymer-beton-als-beitrag-zum-klimaschutz/ (accessed 19.03.2023)
[4] Hillebrandt, Annette et al.: Atlas Recycling: Gebäude als Materialressource. Munich 2018, p. 70
[5] Moffatt, Jack; Haist, Michael: Konzepte zur Herstellung von ressourceneffizienten Betonen am Beispiel der Granulometrie. In: Nolting, Ulrich et al.: Ressourceneffizienter Beton: Zukunftsstrategien für Baustoffe und Baupraxis. 15th Symposium, "Baustoffe und Bauwerkserhaltung". Karlsruhe Institute of Technology (KIT). Karlsruhe 2019, p. 34
[6] Zeumer, Martin; El Khouli, Sebastian; John, Viola: Nachhaltig konstruieren: Vom Tragwerksentwurf bis zur Materialwahl – Gebäude ökologisch bilanzieren und optimieren. Munich 2014, p. 10
[7] Meyser, Johannes: Ressourcenschonung durch Wiederverwendung von Betonfertigbauteilen: Die Lehrbaustelle Plattenvereinigung. In: Baabe-Meijer, Sabine; Kuhlmeier, Werner; Meyser, Johannes: bwp@ Spezial 5. Hochschultage Berufliche Bildung 2011. Trade convention 03. Hamburg 2011, p. 1–15. www.bwpat.de/ht2011/ft03/meyser_ft03-ht2011.pdf (accessed 13.10.2022)
[8] Asam, Claus: Die Wiederverwendung von Betonfertigteilen als Beitrag zum nachhaltigen Bauen. IEMB-Info 2/2007. Institute for Conservation and Modernization of Buildings at the TU Berlin. Berlin 2007, p. 4

Focus on Steel

[1] Moro, José Luis: Baukonstruktion – vom Prinzip zum Detail, Volume 1 "Grundlagen", 3rd edition. Berlin 2021, p. 294
[2] Eggen, Arne Petter; Sandaker, Bjørn Normann: Stahl in der Architektur. Konstruktive und Gestalterische Verwendung. Stuttgart 1996, p. 30–37
[3] Helmus, Manfred; Randel, Anne: Sachstandsbericht zum Stahlrecycling im Bauwesen. Bauforum Stahl. Wuppertal 2014. https://bauforumstahl.de/upload/documents/nachhaltigkeit/Sachstandsbericht.pdf, p. 5 (accessed 15.9.2022)
[4] ibid., p. 1
[5] Hestermann, Ulf; Rongen, Ludwig: Frick/Knöll Baukonstruktionslehre 1. 36th edition. Wiesbaden 2015, p. 262
[6] ibid.
[7] Belz, Walter: Zusammenhänge: Bemerkungen zur Baukonstruktion und dergleichen. Cologne 1999, p. 54
[8] ibid., p. 55 and see note 2, p. 35
[9] see note 2, p. 99 and see note 5, p. 264
[10] ibid., p. 264
[11] ibid., p. 264
[12] Hillebrandt, Annette et al.: Atlas Recycling: Gebäude als Materialressource. Munich 2018, p. 63
[13] see note 3, p. 4
[14] Kuhnhenne, Markus et al.: Wiederverwendung im Stahlbau und Metallleichtbau in Europa. In: Hauke, Bernhard; Institute Building and Environment; German Sustainable Building Council (eds.): Nachhaltigkeit, Ressourceneffizienz und Klimaschutz: Konstruktive Lösungen für das Planen und Bauen – Aktueller Stand der Technik. Berlin 2018, p. 215

Focus on Loam

[1] Marsh, Alastair T. M.; Kulshreshtha, Yask: The State of Earthen Housing Worldwide: How Development Affects Attitudes and Adoption, In: Building Research & Information, 50/2022, p. 485–501. https://doi.org/10.1080/09613218.2021.1953369 (accessed 12.05.2023)
[2] Hebel, Dirk E.; Moges, Melakeselam; Gray, Zara: SUDU: the Sustainable Urban Dwelling Unit. Berlin 2015
[3] ibid.
[4] Sauer, Marko (ed.): Martin Rauch: Gebaute Erde. Munich 2015

Index

3D print → 92ff.

aggregate → 15, 77, 174
aircraft construction → 50f.
aluminium → 57, 59, 79, 135ff.
Anthropocene → 12ff.
antiquity → 30ff., 45, 158
assembly/joinery → 27, 66, 105ff., 114ff., 158
asservatia → 32ff.
atmosphere → 14, 17, 68

back-ventilated facade → 42
BIM → 94
biosphere → 12ff., 25
boundary value → 20, 82ff., 90, 97, 101
building class → 6, 26, 66, 169
building component activation → 101
building component level → 54ff.
building directive → 68
building elements → 15, 59, 71, 97, 119, 121, 209
building envelope → 44, 62, 101, 120f., 177, 192
building material cycle → 33
building simply → 26, 98ff., 119
building stock → 7, 24, 34ff., 67ff., 74, 81, 98
building substance → 30, 34ff.
building system → 51ff., 177

CAD → 97, 132
cascading use → 20f., 25, 58, 67, 133, 161, 177
ceiling construction → 72, 74, 78, 93, 161, 209, 220
certification → 22, 87f., 91, 95
chemicals → 18, 81ff.
circular construction → 25, 66
Circular Economy Act → 40, 46, 177
circular system → 12, 14, 23f., 62
circularity indicator (CI) → 96
climate balance/climate footprint → 57, 99, 177
climate change → 13, 33, 69
climate zone → 50, 53, 209
CMR substances → 84ff.
CNC → 42, 62, 132
CO$_2$ equivalent → 68
comfort/wellbeing → 23, 42, 45, 100f., 103, 121
composite material → 15, 20, 26, 58, 67, 84, 174
composting → 12, 21, 25, 70ff., 133
connection methods → 46, 54, 104, 110
connector → 47, 52, 54, 67, 104ff., 133, 191
contamination/impurity → 20ff., 70f., 79, 88, 105
conversion → 14ff., 30ff., 51, 58, 69, 100, 103
copper → 18f., 61, 70, 81, 137, 190ff.
Cradle to Cradle → 23ff., 87f.
craft/trade → 27, 38, 41ff., 60, 66, 84, 90f., 95, 105, 108ff.

database → 24, 68f., 83, 95, 185, 197
demolition material → 161
demolition site → 61
digitisation → 22, 67, 91ff.
DIN standard → 25, 66, 169
downcycling → 15, 21ff., 35
dry construction → 27, 78, 122, 124, 126f., 209

eco efficiency → 23
economic efficiency → 48, 55, 100, 177
ecosystem → 13f., 16, 80, 96
efficiency → 23f., 41, 54, 100
energy carrier → 18, 24, 45, 67, 73
energy consumption → 102, 119
energy efficiency → 98ff., 103, 141
energy source → 18, 62, 67
environmental impact → 17, 70, 89
environmental justice movement → 17
EU taxonomy → 69
European Green Deal → 61

fauna → 14, 18, 80
filigree construction → 43, 122, 126, 192
fire-retardant → 80ff.
fire safety/fire protection → 6, 54, 85, 107
flexibility → 54f., 99f., 104, 106
flora → 14, 18, 80
footprint → 42, 73, 164, 173, 201, 205f.
force-fit connection → 106ff., 113ff., 172, 193
form-fitting connection → 46, 106ff., 131ff., 149, 160
frame construction → 177
full declaration → 84ff., 90f.
functional separation → 54

gabion → 63
German Sustainable Building Council → 91
global warming potential (GWP) → 68, 77
globalisation → 41
gothic → 31
greenhouse gas → 13, 58, 60, 68, 87, 98, 118, 175, 189
greenwashing → 69, 88

half-timbered construction → 125, 138, 214
health → 17, 67, 80ff., 130, 132, 207
historic preservation → 34ff., 100
Holocene → 13
hunting for building parts → 59
hybrid → 44, 107, 127
hydrocarbon → 68, 81f.
Hydrosphere → 17

industrial construction → 46
Industrial Revolution → 41
industrialisation → 13, 40f., 46, 48, 130, 206
infill → 43ff., 133, 209
insulation material → 62, 72f.
intervention → 39

land degradation → 18
land reclamation project → 16
layer composition → 57, 122
life cycle → 20, 27, 34, 79, 102, 121
life cycle analysis → 68, 99
life cycle costs → 24, 96
lignification → 42
loam (construction) → 43ff., 61f., 78, 166, 206ff.
longevity → 25, 46, 69, 105
Lustron house → 49f., 54f.

Maison Tropicale → 50
masonry construction → 158, 160ff.
material declaration → 80, 90f.
material passport → 22, 91, 96, 119
material store → 20ff., 27, 63, 67ff., 95f., 119
materials cycle → 21, 33, 35, 63
materials library → 69
metabolism → 12, 15, 22, 81
methane → 68
modular system → 27
modularity → 55

(**n**early) zero energy building standard → 99
node → 43, 52, 54
non-residential buildings → 21, 35, 98f.

obsolescence → 17
ozone → 68

paper → 51, 55, 57, 74, 121, 150f.
Paper Log House → 51f., 55
participatory process → 60
People's Pavilion → 52ff.
performance gap → 99
photosynthesis → 12
plants → 12, 42, 49, 72, 78, 80, 157
pollutants → 17, 67, 70, 78, 80ff., 85ff., 208
prebound effect → 102
prefabricated concrete element → 27, 52, 115, 123, 168, 176
primary materials → 20, 23, 58, 60, 67, 69f., 76
product as a service → 24, 54, 63
product label → 85, 87

quality control → 88f., 91, 108

range → 17f., 70
raw material reserve → 17
REACH Regulation → 80ff.

recovery/reclamation → 20, 24, 75, 133
recyclate → 9, 69, 161, 177
recycling → 18, 30, 34, 71, 96f., 176ff., 206
renovation/rehabilitation → 7, 34, 36, 53, 98ff., 102f.
residential area → 34, 40
resource consumption → 9, 17, 23, 42, 71
(rooftop) extension → 59ff.

screw connection → 50, 62, 63, 143, 147, 183ff., 193
secondary material → 22, 58, 68, 70, 79
secondary raw material → 19, 22, 66, 71, 104
semi-finished product → 54, 95, 190ff.
sharing economy → 24
sociology → 60
soil erosion → 18
Solar Decathlon → 60f.
solid construction → 122ff., 127, 146ff., 154, 162, 182, 210, 214
solid timber construction → 150
soundproofing → 6, 25, 55, 118, 126f.
spolia → 30ff.
standardisation → 22, 25, 66, 208
steel construction → 46, 190ff.
straw → 59, 77, 154ff., 207ff.
sufficiency → 23, 41, 47, 63, 92
supply chain → 69, 87, 95
sustainability → 15, 26f., 33, 41, 76, 92, 104

thermal envelope → 25, 42
thermal insulation → 25, 100, 102, 132
thermal insulation composite system → 42
timber construction → 42f., 130f.
timber frame construction/ wall frame construction → 133, 135, 141, 143, 148, 185
timber species → 70, 76, 132
tiny houses → 53
typology → 43, 60, 103, 206

upcycling → 21, 57
urban mining → 20, 30, 35, 97
use phase/operations phase → 17, 26, 53, 58, 73, 84, 95f., 98, 105f., 108, 119

value creation → 24, 47, 56, 60, 69, 95
value retention → 68
vernacular architecture → 40ff.

warranty → 22, 25, 76
waste glass → 62f., 71, 73
Wasteland Exhibition → 56
weather protection → 62, 76, 133
wood-based material → 76

Zollinger roof → 93

231

Imprint

Editors: Dirk E. Hebel, Ludwig Wappner, Katharina Blümke, Valerio Calavetta, Steffen Bytomski, Lisa Häberle, Peter Hoffmann, Paula Holtmann, Hanna Hoss, Daniel Lenz, Falk Schneemann

Authors: Dirk E. Hebel, Ludwig Wappner, Werner Sobek, Thomas Auer, Katharina Blümke, Elena Boerman, Lisa Häberle, Andreas Hild, Peter Hoffmann, Christian Holl, Hauke Horn, Hanna Hoss, Daniel Lenz, Falk Schneemann, Daniela Schneider

Project editing: Katja Pfeiffer (project management), Cosima Frohnmaier (detail catalogue), Jana Rackwitz (theory chapters), Barbara Kissinger (CAD)

Translation into English: Mark Kammerbauer, DE–Munich

Copy editing: Stefan Widdess, DE–Berlin

Proofreading: Meriel Clemett, GB–Bromborough

Cover design: Wiegand von Hartmann, DE–Munich

Drawings: Katharina Blümke, Robina Behrendt, Patrick Bundschuh, Valerio Calavetta, Luca Diefenbacher, Mattis Epp, Lisa Häberle, Hanna Hoss, Felix Caspar Jörgens, Sebastian Kreiter, Salesia Trenker, Amelie Vierhub-Lorenz

Production and DTP: Simone Soesters

Reproduction: ludwig:media, AT–Zell am See

Printing and binding: Gutenberg Beuys Feindruckerei, DE–Langenhagen

Paper: Materica Clay 120 g (cover), Magno Volume 135 g (content)

© 2024, first edition
DETAIL Architecture GmbH, DE–Munich
detail.de

ISBN 978-3-95553-636-7 (Print)
ISBN 978-3-95553-637-4 (E-book)

This product was manufactured from materials originating from reputably managed, FSC®-certified forests and other monitored sources.

This work is subject to copyright. All rights reserved. These rights specifically include the rights of translation, reprinting and presentation, the reuse of illustrations and diagrams, the reproduction on microfilm or on any other media and storage in data processing systems. Furthermore, these rights pertain to any and all parts of the material. Any reproduction of this work, whether whole or in part, even in individual cases, is only permitted within the scope specified by the applicable copyright law. Any reproduction is subject to remuneration. Any infringement will be subject to the penalty clauses of copyright law.

Bibliographic information from the German National Library: The German National Library lists this publication in the Deutsche Nationalbibliografie (German National Bibliography); detailed bibliographic data is available on the Internet at http://dnb.d-nb.de.

The contents of this textbook were researched and developed with great diligence and a conscientious effort to reflect the best available knowledge. We assume no liability for any errors or omissions. No legal claims may be derived from the contents of this book.